CHIMPANZEE CULT

Chimpanzee Culture Wars

RETHINKING HUMAN NATURE ALONGSIDE JAPANESE, EUROPEAN, AND AMERICAN CULTURAL PRIMATOLOGISTS

NICOLAS LANGLITZ

PRINCETON UNIVERSITY PRESS
PRINCETON & OXFORD

Published by Princeton University Press
41 William Street, Princeton, New Jersey 08540
6 Oxford Street, Woodstock, Oxfordshire OX20 1TR

press.princeton.edu

ISBN 978-0-691-20427-7
ISBN (pbk) 978-0-691-20428-4
ISBN (e-book) 978-0-691-20426-0

British Library Cataloging-in-Publication Data is available

Editorial: Fred Appel and Jenny Tan
Production Editorial: Jenny Wolkowicki
Cover design: Chris Ferrante
Production: Erin Suydam
Publicity: Kathryn Stevens and Kate Hensley
Copyeditor: Laurel Anderton

Cover image: From illustration for *The Graphic*, Nov. 28, 1885. Look and Learn / Illustrated Papers Collection / Bridgeman Images

This book has been composed in Arno Pro

Printed on acid-free paper. ∞

Printed in the United States of America

10 9 8 7 6 5 4 3 2 1

For my familoid, Donya, Kiki, and Loretta

CONTENTS

PREFACE

WORK ON THIS BOOK began in a Nicaraguan hammock. Having just sent off a manuscript on the revival of psychedelic research, I fantasized about a new project that would give me a break from all things neuro.[1] To the bellowing cries of howler monkeys, I read Roger Fouts's *Next of Kin: My Conversations with Chimpanzees*, a memoire from one of the ape language projects.[2] In the late 1960s, at a time when LSD and other hallucinogenic drugs fostered hopes that scientists could push human consciousness beyond its present limitations toward capacities not yet realized, primatologists began to explore and expand the minds of great apes by teaching them natural and artificial languages to communicate with humans. Just as psychedelics had endowed psychologists, psychiatrists, and psychopharmacologists with the sense that we don't know what even the human mind can do, the ape language projects conjured up an undreamed of primate potential.

Since the accidents of life had made me a professor of anthropology, a historical and ethnographic study of primatology appealed to me because it promised to reconnect my humanistic interest in the sciences with the constitutive concerns of anthropology. At a time when humans appeared a curiouser and curiouser species of hominoid, most of my colleagues in American cultural anthropology had abandoned *anthropos* as the epistemic object that had originally defined their discipline. When, fresh out of German medical school, I entered the joint medical anthropology program of the University of California at Berkeley and San Francisco in 2003, it already struck me how eagerly cultural anthropologists embraced the science studies doctrine that the ontological divide between nature and culture had collapsed while no rapprochement with evolutionary anthropology ensued. Shifting my research on the naturalization of the human from neuroscientists to evolutionary anthropologists and primatologists seemed a good way to recolonize the heartland of what had become my discipline.

But, as has happened so often in the history of anthropology, at the exact time I grew interested in the ape language projects, the last of these *Pan/Homo* cultures went extinct. In 2012, Sue Savage-Rumbaugh was put on leave after staff accused her of mistreating the bonobos in her care at the Iowa Primate

Learning Sanctuary—a disenchanting but consistent end for a research tradition that had brought much misery to both primates and primatologists.

Meanwhile, the idea that humans were not the only indeterminate animals had stirred up controversy in another—extant—subfield of evolutionary anthropology. The origins of cultural primatology can be traced back to 1950s Japan, where Kinji Imanishi proposed that monkeys had culture, too. Yet European and American primatologists really adopted the culture concept only in the 1980s, just as many American cultural anthropologists dropped the analytic category that had delineated their field. In the thick of the culture wars, culture itself became a bone of contention. And so did the plea of cultural primatologists to extend multiculturalism beyond the human. The resulting chimpanzee culture wars have become the topic of this book.

The reader will have picked up this book for his or her own purposes and will make good use of it in ways that I cannot foresee. One I do anticipate, though: my most immediate colleagues, cultural anthropologists and science studies scholars, might read *Chimpanzee Culture Wars* to better understand how evolutionary anthropologists think about culture. I would love to see my observations serve as a basis for both sides to make up for the exchange that could have happened when cultural primatology went from Japan to Europe and North America in the 1970s, if it hadn't been for the culture wars.

Yet the world has moved on since then. Writing in the late 2010s, I would like to push into the center of this hoped-for conversation the question of whether the comparison of chimpanzee and human culture can help us understand why we are living in a world disfigured and remade by *Homo sapiens* rather than some other primate species. Why are we living in the Anthropocene rather than a Chimpocene? As we're finding once again that we have no idea what human and ape minds can do and as our focus shifts from consciousness to ecological expansion, the chimpanzee culture controversy forces us to reformulate the question of human potential: What are we to make of our distressed ascent to a superdominant species?

———

Cultural primatologists took over not only the culture concept from cultural anthropologists but also the custom of acknowledging their research subjects. I would like to imitate their expressions of gratitude to nonhumans—although even the subjects of the long-discontinued Japanese ape language project I visited will never read this book or any other. But I do feel privileged that I had the chance to see for myself how the chimpanzees of Taï Forest in Côte d'Ivoire, Loango National Park in Gabon, Bossou in Guinea, Kyoto University

Primate Research Institute and Kumamoto Sanctuary in Japan, and Leipzig Zoo in Germany related to the human beings who studied them.

Following my own field's tradition, I wish to thank Christophe Boesch, Tetsuro Matsuzawa, and their coworkers for allowing me into their territory and for serving as both subjects and teachers. I am equally grateful to my interlocutors in cultural anthropology, sociology, the history of science, science and animal studies, and philosophy who have helped me make sense of the chimpanzee culture wars. For the sake of keeping a long list short, I will lump together everybody I am indebted to: Ikuma Adachi, Lys Alcayna-Stevens, Pamela Asquith, Clément Ba, Ulrich Moussouami Bora, Cameron Brinitzer, Sebastian Conrad, Alice Crary, Gabriela Bezerra de Melo Daly, Talia Dan-Cohen, Tobias Deschner, Jan Engelmann, Abou Farman, Didier Fassin, Soumah Aly Gaspard, Yuko Hattori, Misato Hayashi, Satoshi Hirata, Larry Hirschfeld, Cat Hobaiter, Kimberley Hockings, Nene Kpazahi Honora, Michael Huffman, Louis-Bernard Ibally, Jessica Junker, Ammie Kalan, Fumihiro Kano, Miriam Kingsberg Kadia, Hjalmar Kühl, Kevin Langergraber, Sylvain Lemoine, Dana Leon, Resi Löhrich, Lydia Luncz, William McGrew, Alexander Mielke, Erika Milam, Richard Moore, Naruki Morimura, Roger Mundry, Michio Nakamura, Daniel Povinelli, Anna Preis, Sindhu Radhakrishna, Hugh Raffles, Tobias Rees, Julian Rohrhuber, Janet Roitman, Esther Rottenburg, Osamu Sakura, Liran Samuni, Stephanie Schiavenato, Jutta Schickore, Sylvia Sebastiani, Michael Seres, Giulia Sirianni, Ann Stoler, Shirley Strum, Wakana Suzuki, Peter Thomas, Miriam Ticktin, Michael Tomasello, Carel van Schaik, David Watts, Andrew Whiten, and Roman Wittig. I completed this book as a Deborah Lunder and Alan Ezekowitz Founders Circle member during a sabbatical at the Institute for Advanced Study in Princeton—one of the best years of my life.

Permission to publish revised and expanded versions of the following journal articles is hereby gratefully acknowledged: "Synthetic Primatology: What Humans and Chimpanzees Do in a Japanese Laboratory and the African Field," *British Journal for the History of Science Themes* 2 (2017): 1–25; "Salvage and Self-Loathing: Cultural Primatology and the Spiritual Malaise of the Anthropocene," *Anthropology Today* 34, no. 6 (2018): 16–20.

Prologue

SITTING ON A NARROW STRIP of savanna, which the Gabonese maintained between the coastal forest swamps to our left and the deep somber rainforest to our right, I thought of Andrew Battell. During the two and a half years he had lived in Loango, the seventeenth-century British merchant had maintained a healthy respect, even wariness, of these woods so "covered with baboons, monkeys, apes and parrots, that it will fear any man to travel in them alone."[1] But philosophers should undertake the arduous journey to the secluded kingdom, Jean-Jacques Rousseau suggested from a comfortable fauteuil in Paris one and a half centuries later. He had read about this wondrous place in Africa in a book from 1617, *The Strange Adventures of Andrew Battell.*[2]

Two kinds of monsters lived in the forest surrounding the small port where the Englishman stayed. The people of Loango called them Pongoes and Engecoes.[3] While Battell unfortunately forgot to describe the Engecoes, he claimed that the Pongoes' hairless faces and the proportions of their furry bodies resembled those of human beings. They built shelters against the rain and buried their dead under heaps of branches. In the morning, they sat around the dying campfires next to which the locals had slept during a night in the forest. The Pongoes did not know how to sustain these fires, nor could they speak. Yet this description of them reminded Rousseau so much of humans that he wondered whether Battell had not stumbled upon "genuine Savage men . . . in the primitive state of Nature" who—unlike Europeans—had not yet developed their virtual faculties toward sorrowful perfection.[4]

In 1755, the questions of who we were, where we had come from, and where we were going, as humans and especially as modern humans, appeared to be of eminent political importance. The French Revolution was fast approaching. Rousseau's *Discourse on the Origins and the Foundations of Inequality among Men* asked what form of government would be most appropriate to human nature—and it was not Louis XVI's absolutist monarchy. The philosopher speculated that originally, humans had enjoyed a state of freedom and equality that contrasted sharply with the subjection to strictly hierarchical social order under the Bourbons. If the Pongoes and Engecoes could shed light on how humans had lived before they formed societies fostering dependence and inequality, they might also provide inspiration as to how humankind might overcome its modern predicament.

According to Claude Lévi-Strauss, Rousseau's treatise marked the beginning of anthropology.[5] But it took more than a century before anthropologists followed Rousseau's advice and set out for the field. It took another century before they would go to what is now known as Loango National Park in Gabon to study Battell's monsters. As primate folklore gave way to primate science, these preternatural creatures turned out to be western lowland gorillas (*Gorilla gorilla*) and Central African chimpanzees (*Pan troglodytes troglodytes*).[6] Although neither species built shelters or buried their dead, the evolutionary anthropologists documenting their forms of life had grown convinced that the great apes had developed their own cultures, which could be studied ethnographically.

But these cultures were under threat everywhere. In Loango, park authorities insisted on the principle of *nature et culture*, that is, the peaceful coexistence of humans and the forest's flora and fauna, while local poachers and loggers as well as Chinese oil companies pressed into the protected area to extract valuable resources. Just as cultural anthropology had quickly turned into a form of "salvage anthropology," hurriedly documenting the last gasps of newly encountered ethnic groups to preserve some knowledge of their languages and cultures for the ethnographic archive, the budding field of cultural primatology sought to record and understand nonhuman cultures that were frequently facing imminent extinction.[7]

In recent years, geologists, biologists, and humanities scholars have associated this rapid loss of biodiversity—which, at least in the case of the apes, arguably also entails a loss of cultural diversity—with the dawn of a new natural historical epoch, the Anthropocene.[8] Many anthropologists and posthumanities scholars have expressed anger about the term because it attributes this massive planetary transformation to humankind as a whole rather than "situated peoples and their apparatuses, including their agricultural critters."[9] Instead these critics prefer to talk about the Capitalocene or the Plantationocene.[10] Both terms suggest that only particular and historically rather recent human cultures are to blame for the "outrage" of what capitalists and colonial plantation owners have done to "a vulnerable planet that is not yet murdered."[11] Moralist undertones aside, this is an important debate about the causes of the dramatic natural historical transformation we are experiencing.

While the beginnings of anthropogenic climate change appear to coincide with the industrial revolution, paleontologists like Paul Martin have argued that the current mass extinction event can be traced back all the way to early humans who began to kill off megafauna wherever they migrated from Africa.[12] "While we appear rather unintimidating, and perhaps easy prey given our lack of claws, canines, venom, and speed, we come with a dangerous bag of tricks, including projectiles, spears, poisons, snares, fire, and cooperative social norms that make us a top predator," noted evolutionary anthropologist Joseph Henrich.[13] "It's not just the fault of industrialized societies; our species' ecological impacts have a deep history."[14] The ability of *Homo sapiens* to change the face of the Earth and even the planet's climate has brought back with a vengeance the eighteenth-century question of human nature: What sets humans apart from other species?

But the political context in which this question gains its significance has radically changed. It is no longer the birth pains of the now aging European and American democracies that make this philosophical and zoological puzzle a pressing concern. It is the accelerating transformation of ecosystems and the global climate to which we and the biota alongside which we evolved have adapted over hundreds of thousands of years. When nineteenth-century naturalist Alexander von Humboldt visited colonial plantations in South America, he already realized that the irrevocable loss of life-forms was the outcome of things we do and things we often decide to do collectively.[15] As we are extending his insight from the *Tristes tropiques* to an equally mournful Arctic, we have begun to see recent natural history as a guilt-ridden political process.[16] Today, the question of human nature is the question of why it is we who are responsible for this natural historical tragedy—or should we think of it as a black primate comedy? What happened in the evolution of *Homo sapiens*, originally just another African ape, that enabled this inconspicuous species of hominoid to dominate basically all ecosystems across the globe?

The advent of culture has been an obvious contender because it has allowed us to populate habitats radically different from the one in which our ancestors evolved. But if culture is actually not a uniquely human trait, as cultural primatologists claim, then we need to reconsider the question of what made us so exceptional that we came to conquer the planet while our closest relatives continue to be confined to equatorial Africa. And why are their numbers dwindling while ours are rising exponentially? In other words, what brought about the Anthropocene, if other primates have culture, too? Against this new horizon, the chimpanzee culture controversy poses anew the original question of anthropology: Who are we as a species and what may we expect of ourselves?

That evening on the savanna I was part of a conversation that colors every page of this book. I had followed French Swiss primatologist Christophe Boesch to

FIGURE 1. An evolutionary memento mori: open-air natural history museum, Loango National Park, Gabon. Photo by author.

his group's field site in Loango. Sitting at the edge of the research camp, overlooking an artificial savanna roamed by red river hogs, buffalos, and elephants, we recovered from a day in the forest where we had looked for underground beehives, from which the chimpanzees extracted honey with the help of multipart tool sets. In this paradisiacal ambience, the researchers had built an open-air natural history museum: a table heaped with primate, ungulate, and elephant skulls. At the beach, they had even found the vertebrae of a whale. The display appeared like an evolutionary memento mori. An American research assistant told us about the book she was reading. In the face of ecological catastrophe, it celebrated the power of ideas as humanity's infinite resource on a finite planet.[17] In response to this can-do optimism straight out of Silicon Valley, Boesch scoffed: "From a biological point of view, resource depletion is a sign of success. We multiply magnificently. Soon we will be eight billion humans with only 180,000 chimpanzees left. Isn't that an achievement?"

A few days earlier, when we had met at a beach hotel in Gabon's capital, Libreville, Boesch had explained to me that I would see him wearing two hats, that of director of the Max Planck Institute for Evolutionary Anthropology and that of president of the Wild Chimpanzee Foundation. He had founded this NGO to protect the livelihoods of the apes to whom he owed his prolific career. He was a scientist and an activist, a naturalist and a conservationist. At the beginning of the twenty-first century almost all field primatologists breasted the displacement and extinction of their research subjects. Of course they did: without their primates primatologists would become an endangered species, too. The existence of chimpanzee ethnographers like Boesch was even more precarious. As students of cultural differences, they couldn't do with just a few ape populations, maybe even corralled into sanctuaries jumbling individuals of different origins. They needed a plurality of grown communities to advance their intellectual project.

The sarcasm of Boesch's remark concerning the evolutionary success of *Homo sapiens* revealed the cognitive dissonance between affirming evolution and practicing conservation. The Darwinian in Boesch recognized that no species would last forever and that better-adapted organisms would drive their competitors extinct. There had never been a steady and harmonious state of nature, which modern humans had come to destroy. And yet the fieldworker intimately acquainted with the lives of so many wild chimpanzees, who had put up in his Leipzig office a large portrait of Brutus, the alpha male of the Taï community Boesch had first habituated, resisted the course of natural history and struggled to keep things as they were for as long as possible. Certainly not for the lifetime of the world, but maybe for his own lifetime, he paradoxically sought to preserve nature to correct the course of nature.

Toward the end of this field trip—we had meanwhile moved on to Côte d'Ivoire—I observed the original, but by now seriously decimated chimpanzee group that Boesch had habituated in Taï Forest in the early 1980s. While the apes had disappeared into a *Treculia africana* tree to feed on its giant fruit, I spoke with one of the most senior Ivorian field assistants of the Taï Chimpanzee Project. Louis-Bernard Ibally regretted this loss of a world he had come to know so intimately. But he also believed that you had to destroy to create: "Europe has no more forest, but it's developed. In Africa, we have forest, but no development. Isn't cutting down the trees for the creation of agricultural land a precondition of economic development?" When I responded that deforestation was not simply a consequence of economic growth but also of exploding human populations, first in Europe and now in Africa, Ibally agreed. But, he added, we still needed to have children to survive, for there to be more *petits Nicolas*.

I felt ambivalent about his remark. With many of my European and American contemporaries I shared a sense of misanthropy. But I also recognized such human self-loathing as the spiritual malaise of the Anthropocene, probably more prevalent in North America and western Europe than in West Africa. A recent Swedish Canadian study suggested that by far the most effective measure to combat climate change was to have fewer children.[18] Recognizing how destructive our presence on this planet had become, the Oregon-based Voluntary Human Extinction Movement advocated that we sterilize ourselves. In his essay on such "misanthropology," Abou Farman cites the movement's vasectomized founder as saying: "The creation of one more human by anyone anywhere cannot be justified in light of the number of us dying every day and of the damage we are doing to the planet, causing other extinctions."[19] Hoping not for our total annihilation but for a reduction of the world population from currently over seven billion to two or three billion, Californian science studies scholar Donna Haraway offered her own slogan to such experiments of deliberately infertile living: "Make kin, not babies!"[20]

Unscathed by anthropocenic autoaggression, Ibally had done both: he had children and cultivated kinship with another species, spending more time with the chimpanzees than with his own family. The Guéré around Taï Forest had considered the apes descendants of their own human forebears all along. I had to think of Ibally when, not long after my return from Africa, my wife and I had two daughters (no *petit Nicolas* after all) while I was writing this book about how cultural primatologists conceive of the kinship between humans and chimpanzees, a species with whom we share a last common ancestor, presumably already endowed with the capacity for culture.

What I have learned from Christophe Boesch's brand of naturalism is that culture does not provide the "freedom from biology" that anthropologist Marshall Sahlins once hoped for.[21] "Doesn't culture restrict our freedom as much (or as little) as biology?" asked Boesch's fellow cultural primatologist Frans de Waal.[22] This raises the question of human agency in natural history, which British philosopher John Gray posed succinctly: "We do not speak of a time when whales or gorillas will be masters of their destinies. Why then humans?"[23] The faith Haraway and the Voluntary Human Extinction Movement put in the power of political action betrays a persistent humanism, which imagines us, now including our gut bacteria, if you will, at the helm of natural history. That a determined but tiny group of childless West Coast Americans will revert the global population explosion seems even less likely than that World Climate Conferences will enable our more divided than united nations to achieve the two-degree warming goal. My interest in revisiting the question of human nature, which sets us apart from the apes—here I might be closer to Boesch's adversary Michael Tomasello—is driven by a sense of wonder about the process through which modern humans have become the cause (not the agents) of an inexorable ecological upheaval. This dramatic planetary transformation appears to be the latest, but I hope not the last, chapter in the natural history of culture.

Introduction

WE UNDERSTAND our time as one in which human culture remakes nature. But Japanese and Euro-American primatologists have come to question whether humans are the only primates capable of culture—that is, whether culture amounts to human nature. This book examines the ensuing controversy over chimpanzee culture.

In the 1950s, Japanese primatologists around Kinji Imanishi proposed to attribute "subhuman culture"—or *kaluchua,* as they called it—to nonhuman primates.[1] Their discovery of behavioral differences between macaque troops based on the social transmission of newly invented ways of doing things challenged one of the tenets of modern cosmology. It called into question whether, as American anthropologist Marshall Sahlins put it, "culture is the human nature."[2] French philosopher Dominique Lestel declared the Japanese discovery of cultures beyond the human to be as important as the quantum revolution.[3] In a reductionist zeitgeist, contemporaneous breakthroughs in molecular biology had stolen the limelight from behavioral researchers. In a curious mixture of metaphors, jumbling the hushed and the explosive, science writer Michel de Pracontal spoke of a "clandestine" Copernican Revolution of the life sciences, "dynamiting" the barriers that Western thought had erected between nature and culture, humanity and animality.[4]

This talk about scientific revolutions—really an invention of the mid-twentieth-century history of science—suggested that whoever doubted the double movement of anthropomorphizing monkeys and zoomorphizing humans had to be both a scientific and a metaphysical reactionary.[5] Needless to say, such reactionaries soon raised their voices. A first wave expressed skepticism toward the Japanese application of categories previously reserved for humans to other primate species. Criticism got harsher when, in the course of the 1970s and 1980s, a growing number of European and American primatologists and evolutionary anthropologists chimed in with Japanese anthropomorphism and wondered how unique the cultural nature of *Homo sapiens* really was.[6] Anthropologists resisted what they perceived as an attack on the political ontology not just of their field but of the postwar era, carefully crafted to ward off the murderous excesses of nineteenth-century racism and Nazi biologism.[7] Theory of knowledge became another key site of contention in the debate: comparative psychologists defended their use of controlled laboratory experiments—as opposed to the field observations by a new breed of chimpanzee ethnographers—to explain the cognitive capacities that had set humans on such an exceptional evolutionary path.[8] It was far from clear whether the revolutionaries would come out winners and what such a victory would entail—epistemologically, ontologically, and politically.

US evolutionary anthropologist William McGrew dubbed the resulting controversy the chimpanzee culture wars.[9] The expression alluded to the culture wars over progressive and conservative values that began to polarize American society at about the same time as Western primatologists adopted the culture concept.[10] In McGrew's eyes, theirs was a battle over extending multiculturalism—one of the most contentious progressive causes in the culture wars—to the apes.

Like many cultural anthropologists, cultural primatologists fought for their subjects' inclusion. Just then, however, a new generation of cultural anthropologists dismissed the culture concept because it fostered an image of human groups as bounded and homogeneous. It did not conform with their own vision of an open society that allowed everyone to cultivate a different hybrid identity.[11] Thus, cultural anthropology and cultural primatology were like ships passing in the night. Where they did get into shouting distance, their representatives hurled accusations of racism and lack of scientificity at each other.[12]

Chimpanzee Culture Wars argues that cultural primatology recapitulates cultural anthropology in a dissonant key. Genealogically, both fields can be traced back to philosophical reflections on human nature. In the course of colonial conquest, the discovery that different peoples conducted their lives differently thwarted any simple answer to the question of what distinguished all humans from all other animals. When it began to dawn on primatologists that nonhuman primates also showed significant behavioral variation within their species and that this variation might be the product of social learning, the answer became more complicated still. Now all claims about human and, say, chimpanzee nature had to pass through the eye of cultural diversity.

This led to serious disagreements between chimpanzee ethnographers and comparative psychologists: whereas the former sought to explain local behaviors by comparing different field sites, the latter remained committed to extracting species universals from controlled experiments in their laboratories' "culture of no culture."[13] Yet a new generation of Japanese primatologists had left behind this opposition of laboratory and field research. They conducted fieldwork in the laboratory and experiments in the field.

Unfortunately, border crossings between human and chimpanzee life around an outdoor lab in Guinea fostered no flourishing multispecies society. All over Africa, chimpanzee communities vanished under the pressure of accelerating human population growth. Thus cultural primatology once again followed in the footsteps of cultural anthropology and became a salvage operation, frantically archiving the remaining chimpanzee cultures in the face of an anthropogenic mass extinction event. Just as cultural anthropologists have struggled to account for the loss of cultural diversity during five centuries of Euro-American domination (currently on the wane), cultural primatology is now confronted with the question of how to make sense of the eradication of nonhuman cultural and biological diversity in light of modern humans' savage success.

Contingency Table

This book is based on eight months of anthropological fieldwork among primatologists and their primates. Not all of these scientists would speak of themselves as *cultural* primatologists—especially some of the comparative psychologists featured in this book will appear as critics of cultural primatology. But they all made significant contributions to the chimpanzee culture controversy. Originally, I had wanted to confine the project to a controversy over what made us human between field primatologist Christophe Boesch and comparative psychologist Michael Tomasello, two codirectors of the Max Planck Institute for Evolutionary Anthropology in Leipzig, Germany. Their dispute quickly turned from the zoological and potentially ontological question of what distinguished our species from other animals into a bitter argument over the respective epistemological value of experiments and fieldwork.[14] From psychological tests of human children and chimpanzees, Tomasello inferred fundamental differences between *Homo sapiens* and *Pan troglodytes*. By contrast, Boesch saw in Tomasello's subjects only very young, white, middle-class Germans, who could hardly represent all of humankind, and apes whose history, captive environment, and behavior were so abnormal that they could not possibly stand in for their wild conspecifics. Maybe no group of chimpanzees could typify all others if Boesch was right that the single most important finding of the past two decades had been a pronounced behavioral diversity among chimpanzees.[15] This diversity, arguably cultural, had become an almost obligatory passage point for scientific claims about chimpanzee and human nature.[16] Although their positions and approaches were too idiosyncratic for Boesch to represent naturalistic observers and for Tomasello to represent laboratory workers tout court, their disagreement brought into relief key epistemological and ontological points of contention within the

Euro-American field: tensions between fieldworkers and experimenters looking at wild and captive apes and emphasizing human-animal continuity and discontinuity, respectively.

Against the background of a growing body of historical and sociological studies of primatology in the laboratory and in the field, my account of the chimpanzee culture controversy raises a new question.[17] Although cultural primatologists had made it their mission to document and understand geographical differences in behavior between populations of wild chimpanzees, they did not extend their fieldwork to the laboratories of comparative psychologists. But wouldn't it be possible that laboratories, just like field sites, fostered their own chimpanzee cultures, or, perhaps, *chimpanzee-human* cultures? And if so, wouldn't they require new forms of laboratory ethnography as well as ethnological comparisons between labs?

I soon added a second axis to my research design and included Japanese primatologists. For the sake of comparison with Euro-American field and laboratory research, I planned to look at both field and laboratory research in Japan. Eventually, I worked with Tetsuro Matsuzawa, who is one of the few primatologists doing both. "Matsuzawa's way is unique because it is a holistic approach. We Japanese love to approach things holistically," Matsuzawa told me in our very first Skype conversation. "I don't like to see broken pieces of chimpanzee but want to know the chimpanzee as a whole. That is why I'm doing captive *and* field studies." In contrast to the situation at Leipzig, no epistemological divide set up chimpanzee ethnography against laboratory experiments at the Kyoto University Primate Research Institute (KUPRI). Instead, Matsuzawa had crafted a chain of translations between rigidly controlled experiments, field observations and participant observations in his indoor laboratory, field experiments in an outdoor laboratory in Guinea, and field observations in the West African forest.[18]

Although, methodologically, Matsuzawa's research on chimpanzee culture and cognition remained thoroughly in the realm of the natural sciences, his synthesis of benchwork and fieldwork explored an ontological territory beyond nature and culture, at least as Europeans and Americans had understood these categories in the nineteenth and twentieth centuries. This ontology materialized in the microcosms of his laboratory in Inuyama, Japan, and his outdoor laboratory in Bossou, Guinea. The doyen of Japanese primatology after Imanishi made no effort to tease apart nature and human culture. In the basement of KUPRI, captive chimpanzees interacted with touchscreens, using Japanese kanji and Arabic numerals, while their wild cousins sat right behind a Manon village under the watchful eyes of Japanese primatologists and their camcorders, cracking oil palm nuts with hammers and anvils. Whereas Boesch believed that chimpanzee nut cracking belonged to a genuinely wild culture, which the apes had developed on their own account, Matsuzawa speculated that the Bossou community might have originally learned this use of tools from the local human population. Time and again, the relationship of nature, culture, and varieties of the hybrid spaces Haraway dubbed "natureculture" proved a divisive issue in the chimpanzee culture controversy.[19]

As a first approximation, this study of cultural primatology amounts to a double comparison between field and laboratory as well as between Japan and Europe. Matsuzawa loved such 2 × 2 contingency tables, and my research design imitates cultural primatologists' controlled comparisons between primate cultures. It sets side by side Euro-American fieldwork and Euro-American laboratory work, Japanese fieldwork and Japanese laboratory work, Japanese laboratory work and Euro-American laboratory work, and Japanese fieldwork and Euro-American fieldwork. Social anthropologist Fred Eggan explained the rationale behind controlled comparisons: in a field science that does not allow for experimental controls, juxtaposing geographically and ecologically proximate cultures sheds more light on the few ways in which they differ than juxtaposing far-apart cultures that differ in almost every respect.[20] The method of controlled comparison diverges sharply from the Romantic quest for radical otherness that inspired anthropologists to contrast Western culture with Amazonian or Melanesian cultures.[21] Pamela Asquith's original comparison of Japanese and Western primatology could be read as presenting such stark cultural alterity in the realm of science.[22] But she also looked at two adjacent knowledge cultures within primatology: belonging to the same scientific discipline, Japanese and Euro-American monkey and ape researchers were located in what Eggan would have called the same culture area. Of course, the researchers' ethnicity was hardly the only pertinent difference. For example, Imanishi's Kyoto School focused on fieldwork, which makes it hard to tell whether dissimilarities with Euro-American laboratory research are due to national culture or methodology. That's what the second axis of comparison might help us understand.

Of course, all control is relative. There are always more axes of comparison to add. If we aspired to halfway certain knowledge, we would also have to control for the scientists' disciplinary training in biology or psychology, the ontological commitments informing their research questions, the relations they developed with their nonhuman subjects, and so forth. Some cultural anthropologists inferred from the uncontrollable complexity of the field that they had better abandon comparative approaches altogether. That's why we currently see an abundance of ethnographies and very little ethnology systematically surveying this rich body of case studies.

As far as *Chimpanzee Culture Wars* is concerned, organizing the book in the form of a controlled comparison serves primarily as an experiment in reflexivity, which probes the relationship between my own particularist tradition of anthropology and the history of science and the more systematic and generalizing tradition of cultural primatology. I adopted the method of comparison not to extract law-like regularities from a number of case studies but to map a space of possibilities. As we look at different actors, the question is how and why they realized the possibilities they realized and how these possibilities could be recombined to allow for new knowledge cultures, maybe even new human-chimpanzee cultures.

Yet necessity always casts its shadow over both human and chimpanzee potentials. The possibilities of any epistemic culture are limited by its objects. Their firmness delimits the scope for alternative conceptualizations. Chimpanzee behavior might be cultural but it is not infinitely plastic, and it determines much of what researchers can and cannot do, especially in participant observation. Finally, the course of history threatens to foreclose much of what became possible in the late twentieth and early twenty-first centuries. At least in the wild, cultural primatology will vanish with the last primate cultures.

What This Book Is and Isn't

My ethnographic approach to chimpanzee culture research provides a liveliness and detail that literature- and interview-based studies in the history and sociology of science cannot deliver. This quality comes at a price, though. It was not possible to represent the work of scientists at this high level of granularity unless they cooperated and invited me to their laboratories and field stations. Thus, *Chimpanzee Culture Wars* provides an uneven account of the controversy around ape cultures, very much skewed toward the researchers who allowed me into their groups' professional lives. Key figures such as Andrew Whiten, Frans de Waal, and Michio Nakamura will not receive the space they would deserve in a controversy study that treated all actors equally, or at least relative to the weight of their scientific contributions, because they did not grant access to their research facilities. Of course, I will discuss their work where appropriate, especially in the predominantly historical chapters 1 and 2, but it will be for future scholars to take a closer look at these players.

Just as *Chimpanzee Culture Wars* is no sociological controversy study, it is no primatological review of the literature on chimpanzee and other primate cultures, either. It will not survey the broad and ramified array of questions that cultural primatologists and comparative psychologists have debated: how to define culture; whether to distinguish culture from preculture; whether to dismiss the attribution of culture to nonhuman primates as anthropomorphism; whether to define culture in terms of geographical differences in behavior, biological function, or psychological mechanism; whether all kinds of social learning or only true imitation can produce culture; whether culture has to be cumulative; whether it has to be symbolically and even linguistically mediated, or whether symbolic culture and language are only special cases of culture; whether culture requires social norms and how to define a social norm; whether culture is always adaptive and, if so, what adaptive value social learning has; whether culture constitutes a realm of freedom from biological necessity; whether it is about survival or a sense of belonging; whether culture can or even has to be distinguished from genetics and ecology; whether socially learned behaviors might alter the environment in such a way that the modified environment exerts selective pressure on genetically inherited traits; whether social structure and group character influence social transmission; whether personality, emotions, and the quality of social relationships affect the probability of social learning; whether age, rank, or sex determine whose newly acquired behaviors will spread through a group; whether female migrants enable cultural exchange and diffusion of new traits between chimpanzee communities; whether wild chimpanzees learned certain traits such as nut cracking from observing humans; how cultural traits are formed and maintained; how to demarcate one cultural trait against another; whether human and chimpanzee culture evolved independently or from an already cultured common ancestor; whether culture is uniquely human, limited to primates, or a behavioral feature widely shared across the animal kingdom; whether field observations provide evidence about the learning mechanisms through which newly acquired behaviors are passed on in a group; whether new statistical methods allow causal claims to be derived from field observations; whether experiments on captive animals can prove or disprove the cultural capacity of their entire species; whether experimenters can expect chimpanzees to socially learn from human models, or whether chimpanzees have to be presented with conspecific models; whether humans and chimpanzees have to be tested under the same conditions; whether better experimental designs or the integration of laboratory and field research can put an end to the chimpanzee culture controversy, and so forth. Most of these questions will be addressed in the course of this book, some in passing, some in great detail. But I will not provide a systematic overview. Any reader looking for this can consult numerous monographs, edited volumes, book chapters, and review articles on the subject.[23]

Chimpanzee Culture Wars does not belong to this vast body of literature. It is first and foremost an ethnographic essay about alternative ways of looking at human nature and primate cultures. As an ethnography in the interpretive tradition, it is subject to the limitations of this peculiar way of writing culture, human and otherwise. But it also takes advantage of the genre's unrivaled possibilities, providing a firsthand account of some of the main characters, research sites, and scientific practices of cultural primatology. As an essay, this book attempts to understand the vitriolic war of words over chimpanzee cultures through its protagonists' many attempts to determine what distinguishes us from our nonhuman cousins. It seeks to extend the ethnographic material, including the primatologists' findings, in directions not developed in the scientific literature on chimpanzee cultures. The goal is to tear this material out of its customary frames of reference and to look at it in unexpected contexts. This recontextualization does not aim at synthesis but at exploring and exploiting perspectival differences. Instead of surveying the entire field of cultural primatology, I sought ethnographic interlocutors to articulate and work through questions of my own. Some of these questions were shared by the people I worked with, while others allowed me to look at their research from new angles.

I develop philosophical ideas more through stories than arguments because I'm interested in how they translate and are translated into experiences. Occasionally, these stories invert or at least query the moral or epistemic value of an idea to test the consequences. Austrian writer Robert Musil remarked that, unlike scientific publications that aim at knowledge, the essay seeks to transform human beings through a "reforging of a great complex of feeling (most penetratingly imaged in Saul's becoming Paul) . . . , so that one suddenly understands the world and oneself differently."[24] This brand of essayism can thrill readers but it also risks irritating them. I expect that cultural anthropologists, especially, will take issue with the subtle and not so subtle challenges to some of the dominant value judgments that pervade much contemporary humanities and posthumanities scholarship such as the condemnation of positivism or human exceptionalism. Cultural primatologists might join them in taking umbrage at views like my plea for fatalism in the face of an unprecedented anthropogenic mass extinction event. But first and foremost, the scientists in the readership will notice that the ethnographic essay represents a humanist style of thought that is very different from their own ways of writing. I do not expect readers to agree with any of my efforts but hope that whatever unease and opposition this book provokes, it will help them confront their stakes in the problem space we are about to explore together.

Two Cultures Still

Here is a conundrum for future historians of science: in the late twentieth century, humanities scholars, social researchers, and natural scientists began to experience and promote the collapse of the dichotomies of nature and culture, nature and society, nature and mind, and nature and the human. Since the late nineteenth century, this series of ontological oppositions had organized the disciplinary landscape of universities across the globe, breaking up academic knowledge production into natural and social sciences, *Geistes-* and *Naturwissenschaften, sciences humaines* and *sciences naturelles*. It is true that even in the heyday of this ontologically based division of academic labor, hybrid research fields such as anthropology escaped the clear-cut separation of the two cultures without resolving the problem of how to categorize them. In the twentieth century, however, a countermovement challenged the trend toward disciplinary differentiation and instead propagated interdisciplinarity.[25] Researchers from both sides of the great divide rushed in, trying to occupy what could, in principle, have become a new common ground.

In the 1980s, cultural anthropologists and science studies scholars distanced themselves from culture and society and attributed the opposition of their disciplines' organizing concepts to nature to a historically and culturally contingent cosmology now on the wane.[26] Some even considered this dualist worldview an illusion to be replaced by a more accurate ontological vision of the world as teeming with naturecultural, biosocial, and human/material hybrids.[27] At the same time, the ascending neurosciences advocated the reduction of mind to brain more powerfully than ever, and their philosophical allies envisioned a "unified science of the mind-brain."[28] Euro-American sociobiologists explained the social life of humans and other animals biologically and advocated the integration of the human sciences in a new synthesis of genetics and evolutionary theory.[29] Biological anthropologists expanded the realm of culture far beyond the human, from apes, whales, and dolphins all the way to crows and guppies, and proposed a "unified science of cultural evolution."[30] "It was probably under the influence of ethology, in particular that of the great apes, that modern ontology began to waver once one of its most generally recognized principles was called into question: namely the absolute uniqueness of humans as a species capable of producing cultural differences," noted anthropologist Philippe Descola.[31]

What's puzzling about this historically and geographically protracted metaphysical transformation is that although natural scientists and humanities-oriented social scientists have begun to occupy the new ontological borderland, little epistemological reconciliation or cross-fertilization has occurred. To be sure, the gap between the "two cultures," which Charles Percy Snow had already lamented a few decades earlier, was no longer the same, but it had not become any less divisive.[32] Of course, we could work out a more fine-grained taxonomy of epistemic cultures, which would map many more than two. But most actors in this intricate field continued to understand themselves as humanities scholars, social researchers, or natural scientists. Although the primatologists at the ethnographic heart of this book studied culture, they self-identified as natural scientists. Boesch referred to Tomasello's group of experimental psychologists as *Geisteswissenschaftler*, but that was to explain why cooperation with them had turned out to be so difficult. It reflected the sense that lack of mutual understanding and appreciation remained especially pronounced between those descending from the natural sciences and those descending from the humanities. While social anthropologist Tim Ingold said that primatologists likening their field studies to ethnography did not know what they were talking about, cultural evolutionist Jamshid Tehrani dismissed Ingold and Palsson's volume *Biosocial Becomings: Integrating Social and Biological Anthropology* as "written by social anthropologists for social anthropologists."[33] A prominent cultural primatologist to whom I mentioned a special journal issue on multispecies studies snapped: "Please tell me the name of this journal, so that I may avoid it." Although the filling of the ontological chasm did engender new epistemologies and methods, it did not bring peace and fruitful collaborations to those who had endured or incited the sociobiology wars, the culture wars, and the science wars over the course of four decades. Why are we still living in two epistemic cultures?

My own response to this puzzle has been to engage ethnographically with primatologists who occupy positions at the very center of their field. They have kept their distance from posthumanist attempts at integrating animal minds, societies, and cultures under the terms of the humanities and interpretive social sciences.[34] Instead of denouncing the biologists' approach as not being "genuine" ethnography and accusing them of "abusing" the term, as Ingold did, I followed Christophe Boesch and his coworkers to Taï National Park to familiarize myself with how they actually studied chimpanzee lifeways.[35] In the same spirit, I observed how Tetsuro Matsuzawa and the researchers around him studied chimpanzee culture by integrating laboratory and field experiments as well as field observations and what they called participation observation. In the face of widespread condemnation of the naturalist and cognitivist traditions these scientists represent, I set out to understand their knowledge cultures. While interpretive cultural anthropologists have long spurned positivism as a naive misconception of science as steadily progressing on the basis of theory- and value-free observations, cultural primatologists continue to aim for just that. Although I share many philosophical objections to this philosophy of science, its persistence awakened my ethnographic curiosity. I will play devil's advocate in the humanities and posthumanities and defend the scientists' pursuit of such regulative ideals, even if they will never fulfill their own aspirations.

As I peer across the two-culture divide, I will frequently compare cultural primatology with my own home discipline, although I have never conducted a formal study of the knowledge culture of cultural anthropology. Just as many anthropologists contrast a non-Western culture they studied ethnographically to "the West," which they did not study but know intimately as natives of

Europe or North America, I compare chimpanzee ethnography, Japanese participation observation, and many other primatological research practices with research practices in my own field on the basis of my experience as faculty in an American cultural anthropology department. Matei Candea distinguishes such "frontal comparisons" between us and them from "lateral comparisons" between them and them.[36] My comparisons between Euro-American and Japanese primatologists and between primatological laboratory workers and fieldworkers are lateral, and my comparisons of these ethnographic subjects with my own humanist hinterland go full frontal. For obvious reasons, I have more stakes in the frontal comparison than in the lateral ones, and readers will note polemic undertones—aimed mostly at my own field (disciplinary chauvinism is not among my epistemic vices). I hope that bringing the often ill-articulated norms and forms of our respective subfields into relief will reinvigorate conversations between cultural and evolutionary anthropologists, largely abandoned since the 1980s.

From Second-Order Primatology to the Hominoid Condition

This book is about primate culture in the culture of primatology. But the term *culture* might be used differently when applied to primates and primatologists, respectively. Swiss comparative psychologist Thibaud Gruber and others suggested that "apes have culture but do not know that they do."[37] Lacking the cognitive capacity for representations, they may neither understand that they or others hold beliefs about their cultures nor notice that they do and see things one way while other groups do and see them another way. By contrast, humans, especially since the late eighteenth century, have grown exceedingly aware of such differences. Far beyond Europe have they come to talk about these differences in terms of culture and its translations.

Cultural primatologists are a product of both the evolution and history of cultural thinking. Not only did they apply the culture concept to nonhuman primates; they also applied it to themselves. Dutch ethologist Frans de Waal, for instance, compared Western and Japanese primatology to explore the role of "cultural bias" in science: "What we discover in nature is often what we put into it in the first place."[38] Consequently, de Waal maintained, "whether we grant animals culture is ultimately a human cultural question."[39] When science studies scholar Donna Haraway presented primatology in such constructivist terms a decade earlier, de Waal's colleagues took her to task.[40] Had the culture concept since served as a Trojan horse, smuggling cultural relativism into the predominantly realist citadels of science?

Culture, at least in its modern form, provided "a perspective for the observation of observers," argued German sociologist Niklas Luhmann.[41] He called such observations of observations second-order observations. While first-order observers attend to *what* is (or should be) the case, second-order observers attend to *how* first-order observers arrive at such determinations. Instead of observing the world, second-order observers reveal the contingency of any observation of the world: a different perspective is always conceivable.[42] If Gruber and colleagues were right about the absence of metarepresentations in apes, culture as second-order observation would be the prerogative of *Homo sapiens*.

Although cultural anthropologists, sociologists, and historians institutionalized second-order observations, natural scientists also occasionally raise their eyes from their scientific objects to observe how colleagues have come to see the same objects differently. They often think that "differently" means "wrongly"—and sometimes they are right. At other times, differences in perspective result from the fact that research findings can't be untangled from research practices or depend on contingent definitions of concepts such as culture. For instance, there is no truth of the matter of whether culture must comprise metarepresentations. Thus disagreements arise and second-order observations abound. And so it happened in the controversy over primate culture.

Paul Rabinow declared the distress and dispute growing out of the inevitable plurality of positions in reasoned discourse—he spoke of "the apparently unavoidable fact that *anthropos* is that being who suffers from too many *logoi*"—the starting point of an anthropology of reason.[43] While nonhuman primates might not share this predicament, primatologists certainly do. As primates passionately interested in how other living beings, including our conspecifics, do and see things, even the nonprimatologists among us might profit from applying the estrangement effect produced by second-order observations to primatological knowledge. Through the medium of the history and ethnography of science, *Chimpanzee Culture Wars* raises the question of a second-order primatology that reflects on the vexing profusion of Logoi through which we have come to understand ourselves as cultured apes.

Yet my account won't take the form of a traditional controversy study. In contrast to Amanda Rees's *The Infanticide Controversy: Primatology and the Art of Field Science*, it does not maintain a strict separation of sociological and primatological truth claims. As a sociologist of scientific knowledge committed to the principle of empirical relativism, she remains agnostic about how primatologists describe primate behavior.[44] When Rees opens her book with the admission "I have never been to the field and have never seen wild primates with my own eyes," she asserts herself as a highly disciplined

FIGURE 2. Second-order primatology: observing observers as they observe chimpanzees. Photo by author.

second-order observer who kept her eyes on those first-order observers she read and interviewed without getting distracted by their monkeys.[45] The same could be said about the methodology of historians of primatology like Gregory Radick, Marion Thomas, and Georgina Montgomery.[46] My own approach is situated in anthropology as a field that has always studied human cultures and nonhuman primates. In science studies, it owes more to laboratory ethnography than to controversy studies in that it pays attention to both humans and nonhumans.[47] In the field, you can't help seeing and even interacting with both. Considering that primate societies are also societies and that primatologists are primates, too, some of the most challenging philosophical questions arise where the disciplinary boundaries between primatology and the human sciences become porous.[48]

Boesch once responded to my overbearing questioning: "You're listening to me and believe any rubbish I'm telling you. You are a type of person who has learned to learn conspecifically rather than socially." He took my trusting readiness to learn from other humans in general, rather than from members of my own social group in particular, as a behavioral trait that distinguished *Homo academicus* from both wild chimpanzees and the Hadza foragers he had observed in Tanzania. I had never met any Hadza people, and as far as *Pan troglodytes* was concerned, Boesch's claim was a matter of controversy, but he had a point about me.

For I agree with Tim Ingold—who happens to be one of the fiercest critics of chimpanzee ethnography—that anthropology is not a study *of* people but a study *with* people that aims at learning to see things in the ways our interlocutors do.[49] Of course, unless anthropologists are prepared to adopt contradictory beliefs, they can't possibly espouse everything that people tell them, especially if their interlocutors disagree with each other—and disagree they did in the chimpanzee culture controversy. Controversies shift the focus from *what* is the case to *how* different people determine what is the case. Sociologists of scientific knowledge follow this transition from first- to second-order observations on a one-way ticket. By contrast, I am interested in the back-and-forth between observing the world and observing (and listening to) other observers as they are observing the world. I set out to think about chimpanzee cultures in the Anthropocene alongside people who had departed from different quarters of evolutionary theory to study the inhabitants of a cosmological terra incognita that lies in front of all of us. My brand of anthropology is philosophical but in a very different way than philosophical anthropology. It uses the tools of ethnographic fieldwork to the end of reflecting from different angles on the human, nay hominoid condition today.

Bird's-Eye View

Cultural primatology was born in mid-twentieth-century Japan. Chapter 1 examines this prehistory of the chimpanzee culture wars. The story begins in 1948 with the observation of a troop of Japanese macaques on a subtropical islet. As Imanishi and his students fed them sweet potatoes on the beach, the monkeys invented a way of washing off the sand in the sea. Subsequently, they passed on the new behavior from generation to generation. The Japanese primatologists conceived of this social transmission as preculture. They framed their anthropomorphic conceptualization, as well as a research practice that made no effort to minimize human interference, in terms of Japanese culture. Soon Imanishi's anti-Darwinian evolutionary theory became engulfed in national and international controversy over its association with nationalist politics and its breach of the divide between science and the humanities. Ironically, this self-consciously Japanese brand of scholarship had not only appropriated European and American elements of evolutionist thought, including the idea of animal traditions, but also stirred up a controversy over primate cultures that polarized primatology far beyond the boundaries of national cultures.

The second chapter offers a historical account of the chimpanzee culture wars. In 1978, William McGrew's description of a cultural difference between the chimpanzees of Jane Goodall's field site at Gombe and Junichiro Itani's field site at Mahale sparked a scientific controversy with shrill political overtones. Scientifically, the debate pitted field biologists against experimental psychologists, and proponents of human-animal continuity against those interested primarily in what set *Homo sapiens* apart from other primate species. Chimpanzee ethnographers struggled with the question of whether they had to exclude genetic and ecological explanations to determine the cultural nature of behavioral differences. Politically, challenges to the dualist ontology of one nature and many human cultures became part of a much larger battle over biological humanism that had emerged after World War II to rein in the excesses of Nazi racial ideology. In the culture wars, a new ethnos-centered left broke with this false universalism in the name of multiculturalism. In the chimpanzee culture wars, cultural primatologists suggested extending multiculturalism beyond the human. But just as they adopted the culture concept, many cultural anthropologists decided in the late 1980s to write against culture. They now suspected that talk about cultural differences served the very racist "othering" that it had been meant to replace. Alienated by the moralism of their colleagues in the humanities, cultural primatologists continued their exploration of chimpanzee cultures in a positivist vein.

Chapter 3 takes a close look at how chimpanzee ethnographer Christophe Boesch and his group studied the social transmission of cultural traits in the chimpanzee communities of Taï Forest, Côte d'Ivoire. This ethnographic account of primatological fieldwork in the mid 2010s measures the historical distance to the 1960s when Goodall and others sought to take part in the social life of great apes. In contemporary Taï, by contrast, disengaged observations of habituated chimpanzees served to protect both *Pan* and *Homo*. Despite the researchers' efforts to keep human-animal relations as neutral as possible, different chimpanzee communities related to their observers differently. Whether behavioral differences were considered cultural and whether their observation was ethnographic would be revealed only retrospectively as data analysis ruled out alternative accounts. In the forest, chimpanzee ethnography could hardly be distinguished from other forms of fieldwork. The collective empiricism of cultural primatologists aspired to a separation of observation from theory and interpretation. Humanities-oriented anthropologists and even some primatologists who were committed to the epistemic virtues of thick description dismissed such an ethnography by ethogram as an abuse of the term *ethnography*. But Boesch's approach to writing wild cultures turned out to share another important feature with humanities scholarship: references to philosophical classics gave it an intensely polemic bent rarely found in the scientific literature.

From the Ivorian rainforest, the reader will accompany the primatologists back to the Max Planck Institute for Evolutionary Anthropology in Leipzig, which had reintroduced a new biological anthropology to a country that still associated the discipline with Nazi racial theory. Chapter 4 examines how Boesch's colleague and codirector Michael Tomasello derived truth claims about the anthropological difference between *Homo sapiens* and *Pan troglodytes* from controlled experiments comparing the social cognition of human children with that of grown chimpanzees. Tomasello's claim that humans were the only primates capable of culture and cooperation received an enthusiastic reception by German philosophers. The Frankfurt School and a new generation of philosophical anthropologists had finally found an ally in their battle against neuroreductionists and genetic determinists. Yet Boesch called into question the validity of Tomasello's findings by pointing out that the social behavior of both humans and apes was too contingent on local circumstances for Leipzig kindergarten children and zoo chimpanzees rescued from a Dutch pharmaceutical company to represent all of humanity and chimpanzeehood. He accused Tomasello of not controlling for the different conditions under which Tomasello tested humans and apes. The ensuing controversy over the relationship between laboratory work and fieldwork happened at a time when new statistical methods were opening up vast new possibilities for chimpanzee ethnography, even fostering hopes that experimentation with captive animals would become superfluous because uncontrolled observations in the wild would allow the establishment of causal relations. Could Boesch's cultural primatology inform a different philosophical anthropology than the one drawing from Tomasello's comparative psychology?

In the debate with Tomasello, Boesch pointed to Tetsuro Matsuzawa's experiments at the Kyoto University Primate Research Institute as a model for testing humans and chimpanzees under the same conditions. The fifth chapter explores how *Pan* and *Homo* came to share a life in this Japanese laboratory. Matsuzawa's Ai Project, named after his most famous nonhuman "research partner," began as the Japanese ape language project in 1976. But it soon morphed into a much broader comparative cognitive science program that provided a new face to Japanese primatology in the post-Imanishi era. For both methodological and ethical reasons, he sought to square tight experimental control with maximizing the captive chimpanzees' freedom to show spontaneous behavior. Field observations in the laboratory captured such unexpected actions, while participant observation enabled social learning across species boundaries. Researchers introduced the laboratory animals to stone tools from their wild relatives to study chimpanzee cultural cognition under controlled laboratory conditions. But these unconstrained face-to-face interactions between primates and primatologists in an experimental booth also provided a so-called cultural correction device that put the scientists' interpretations of ape behavior to the test without experimental controls or statistical analyses: if they misread chimpanzee minds, they risked serious injury. The often violent nature of chimpanzee social life strictly limited how much humans could take part. At Kumamoto Sanctuary, however, younger researchers from Matsuzawa's lineage had repurposed participant observation to apply delicate measuring instruments such as EEG caps or head-mounted eye tracker goggles to otherwise unruly apes. Thus a new generation of Japanese primatologists integrated high-tech laboratory experiments, field observations in the laboratory, and participant observation—and eventually extended this synthetic primatology to the field.

Chapter 6 follows Matsuzawa and his coworkers to their outdoor laboratory in Bossou, Guinea. Revered as the totem animal of the Manon and deprived of almost all primary rainforest, the Bossou chimpanzees had learned to live on human crops in an agricultural landscape. In contrast to Boesch's emphasis on so-called wild cultures, Matsuzawa speculated that historically, this chimpanzee community might have learned from the human population how to crack the oil palm nuts that local farmers cultivated. Field experiments allowed the primatologists to study how female immigrants passed on their knowledge of how to crack other kinds of nuts within the group. At this point, Japanese cultural primatology contradicted the Manon's mythological understanding of "their" apes as a bounded community of nonnatural animals. Chimpanzee road crossings provided an opportunity for a natural—or really "naturecultural"—experiment in an anthropogenic environment. Ethnoprimatologists collaborating with Matsuzawa studied the ecological interface between humans and primates and used their insights for conservationist ends. After a political conflict over the protection of a small patch of primary forest on a sacred hill, the Japanese primatologists took over the Manon's position that the livelihood of the Bossou chimpanzees was better served by plantations than by a nature reserve. And yet amid aggravating conflicts between humans and chimpanzees, the numbers of the latter were in free fall.

The extermination of each chimpanzee community not only endangered biodiversity; it also diminished chimpanzee cultural diversity.[50] In the footsteps of cultural anthropology, which had sought to catalog disappearing human cultures almost since its inception, chimpanzee ethnography began to do the same for chimpanzee cultures. The last chapter examines how cultural primatologists fought for the conservation and documentation of quickly dwindling chimpanzee communities. As *Homo sapiens* outcompeted *Pan troglodytes* in sub-Saharan Africa, Boesch, Matsuzawa, and the first generation of African primatologists forged alliances and made enemies in their attempts to prolong the coexistence of humans and apes for a few more years. In West Africa, the chimpanzee culture wars entwined scientific controversy with political crises and ethnic strife. In the face of the sixth mass extinction event in natural history, cultural primatologists set out to collect as much data about the lives of as many chimpanzee communities as possible. Boesch's Pan African Project almost quadrupled the number of documented chimpanzee cultures by switching from close-up ethnographic observations of habituated groups to big data collection of camera trap recordings, fecal samples, and material artifacts. This collective effort at over thirty-five field sites built up an archive for future primatologists who might no longer have a chance to experience chimpanzee cultures firsthand. Epistemologically, the effort to build such an archive was based on an elegiac positivism, which minimized the theory-laden nature of recorded observations. For nobody knew what theoretical commitments would orient the work of its users. A sense of human guilt and anticipatory grief for the demise of our own species marked this salvage primatology.

The conclusion revisits the questions raised in this introduction in light of the ethnographic chapters. It compares the knowledge cultures of Boesch's field station, Tomasello's laboratory, Matsuzawa's laboratory, and Matsuzawa's field station with each other to map a space of no-longer-available possibilities. By the time of this book's publication, the scientists who created these sites of learning with their unique styles of inquiry will all be retired and a new generation will have taken over. In the face of pleas for unifying the so-called two cultures of the sciences and the humanities, I espouse the conservation and promotion of a much broader epistemic diversity, which must include voices that call into question the value of such epistemic diversity.

Not bound by the interpretive modesty that characterizes positivist knowledge making, the epilogue takes inspiration from Imanishi's speculative natural history to imagine what a joyous primatology, unburdened by the contrite Christian opposition of humanity to nature, might look like. Would it be possible not to cast the extinction of chimpanzee cultures as moral failure?

1

The Birth of Cultural Primatology from the Spirit of Japanese Uniqueness

WHEN ANTHROPOLOGIST OF SCIENCE Sharon Traweek studied American and Japanese physicists in the mid-1970s, her interlocutors represented their world as a "culture of no culture," whereas she noticed culturally distinct styles of research.[1] Like the great apes in the article by Thibaud Gruber and colleagues, these scientists seemed to have cultures without knowing that they did.[2] This chapter presents a very different culture of science, one that is self-consciously cultural.

Japanese primatologists who understood their primatology as specifically Japanese have been an object of anthropological study since Pamela Asquith began to observe and engage with the Kyoto School in the late 1970s.[3] She belonged to the first generation of cultural anthropologists who turned their ethnographic gaze to science. One of their goals was to show that present-day European and American ways of looking at the world weren't the only rational options. They showed that nature did not speak for itself, but different groups of people spoke for it, and these people offered different accounts. The anthropologists' method of cultural and historical comparison served not to extract general historical laws but to discern possibilities.

Japan proved a powerful point of comparison because it was modern and yet radically different from the West. A highly industrialized country with internationally competitive research facilities, it had developed largely apart from Europe and North America. Soon Western anthropologists flocked in to demonstrate that disciplines such as genetics, medicine, or physics took one form in France or the United States and quite another in Japan.[4] In the case of primatology, Asquith found that its Japanese variety differed from the Western branches of the discipline in its acceptance of human-animal continuity and an unrepentant fabrication of anthropomorphisms such as the idea of monkey cultures.[5]

In retrospect, however, it seems safe to say that cross-cultural comparisons of different scientific disciplines have proved national cultures to be a poorly delimited unit of analysis for the study of an enterprise as cosmopolitan as twentieth-century science. Although Traweek wanted to show that American and Japanese high-energy physicists thought very differently, she had to admit that "what I will describe as the dominant 'Japanese' style also exists in the United States; and the prevailing style in the United States most certainly also exists in Japan."[6] As Asquith moved from ethnographic to archival work on Imanishi's evolutionary theory, she discovered that "contrary to a . . . cultural underpinning to his ideas . . . , there was instead an almost seamless fusion of Western and Japanese science."[7] How Japanese is Japanese primatology after all?

This chapter provides an account of the birth of cultural primatology. It examines how self-consciously Japanese primatologists proposed and provided evidence to support the idea that nonhuman primates had culture, too. It departs from my own visit to Koshima, a subtropical islet where this research began in 1948, which will eventually serve to mark the historical distance between Japanese primatology then and now. Yet the bulk of my account is based on a review of the available literature. It is the product of second-order observations, which enabled me to decompose and recompose the narratives surrounding the Japanese origins of cultural primatology. This retelling provides the backstory to the ethnographic chapters that follow.

While the culture concept enabled second-order observations of various sciences and provided a rationale for cultural comparisons between knowledge cultures in Japan and the United States, early anthropology of science approached culture itself in the realist fashion of first-order observations: it assumed that there was such a thing as the culture of Japanese primatology, American physics, or a culture of no culture. By contrast, *Chimpanzee Culture Wars* extends the practice of second-order observation to the concept of culture itself as it traveled between disciplines and countries.

This shift from first- to second-order observations of culture brings out a different story. While cultural primatology was clearly born from the *spirit* of Japanese uniqueness, it was not uniquely Japanese. Instead, cultural primatology emerged as the product of a lively exchange of knowledge across Japan's borders. The culturalist packaging of the discovery of social learning in Japanese macaques and its framing in an anti-Darwinian evolutionary theory rich in nationalist undertones might have provoked more controversy and polemics than the supposedly Japanese penchant for anthropomorphism ever did. We will see how, in the 1980s, a new generation of Japanese primatologists broke with Imanishiism and sought to restore a cosmopolitan conception of science, not least by offering an alternative history of their field that emphasized transnational flows of knowledge. This narrative prepares the ground for chapters 4 and 5, which feature the work of Tetsuro Matsuzawa, who would become the most prominent representative of this generation—even though he chose not to abandon but to reinterpret the spirit of Japanese uniqueness.

Koshima: Myths and Macaques

Behind a curtain of warm rain on the steep forested slopes of Koshima, a bright red gate marked the entrance to a small bay. Such torii gates symbolized a crossing from the world of human dwellings and cultivated lands to the realm of wild animals and untamed spirits—or the other way around, from a nature inhospitable to humans to one made to serve our needs. The Koshima gate, however, admitted the visitor to a place where the opposition of undomesticated nature and human culture itself had been called into question. On the islet, Japanese macaques had provided the first evidence for a nonhuman culture. The idea of a monkey culture would not only shake the Western oppositions of nature and culture, humanity and animality, as Michel de Pracontal proposed, but would also add a new entry to the ontological inventory of Japan.[8] Some commentators considered the observations of social learning in *Macaca fuscata* by Imanishi and his group in 1953 so momentous that they declared Koshima the scene of a scientific revolution.[9]

As a fisherman ferried us over, the postdoc now running the research station related that the monkey troop had descended from two animals that had served a young princess. The princess's father had confined her to this island after his daughter had fallen in love with a man she was not supposed to marry. Since the lonely woman's prayers had cured many villagers on the mainland shore of their afflictions, locals honored her by building a Shinto shrine on this volcanic rock formation. Only three hundred meters separated the islet from the shore. Every decade or so the water receded far enough for a sandbank to emerge, enabling migration from and to Koshima, diversifying the gene pool of its legendary monkey population.

Koshima lay off the coast of Kyushu, the southwesternmost of Japan's four main islands. Here, the boundary between myth and history became blurred, as French anthropologist Claude Lévi-Strauss noted after a visit in 1985.[10] Japan has often been presented in such exoticist terms as a place escaping the supposedly Western oppositions of nature and culture, human and animal, subject and object, or science and spirituality. Danish science studies scholars Casper Bruun Jensen and Anders Blok, for example, described Japanese primatology as a Shinto- and Buddhist-infused form of "techno-animism."[11] The islet of Koshima appeared like a manifestation of this hybridizing cosmology, placing side by side a temple and a research site, an abandoned field once cultivated by the villagers and a wild forest cut across by electric wires. The wires led to a decaying emergency call tower from a time before scientists brought their own cell phones. And then there were the cultivated monkeys. They had not only adopted and passed on new behaviors in response to the primatologists' interventions but, like all animals, they were conceived of by Japanese Buddhists as bearers of souls transmigrating between different species and even between animals and gods.[12]

This metempsychosis doctrine fostered a sense of spiritual connectedness with other forms of life and consequently the acceptance of evolutionist thought in Japan, the postdoctoral researcher explained to me. After all, Buddha had been an elephant before becoming human—a thought that would have horrified Christians who held that only human beings had been endowed with souls. In Europe, anthropomorphism had been a theological sin long before it became a scientific one, as historian of science Lorraine Daston pointed out. Originally, medieval theologians had been less concerned about the application of human categories to animals than to God and the angels.[13] But they already recognized anthropomorphism as both problematic and inevitable. The Buddhist doctrine of rebirth in the form of demigods, humans, animals, and ghosts raised no such concerns. Consequently, the postdoc told me, "researchers treated the macaques like humans."

However, her culturalist account of Japanese primatology left unanswered why, before the scientists' arrival in 1948, hunters who were just as Japanese as the primatologists had shot the Koshima macaques to sell their body parts as Chinese medicine, and why they had trapped live animals as pets for American occupation forces. As both Euro-American and Japanese scientists and scholars attributed the embrace of human-animal continuity in Japanese primatology to Buddhist tradition, the Japanese authorities culled about ten thousand macaques per year as part of pest control programs. Biomedical researchers regularly sacrificed monkeys in their laboratories—followed by a Buddhist memorial service to give thanks and ask the dead animals for forgiveness.[14]

Such rough treatment of creatures with whom we might not just share a last common ancestor some twenty-five million years ago, but who might have provided a new body to the soul of a recently deceased human being, appears paradoxical. Japanese primatologist Masao Kawai, who had studied the Koshima troop under the tutelage of his mentor Kinji Imanishi, sought to resolve the contradiction by comparing the brutalization of animals to domestic violence in Japan. If Japanese children threw stones at an animal and bullied it, he explained to anthropologist Pamela Asquith in the late 1970s, it was not like a naughty boy harassing an outsider but like a parent beating his or her own child. In Japanese culture, he claimed, such behavior toward other species expressed "the feeling of unity and belonging in a family; it is a corporate feeling which includes animals."[15]

Until the day of my visit in 2015, sweet potatoes and wheat had maintained the "friendly relationship" between Japanese primatologists and Koshima macaques that Kawai and his colleagues had forged half a century earlier.[16] As our boat approached the beach of Ōdomari Bay, the field assistant cried out and the macaques responded in kind from the laurel forest. Before we landed at the shore, the monkeys had already run down the steep and rain-slicked cliffs covered by myriad rock lice. The troop greeted us expectantly.

Before the provision of any food, however, a visiting researcher from Takasakiyama Monkey Park prepared a field experiment, comparing how fast the macaques could pick up wheat grains with their hands and mouths, respectively. As the more courageous among the impatient animals tried to steal the food before the experimental setup was complete, the researcher threw stones after them. At the chimpanzee field sites that I had previously visited in Africa, such aggressive gestures would have had very serious consequences for the researcher. Gravely concerned about reactions on the part of the much stronger but also much shyer apes, the European and American fieldworkers I had followed in Gabon and Ivory Coast sanctioned all actions that jeopardized their efforts to keep human-animal relations as neutral as possible.

On Koshima, the cradle of Japanese primatology, zoologist Kinji Imanishi and his students had developed a very different philosophy. They fed the animals, conducted simple experimental interventions, and crafted complex social relations across species boundaries. Naoki Koyama even reported how he laid the foundation for his friendship with the monkeys of Arashiyama who had previously threatened him: one day, he caught the highest-ranking individual off guard and "kicked him in the jaw with my boot."[17] Unconcerned that human meddling could spoil the naturalness of their field site, these Japanese primatologists did not understand themselves as detached onlookers of primate life. Imanishi's approach to the "society of living things" had grown out of an ecological study of mayfly larvae but soon began to trade concepts and practices with the human sciences, especially social and cultural anthropology where participant observation had been constructed as the royal road to the "native's point of view."[18]

Monkeys Have *Kaluchua*

In 1952, Imanishi speculated that not all animal behavior was based on instinct. His paper "The Evolution of Human Nature" contained a fictive dialogue between an evolutionist, a layman, a wasp, and a monkey who discussed whether the latter had culture, too.

EVOLUTIONIST: People say, animals live only by instinct, while humans have culture.

LAYMAN: To be a human means to have a culture.

EVOLUTIONIST: . . . Instinct is inherited through a genetic channel, while culture is transmitted through a nongenetic channel. Culture is acquired through learning and teaching, so that the model and the pedagogy are necessary. Therefore, a group life is inevitably required for the establishment of a culture. Thus, to maintain a culture as culture, the group life must be a perpetual one.

LAYMAN: If the condition for the establishment of a culture is just like you described, then culture may be seen not only in humans but also in other animals that live in a perpetual social group. How is it in monkeys? Do you have culture, monkey?

MONKEY: We live in a perpetual social group. . . . But it is not made clear yet how much of our behavior is determined by instinct and how much is determined by culture.[19]

Imanishi's willingness to admit nonhuman animals to the realm of culture has been attributed to Japanese anthropomorphism.[20] He sought to find primordial germs of humanity in animals.[21] But note that the dialogue begins with the evolutionist referring to the belief—apparently widely shared among Imanishi's countrymen—that animals act on their instincts while only humans have culture. Imanishi's text suggests that he expected his Japanese readers to at least frown on his anthropomorphization of monkeys.

Historian Tessa Morris-Suzuki argued that the concept of culture, or *bunka*, had gained wide currency in Japan only in the 1920s.[22] Historian Miriam Kingsberg Kadia even suggested to me that only under the American occupation of Japan between 1945 and 1952 did culture become a salient variable for the understanding of human difference. Originally, *bunka* and the related word *bunmei* (civilization) had been adopted from the Chinese classics, where they referred to the ordering and improvement of society by the written word, learning, and scholarly rule (*bun*) instead of the sword (*bu*). Analogous to the semantic transformation of the English words *culture* and *civilization, bunka* and *bunmei* also began their careers designating a realm of human activities that distinguished a ruling class from commoners. In the Meiji Era (1868–1912), when Japan opened to European and American Enlightenment philosophy, science, and technology after two centuries of self-imposed isolation, *bunmei* came to be associated with a newly imported scientific rationality and industrial progress. While *bunmei* temporalized ethnic differences in terms of developmental stages, *bunka* spatialized them as differences between contemporary but geographically apart cultures. The idea that *nihon bunka*

(Japanese culture) designated a way of life and a mind-set shared by all Japanese, which distinguished them from other peoples, especially from Westerners, emerged between the 1920s and 1940s as part of nationalist backlash to Meiji cosmopolitanism. This was the time when Kinji Imanishi, born in 1902, came of age.

At the height of the Second World War, worried that he might not come back from his deployment to Japanese-occupied Manchuria, Imanishi laid out his view of the natural world. He considered it so personal that he presented the resulting book as a "self-portrait."[23] He proposed an account of evolution that stood in stark contrast to Darwin's emphasis on the natural selection of the fittest individuals. For Imanishi, the existence of living things was more than an unceasing quest for food and sex: "Is it true that their lives begin and end in wretched brutishness? If living things are like that, then why are flowers and butterflies beautiful?"[24] The spiral shape of fossilized ammonites made Imanishi wonder whether the biological world produced its own kind of art. "Is there not something that could be called culture, although it is of course different from human culture?" he asked.

Imanishi even speculated whether such nonhuman culture—rather than natural selection of random variations—might have been the origin of species. On the last pages of his short book, he proposed that "changes gradually generated in nonessential cultural features, which have no direct relation to their way of surviving," could "unexpectedly develop into the distinctive traits distinguishing each species" after its separation from a common ancestor.[25] Just as differences in climate could not answer the question of why science developed in Europe and not in Asia, different environments alone could not explain the evolution of different forms of life. Each developed according to an unconscious design or destiny. "To best express the flower which blooms on this destiny," Imanishi concluded, "I venture to borrow the term culture."[26]

A decade after the publication of *A Japanese View of Nature* in 1941, the beauty of ammonites no longer served as Imanishi's paradigm of nonhuman culture. He had returned from Manchuria to Kyoto as an unpaid but independently wealthy university lecturer in the Department of Zoology. In 1948, Imanishi and three dedicated students, Junichiro Itani, Shunzo Kawamura, and Masao Kawai, began to observe a troop of Japanese macaques on Koshima at the southernmost tip of Kyushu. They looked at *Macaca fuscata* from a "biosociological" perspective, which would come to distinguish the so-called Kyoto School of primatology from other forms of primate research in Japan.[27] Kazutaka Sugawara, an anthropologist working in this tradition, compared this early Japanese approach to the structural-functionalism of British social anthropology.[28] Imanishi began to wonder whether the monkey society they observed had its own culture.

In the dialogue Imanishi wrote in 1952, he further pursued his idea that nonhuman species could have culture, but that it would be different from human culture. To distinguish between the two, he introduced a neologism usually transcribed as *kaluchua*. Japanese speakers would pronounce it like the English word *culture*. In opposition to the idea that monkeys had culture, Imanishi had the layman repeat the widely accepted view at the time that "we became human, not animals, because we had *bunka*." The fictive evolutionist responded: "Wait a minute. I did not say *bunka*, I said *kaluchua*."[29] Seven years after the US Air Force had dropped two nuclear bombs to force Imanishi's home to surrender, attributing the Japanese notion of *bunka* to people and a homonym of American *culture* to monkeys might well have been a slight. Imanishi later explained the need for this distinction by pointing out that the term *bunka* had been too human centered, connoting the noble, intentional, sophisticated, and complex aspects of our intellectual activities. Attributing it to nonhuman primates evoked the image of "an ape wearing a costume."[30]

Presumably, Imanishi's remark alluded to traditional Japanese monkey performances, which since medieval times had culturalized macaques by dressing them in human clothes. In her historical study of these shows, anthropologist Emiko Ohnuki-Tierney noted the ambiguous nature of this anthropomorphism: "Seeing a disconcerting likeness between themselves and the monkey, the Japanese also attempt to create distance by projecting their negative side onto the monkey and turning it into a scapegoat, a laughable animal, who in vain imitates humans."[31] In folktales, the monkey who tried to be a human served as a mirror for those ridiculous humans who hoped for more than they had.[32] As *burakumin*, members of an outcast group at the very bottom of Japanese society, monkey trainers had good reason to entertain fancies of upward social mobility. Between the 1920s and 1970s, they sought to eradicate all traces of their identity by escaping the stigmatized occupations usually adopted by members of their communities.[33] Thus, monkey performances had virtually disappeared from public life when Imanishi's concept of *kaluchua* broke with the tradition of presenting macaques as if they were human.

Yet de Pracontal proposed that Imanishi coined the term *kaluchua* not because he wanted to avoid anthropomorphism but because he considered *bunka* too constrictive for animals (just as constrictive as a monkey squeezed into a costume).[34] Kyoto primatologist Michio Nakamura told me that Imanishi preferred *kaluchua* because it covered simple collective behaviors and he wanted to dissociate culture from intelligence.[35] It's true what historian of science Lorraine Daston said: "Underneath the smooth word 'anthropomorphism' are hidden a multitude of *anthropoi*, of kinds of humanity . . . as well as the multitude of *morphoi*, of shapes of understanding other minds."[36] The form that Imanishi gave to anthropomorphism was Japanese, distinctly modern, but also highly idiosyncratic.

Considering that many Japanese primatologists as well as their critics and commentators have identified the Kyoto School with its anthropomorphism, I would like to emphasize that Imanishi was equally interested in similarity and difference. He dismissed a mechanistic view of living things as just as subjective and unscientific as an anthropomorphic view.[37] In his first book, he had already pointed out that his attribution of qualities such as society and fine art, which many of his Japanese readers considered uniquely human, to other species did not imply that animals were "placed on the same level as humans nor that humans are being reduced to other living things," but that all life-forms shared an "essential similarity in fundamental characteristics."[38]

First and foremost, Imanishi aspired to a symmetrical perspective that looked at humans and nonhumans through the same optic: "We cannot regard animals as mere automata nor humans as special creations of an omnipotent and omniscient deity. We cannot deny that humans, along with other animals, are the result of the growth and development from one thing, however pre-eminent humans have become. We can recognize this world because we humans are part of it. Therefore, it is not at all surprising that shamans and poets could talk and listen to non-human life such as trees and stones."[39] To disenchanted anthropologists this amalgamation of evolutionary biology and the quintessential myth of communing with nonhumans promised an escape from the cosmic solipsism of modernity.[40]

In 1952, *kaluchua* was an object of aesthetic musings and natural historical speculation. The following year, however, Satsue Mito, a local elementary school teacher who helped Imanishi and his students as a field assistant on Koshima, observed something they had never seen before. At the beach, Mito regularly fed leftovers from her home as well as locally grown sweet potatoes to the monkeys to habituate the animals to human presence. One late summer day, a young female they called Imo used a small stream and, later, seawater to wash the sand off her potatoes before eating them. Initially, only her closest relatives copied the behavior. Then more and more distant group members followed suit, eventually passing it on to new generations, until almost the entire troop had adopted sweet-potato washing. In the following years, Imanishi's team meticulously documented the invention and spread of several new behaviors, including wheat washing and panhandling to researchers. It was the first time that primatologists had observed the nongenetic transmission of a behavior in the wild. What had been fiction in 1952 became science in 1953.

Imanishi presented their discovery of a "subhuman culture"—his students would speak of "subculture" and "pre-culture"—to an anglophone readership in terms highly reminiscent of the debates over culture in nineteenth- and early twentieth-century Europe and America.[41] They conceptualized culture as a "transmission" of behavior from generation to generation. However, in the case of the new food habits, which the monkeys acquired as the researchers fed them unfamiliar items, it was usually very young animals who first invented or adopted them and then passed them to their mothers. Some older individuals, especially the males, would never take them on.[42]

Over the years, Imanishi grew frustrated with the identification of *kaluchua* with sweet-potato washing. This focus on subsistence techniques would eventually lead primatologists to study chimpanzee nut cracking and termite fishing as paradigm cases of ape culture. But food appeared peripheral to Imanishi's concern with group living and collective identity. In 1966, he abandoned the term *kaluchua* and began to use *bunka* in a more extensive sense instead. Most members of the Kyoto School followed suit. Four decades later, a new generation proposed to let the term *kaluchua* rest but to return from research on food acquisition to Imanishi's original idea of studying social customs and organization in terms of culture.[43]

Across the Pacific, members of the Culture and Personality school, often students of Franz Boas, who refused to think of culture in social evolutionist terms, avidly read the works of Sigmund Freud to understand how socialization in different cultures formed different types of personality. Donna Haraway claimed that Imanishi and his group had borrowed the notion of culture from these American anthropologists.[44] Margaret Mead's and Ruth Benedict's works might very well have been among Imanishi's books, still accessible today at Gifu University Library.

Be that as it may, Imanishi also explained cultural transmission in psychoanalytic terms: just as humans developed a superego by way of identification with others, macaque children developed their "culture and personality" by first identifying with their mother, then—not knowing their fathers—with the group leader.[45] This resulting "pre-culture" defined "the social character of the troop such as cautiousness, tameness, and aggressiveness."[46]

This psychological mechanism ensured the reproduction of the group's hierarchy. The children of dominant females would grow up in closer proximity to the leader and would be "well enculturated."[47] Introjecting the leader's "personality" made the simian *jeunesse d'orée* more group oriented and prepared them to behave as would-be leaders under similar conditions.[48] Thus, the elitist conception of culture as a privilege of the ruling class rather than a feature equally shared by all members of a group found its way into zoology.

The role model function of the leader gave Imanishi's monkey culture a conservative bent. It prevented nonconformist and antisocial members of the group (*oikia* in the terminology of Imanishi's evolutionary theory) from causing social unrest: "Those [males] with unsuccessful identification will not go on smoothly with the tradition or the culture of the oikia and leave the oikia more unhesitatingly. . . . Thus the oikia itself is conserved."[49]

Hence, the nineteenth-century conception of culture as collective identity, which had come to define American cultural anthropology, also found its match in Japanese primatology. From the start, the Kyoto School emphasized that culture was not just about social learning of survival skills but provided a sense of belonging. "A group does not split into individuals because of kaluchua," Imanishi wrote in 1952, and shortly afterward, Itani added that "it can be said that kaluchural personality is what makes mutual communication possible and makes bonding between individuals to form an orderly group."[50]

This social character distinguished one troop of *Macaca fuscata* from all others—just as human cultures manifested behavioral diversity in *Homo sapiens*. "Japanese people eat sea cucumbers, but people in Western countries do not," noted Satoshi Hirata, Kunio Watanabe, and Masao Kawai as they looked back at the tradition of sweet-potato washing on Koshima.[51] By the late 1950s, this population was no longer the only group Imanishi and his students were observing. This made possible a "comparative study of oikiae belonging to the same species."[52] "Some groups seem to leave very little freedom to the individual," said Belgian physical anthropologist and Washburn student Jean (or John) Frisch in summarizing the early results of Japanese cross-cultural comparisons for an anglophone audience.[53] "Other populations show much less cohesion and many animals are seen to act and move much more freely." He concluded that it wasn't possible "to study one group and then set out to write a book about 'Japanese monkeys.'"[54] Each group deserved its own monograph. Michio Nakamura, a chimpanzee researcher representing the third generation of the Kyoto School, described their early books about the social worlds of nonhuman primates as "ethnographies."[55]

Provisioning a *Macaca/Homo* Culture

The Japanese researchers knew very well that it had been they who triggered the spread of the newly acquired behaviors they observed in the macaques. Kawai freely admitted: "Pre-cultural behaviors of the Koshima troop were all begun under the conditions of provisionization. It is quite doubtful whether or not such inventive behaviors as [sweet-potato washing] and [wheat washing] should have been developed in the natural life."[56] Only when research assistant Mito started throwing peanuts into the water did the monkeys begin to bathe in the sea. Eventually, they also developed the so-called give-me-some behavior—begging researchers for food—which Kawai interpreted as a symbolic expression of their friendly attitude toward humans.[57] It contrasted with the behavior of groups that Imanishi's students had observed in monkey parks where the animals snatched food from tourists instead of kindly asking for it. At Takasakiyama National Park, Itani had documented how the group learned to eat candy they obtained from visitors and scientists.[58] Observing how different groups of *Macaca fuscata* interacted differently with different groups of *Homo sapiens*, the researchers learned about monkey culture. Or was it really *Macaca/Homo* culture?

"Provisionized monkeys are, in a sense, acculturated monkeys," wrote Imanishi, "in that provisionization means not only a change in their food habits but also a change in their social behavior and personalities through the contact with man."[59] But, he insisted, they were neither captured nor domesticated, but "wild as ever." Whether they could still count as wild after a few generations of acculturations was a different question, though.

Provisionization or provisioning was so key to the early Kyoto School's approach that DeVore and Hall referred to it as "the Japanese method of 'provisionization'"—a subsequently often repeated claim.[60] Handing out food to the monkeys changed their behavior. The main goal was to speed up their habituation to human presence. So far, the Koshima macaques had experienced humans mostly as predators. In 1947, only one year before the scientists arrived, which also happened to be the year when unauthorized hunting of Japanese monkeys was outlawed, many members of the group had been shot.[61] But a bountiful table drew "the forest sprites" out of their wooded cover onto the open beach of Ōdomari Bay where they could be easily observed.[62] At the beginning of the habituation process, the researchers offered the monkeys their natural foods—things they were already familiar with. Over time, however, they introduced products of human agriculture such as sweet potatoes, wheat, and soybeans because these were easier to obtain.[63] While some suggested that provisioning originated from bird-watching, others believed it had grown out of providing food for carp, birds, deer, and monkeys at Shinto shrines where monkeys served as mediators between priests and *kami*, mostly by collecting food offerings left for these spirits of the natural world.[64]

Whatever its genealogy, the intervention into the macaques' feeding ecology led them to develop new behaviors and learn them from each other. This provided an opportunity to observe social transmission in real time and to compare the group's behavior before and after decisive innovations in response to new food items. Once a new behavior had been established and was shared by the whole group, it would become more difficult to identify as cultural because it could just as well be an instinctual response to a particular ecological situation. Of course, comparisons between groups might have provided important clues. But on their own, behavioral differences could always be attributed to environmental differences. In other words, without

FIGURE 3. Japanese macaques picking up provisioned wheat: the first recorded culture of nonhuman primates on Koshima. Photo by author.

provisioning, Imanishi and his students might never have discovered monkey *kaluchua*.

Consequently, an altogether different history of Japanese primatology would have been perfectly possible, Juichi Yamagiwa argued. At the time of my fieldwork, the gorilla researcher was president of the University of Kyoto and had little time to observe primates other than *Homo academicus*. In the 1970s, however, he had managed to habituate a macaque troop on Yakushima Island without any food handouts. Had his Kyoto School antecedents approached the Koshima group differently, "research on Japanese monkeys would probably have taken a completely different course."[65]

The new abundance of food set off an explosion of the Koshima population. When research began in 1948, the small islet was home to 20 macaques; by 1962, there were 59.[66] In Takasakiyama, the population skyrocketed between 1953 and 1979, from 220 to more than 2,000 animals.[67] Considering how dramatically the behavior of *Homo sapiens* had changed after the agricultural revolution as human populations took on a sedentary lifestyle and grew rapidly, the deliberate shrinking of the monkeys' nomadic range and a multiplication of their group size must have had dramatic consequences for their social life as well. Japanese primatologists documented some of these transformations, most prominently the breaking up of oversized groups.[68] The scientific object of primate sociologists changed under their eyes and hands. Otherwise sympathetic Western colleagues like Frisch pointed out that control studies of unprovisioned groups were "badly needed if we want to make sure that the behavior has not been . . . substantially altered by what is, in fact, a semidomestication."[69] Imanishi recognized the problem. He wanted to curb the effect of provisioning on group size by capturing "surplus monkeys" and supplying them to the newly founded Japan Monkey Centre in Inuyama for biomedical experiments.[70]

The literature on Japanese conceptions of nature suggests that such human intervention might have confronted primatologists with practical difficulties, but it did not provoke a metaphysical crisis. *Shizen,* which in the twentieth century came to translate as the English word *nature* and the German *Natur,* never required a phenomenon to be independent of human influence. From the start, Imanishi had not conceived of humans as standing outside of nature.[71] That the macaques responded to human actions did not make their form of life any less natural. Nor did it make it less cultural. In contrast to their Western critics, Japanese primatologists saw no reason why provisioning or other forms of human interference would have called into question their discovery of monkey culture.

The Politics of Japanese Nature and Monkey Culture

Shizen, however, was no perennial feature of Japanese ontology. In the late nineteenth century, some English-Japanese dictionaries still didn't even list it as a possible translation of the English word *nature.* Instead they offered a bouquet of other terms, each carrying its own metaphysical connotations.[72] From his history of nature studies in early modern Japan, Federico Marcon inferred: "There is nothing *natural* in our conceptions of 'nature.'"[73]

What exactly *nature* meant was not just a philosophical but also a political question. If nature roughly amounts to what is predetermined about ourselves or any given situation, then whoever gets to define it also gets to define what we can hope to change. This led historian Julia Adeney Thomas to write a political history of the Japanese notion of nature, which suggests that Imanishi did not inherit an ancient autochthonous understanding but participated in reconstructing the concept of *shizen* in wartime and postwar Japan. Originally, the term had been borrowed from Chinese, where it had meant idleness or purposelessness. By contrast, in twentieth-century Japan it came to be associated with spontaneity and inherent, unmanipulated qualities. "Nature bubbles up from within," noted Thomas.[74]

Imanishi was a child of the Meiji period, which began to historicize nature. In 1877, the Japanese authorities appointed the American Edward Morse as the first professor of zoology at the newly founded Imperial University of Tokyo. He introduced his audience to evolutionary theory. A committed follower of philosopher, biologist, and sociologist Herbert Spencer, Morse liberally applied Darwin's theory of evolution through natural selection to more or less advanced human societies—placing Japan among the already civilized nations.[75] Both supporters of the Japanese emperor and their democratic opponents embraced the social evolutionist idea that the peoples of the world were on the road of progress. That Europeans and Asians were ahead of the game appealed to all political camps.[76]

Toward the end of the Meiji Era, however, Japanese intellectuals began to suspect that in the social evolutionist worldview, they would always lag behind the West. If there was only one modernity, theirs would be belated; if there was only one nature against which to measure progress, it would always favor Europeans. Spencer's brand of universalism undermined their national identity. Moreover, its emphasis on cutthroat competition and constant change threatened to destabilize their social order. The outcome was what Thomas called the "acculturation of Japanese nature": *shizen* became the conceptual "fulcrum to distinguish Japanese nature from Western nature," designating a nature that was "neither universal nor the particular domain of the West."[77]

Doubts about Japan's place in the modern world gave rise to a nationalist discourse that presented the Japanese as a unique people, culturally homogeneous and radically different from all other peoples. In the 1930s and 1940s, *shizen* gradually morphed into what Thomas called "ultranational nature."[78] In contrast to the appreciation of nature in Nazi Germany, Japan's wartime ally, the Japanese understanding of nature was less as a physical environment, which National Socialist *Naturschutz* sought to protect against human meddling, but as a unifier of all aspects of Japanese existence, including human life. After the defeat in World War II, national self-appraisals continued in more self-critical fashion, taking cues from Ruth Benedict's anthropological monograph *The Chrysanthemum and the Sword*.[79] Imanishi's project came to play an important role in a culturalist genre of writing known as *nihonjinron*, literally "discussions of the Japanese," or *nihonbunkaron*, "discussions of Japanese culture." Especially between the 1960s and 1980s, these popular and scholarly publications redefined national identity by contrasting Japan with a monolithic West (not China, India, or Africa). While few explicitly claimed supremacy, they all emphasized Japanese uniqueness.[80]

Many intellectuals demanded that Japan create its own science instead of continuing to imitate overseas methods and philosophies.[81] Following the lead of the Kyoto School of philosophy, with which the Kyoto School of primatology had been entangled socially and intellectually since the interwar years, Imanishi presented his group's approach as the product of an authentic Japanese culture grounded in an animistically based harmony with Japanese nature.[82] Cast in such culturalist terms, their primatology probably received more national and international attention than any other brand of science self-consciously made in Japan.[83]

Anthropologist Tadao Umesao, a student of Imanishi's who had studied the social life of tadpoles before turning to that of nomads at a research center in China, which Imanishi headed during the Second World War, attributed the birth of Japanese primatology to his people's national character: "The development of primatology in Japan is due to the intimacy which subsists here between man and monkey. For Europeans, of course, there is an unbridgeable gap between man and the animal kingdom."[84] Against this essentialist view of Westerners stood an equally essentialist view of the Japanese, who derived a sense of spiritual connectedness with other life-forms from the Buddhist metempsychosis doctrine.

But hadn't Darwin denied any fundamental difference between humans and the higher mammals?[85] Although Imanishi agreed with the Englishman that all extant species, including humans, had one and the same origin, the discovery of macaque *kaluchua* was couched in an explicitly anti-Darwinian evolutionist framework. Life as seen from Kyoto did not amount to a struggle over the survival of the fittest but was a fundamentally harmonious affair based on cooperation rather than competition within and between species.[86] The idea that primate sociality was less about fighting than about learning from each other fit well into this cosmology.

Beyond evolutionary theory, Masao Kawai attributed their group's discovery of monkey *kaluchua* to a methodology also enabled by the Japanese mindset. They studied Imo and the other animals as group-living individuals rather than as generic representatives of their species. For this purpose, they identified every single animal—another feature that supposedly distinguished Japanese from Western primatology at the time. Individual identification provided the basis for Imanishi's sociological approach.[87] Unless researchers recognized each animal, they could not collect data on how an innovation such as Imo's sweet-potato washing was taken up (or not) by other members of the group.

Studying how the monkeys passed on new behaviors to successive generations required primatologists to switch from conducting brief expeditions to building up a continuous research presence. In the case of Koshima, the monkeys have been observed without interruption from 1948 until this day. In their book on long-term field studies of primates, Peter Kappeler and David Watts credited Imanishi and his students with developing the triad of "habituation (facilitated by provisioning), individual identification, and long-term observations," which has often been presented as a hallmark of Japanese primatology but is now considered "the methodological standard for most primatological fieldwork."[88]

Kawai proposed a fourth characteristic of the Japanese approach: they understood nonhuman primates empathically rather than objectively. He doubted that this newly acquired research practice could be socially transmitted across national borders. He believed that his so-called feel-one approach, or *kyōkan*, represented a more instinctive response rooted in the Japanese paleocortex, which also allowed his people to tell apart individual animals.[89] In the 1990s, Kawai invoked Tadanobu Tsunoda's work on "the Japanese brain," another prominent expression of *nihonjinron*, suggesting that more pronounced activity in the left hemisphere enabled Japanese primatologists to recognize the faces of particular monkeys, which all looked the same to Western observers.[90] In a study of scientific nationalisms in East Asia, anthropologist Margaret Sleeboom remarked: "Kawai creates a greater distance between the Japanese and the Ōbei-ans [Westerners] than the one he claims to exist between the Japanese and macaques."[91] In a transvaluation of racist identifications of non-Europeans with nonhuman primates, this first generation of Japanese primatologists turned the insult into a compliment and used it to suggest the superiority of their observational skills over those of their European and American colleagues.

According to Umesao, Imanishi developed his "cultural primatology" in direct opposition to Western cultural anthropology.[92] Kyoto School philosopher Ueyama Shunpei later described Imanishi's antagonism to "the yanks" (*amechan*) in terms of hurt national pride: he wanted to demonstrate that "though we've been beaten in the war, we won't be done in by their scholarship."[93] From this sense of resentment, cultural primatology was born with a polemic bent in postwar Japan.

Nihonjinron left little room for differences *within* national cultures. However, some Japanese scholars had also responded critically to what they perceived as the Kyoto School's anthropomorphisms.[94] Imanishi's suggestion that monkeys had *kaluchua* was not traditionally Japanese, but an intellectual innovation. At the same time, their Christian heritage did not keep European and American Darwinians from applying one and the same lens to *Homo sapiens* and other animals. The confluence of Buddhist and Darwinian emphases on human-animal continuity as well as the confluence of Euro-American and Japanese anti-anthropomorphisms suggests that the front lines did not have to be drawn between Japan and the West. The rifts ran straight through Japan, Europe, and North America, respectively. They divided the scientific field far beyond the borders of nation-states. At the same time, these internal divisions had the potential to build bridges between certain factions in Japan and like-minded factions in Europe and the United States.

Reception and Reaction

On the international stage, the Kyoto School's presentation of a bold alternative to the predominant currents of European and American evolutionary anthropology and biology received a mixed echo. The suggestion that monkeys had culture was not dismissed out of hand. In the first comprehensive review of Japanese primatology, Jesuit father Frisch wrote in the *American Anthropologist*: "To the extent to which culture is equated with learned, traditional behavior, monkeys appear to have indeed much more 'culture' than anthropologists have often thought."[95] Yet he hoped to discover what *Homo sapiens* had "in common with other primates, but raised to a level which makes it human."[96]

Similarly, American anthropologist Irving Hallowell argued that the social transmission of behaviors, which Imanishi and his students had found on Koshima and in Takasakiyama, represented only "a prerequisite of culture and an earmark of an earlier protocultural behavior plateau."[97] What distinguished social learning in nonhuman primates from full-blown culture in humans was the monkeys' conservatism. "Every generation learns the same thing which its parents have learned," noted Hallowell in the face of Imo's invention of sweet-potato and wheat washing.[98] "In only very few instances the evidence is conclusive that the learned behavior can be modified or added to and that the modifications and additions are transmitted to subsequent generations." In the chimpanzee culture controversy, this last criterion—Michael Tomasello would call it "the ratchet effect" of cumulative culture—would become one of the most hotly contested defenses in the battle over human exceptionalism.[99] Defining culture as the social transmission of simple habits "confuses the conceptualization" of hominid evolution, Hallowell maintained.[100] At least until 1966, however, Imanishi and his students did not speak of culture or *bunka* anyway. Just like their American colleague, they spoke of preculture, subculture, subhuman culture, or *kaluchua*.

In 1960, some of the most eminent American and European primatologists of their time—from Ray Carpenter and Sherwood Washburn to Solly Zuckerman and Adolph Schultz—responded to an article titled "The Social Organization of Subhuman Primates in their Natural Habitat," which Imanishi had published in the US journal *Current Anthropology*. They appreciated his observations but took issue with his terminology—although Carpenter acknowledged the problems of translation across languages and cultures.[101] Michael Chance and John Emlen expressed concern about anthropomorphic descriptions of macaque life: Did animals that didn't pair-bond really live in "families"?[102] Emlen and Schultz also questioned the anthropocentrism underlying Imanishi's characterization of monkeys as "subhuman" and suggested that "nonhuman" would be a more objective term.[103] Imanishi responded that he wouldn't apply the term *subhuman* indiscriminately to all nonhuman primates, but "as a social anthropologist, I am interested in the origin of culture, and, in my usage, the 'subhuman' level of primates is nearly equivalent to the 'subcultural' level of primates."[104]

This exchange continued, not only around the question of primate culture but also around Kawai's work on female hierarchies and dependent rank in Japanese macaques as well as Yukimaru Sugiyama's observations of infanticide in langurs.[105] Stuart Altmann, especially, invited and translated Japanese primatologists in the mid-1960s.[106] At the time, lemur researcher Alison Jolly remembered, psychologists continued to taboo anthropomorphism while many primatologists quietly sidestepped their warnings.[107] Pamela Asquith suggested that the Japanese embrace of anthropomorphism and the Western rejection of it set up oriental and occidental primatology in opposition to each other.[108] Her Japanese host Sugiyama worried that this portrayal of Japanese primatology would push Japanese and Western scientists even further apart. Culturalist bias had led American researchers to advise their graduate students not to trust or cite Japanese studies and not to publish their findings in the Japanese journal *Primates*.[109] It is hard to assess how widespread and deep these reservations were because they were rarely expressed in print.

In comparison with their American and European colleagues, Japanese primatologists remained at the periphery of the field, argued Asquith.[110] Western primatology journals hardly cited Japanese studies, even if written in English, and they did not publish any Japanese authors until *Folia Primatologica* accepted an article by Toshisada Nishida on the bark-eating habits of chimpanzees.[111] At the beginning of the twenty-first century, Nakamura still complained that the traditional Japanese approach to primatology, especially ethnography, had been devalued, leading many Japanese primatologists to kowtow to international expectations.[112] Newspaper editorials had occasionally chastised such copying of Western customs as *saru mane*, or monkey imitation.[113]

In an analysis of the marginalization of Japanese anthropology, Takami Kuwayama repined that scholars at the periphery got heard only if they managed to translate their concerns into those of scholars at the center.[114] For example, Indian scholars had developed an intimate familiarity with Western knowledge cultures, which enabled them to carve out powerful positions in postcolonial studies. Japanese anthropologists also worked at the periphery, but their field had been born out of their own experience as colonizers. They had never been subjected to European rule—although the American occupation after the Second World War would leave its mark. Unless they had gone abroad for an academic exchange, which very few members of the Kyoto School did, they had rarely learned how to craft arguments that would be heard by their European and American colleagues, and they often received negative responses from Anglo-American journal editors.[115]

Anti Anti-Darwinism

Maybe it was this marginalization of Japanese primatology at large that initially saved cultural primatology, as one of its centerpieces, from becoming the focus of international controversy. When Imanishi's work finally did become engulfed in a fierce polemic, primates were no longer at the center of his attention. After pulling strings to institutionalize primatology in Japan—the Japan Monkey Centre and the Kyoto University Primate Research Institute both opened in the small town of Inuyama in 1956 and 1967, respectively—Imanishi moved on, providing important impulses to sociocultural anthropology before returning to evolutionary theory. In the 1980s, when his influence was already on the wane, it was especially his anti-Darwinism that came under attack.

International magazines and journals reported about and further fueled the controversy in a rhetoric that could not have been more orientalist and bellicose. In a sensationalist article on this disruption of "the Japanese ethic of perfect harmony," Bob Johnstone, the Japan correspondent of the *New Scientist*, announced: "An 'intercontinental missile' launched from Australia provoked a 'commando raid' in Japan, which in turn led to a 'battle' that raged in Britain for most of last year."[116]

What appeared to be a clash of civilizations had more domestic origins, though. The Australian missile had been launched by a former student of Imanishi's. At Kyoto University, Atuhiro Sibatani had attended Imanishi's 1943 lecture on animal sociology. Later in life, the molecular biologist also approached the world of living things via the human sciences. In collaboration with Kiyohiko Ikeda, he blended Imanishi's view of nature with the respective structural linguistics of the Swiss Ferdinand de Saussure and the American Noam Chomsky into a structuralist biology.[117] They concluded that culture, as Imanishi had described it in the context of his evolutionary theory, was firmly rooted in the structures of human and animal brains. Following the French Greek philosopher Cornelius Castoriadis, Ikeda and Sibatani suggested that "only those social institutions that correspond to the preexisting structure in the brain can be accepted as culture."[118] Thus, Sibatani translated Imanishi's thought into wildly different idioms. On the theoretical level, he was a follower rather than an opponent of his old Kyoto professor. Having emigrated to Sydney, however, Sibatani embodied a cosmopolitan ethic and epistemology. He mixed theorems of occidental and oriental origin just as freely as knowledge from across the divide between the natural sciences and humanities. Japanese culture, however, stifled the kind of critical assessments that such eclectic recombinations required, Sibatani lamented: "In Japan, especially in privileged society, the discussion of any matter in great depth is often considered to be in bad taste—even in science."[119] He decided to shake things up from abroad.

In the face of Imanishi's ardent opposition to Darwin, Sibatani maintained that their evolutionary theories were quite compatible. One of the tenets of Imanishi's view of nature was that the biota did not compete in a struggle for existence but coexisted harmoniously by partitioning their lifestyles and segregating their habitats—a thesis he had developed during the 1930s as he studied the ecology of mayfly larvae in Japanese rivers.[120] In Sibatani's eyes, such habitat segregation was perfectly consonant with Darwin's principle that "the greatest amount of life can be supported by great diversification of structure."[121] What he took issue with was less Imanishi's theoretical position than its nationalist framing.

Sibatani interpreted the popularity of Imanishi's anti-selectionist and anti-individualistic evolutionary theory in Japan against the background of the country's ascent as an industrial power, which fostered "a mood of 'Japan-as-number-one'": "A breed of strong ethnocentrism, chauvinism and xenophobia is now surging in Japan, trying to interpret all things Japanese in a positive light."[122] He pointed out that in contrast to the West, where, most recently in the sociobiology debate, it was mostly the left that opposed Darwinism, in Japan it was right-wingers who wanted Darwinism defeated. Many would have loved to see it replaced by "a truly indigenous science, free from the contamination of 'Eurocentrism'"—and that was what the Kyoto School had on offer.[123] In this political climate, Imanishi was awarded the Order of Culture, the highest honor for scholars in Japan.

As if to confirm Donna Haraway's dictum that "primatology is politics by other means," economic historian Heita Kawakatsu adopted Imanishi's concept of habitat segregation as a model for an identitarian multiculturalism, in which each ethnic group would remain within its own "ecological niche," just as the Japanese had done when they closed off their country during the Edo period.[124] Eventually, even Japan's prime minister Yasuhiro Nakasone accounted for Japan's exceptional development apart from the West in terms of habitat segregation.[125] In his infamous 1986 speech, he claimed that Japan was a much more "intelligent society" than the United States, where "there are quite a few black people, Puerto Ricans and Mexicans."[126]

Soon Sibatani's critique of national chauvinism and anti-Darwinism began to reverberate in the anglophone world. Beverly Halstead, a British professor of geology who continued the lineage of Darwin's bulldogs into the late twentieth century, had read Sibatani's article in the *Journal of Social and Biological Structures* shortly before embarking on a three-month visiting professorship at Kyoto University and decided to operate as the "commando behind enemy lines" about whom the *New Scientist* would report.[127] In the British journal *Nature,* he brought Japanese anti-Darwinism to the attention of the broadest scientific readership possible.[128]

Halstead later characterized his article as "more sociological than scientific."[129] It followed the tradition of an asymmetrical sociology of science, which at the time had come under heavy fire from the budding field of science studies for giving different kinds of explanations for the competing positions in a controversy.[130] While Halstead explained the truth of Darwin's theory in terms of its accurate representation of nature, he attributed the falsity of Imanishi's views to their distortion by Japanese society.

Like Sibatani, Halstead did not see habitat segregation as antithetical to natural selection, nor could he discern any cosmological incommensurability between the Darwinian emphasis on competition and Imanishi's emphasis on cooperation as fundamental principles of life.[131] British evolutionary theorist John Maynard Smith and the sociobiologists had replaced this "false dichotomy" with a "dialectic resolution of the unity of opposites."[132] However, their synthesis left the Darwinian worldview intact in that it affirmed competition as the overarching category while cooperation (within a group) became a special case of competition (between groups).[133]

Halstead provided a social explanation of Imanishi's error. It had not escaped his notice that Japanese anti-Darwinism not only aimed at the West but had grown out of a reckoning with the social Darwinism of the Meiji Era. The new ideology, however, was no better than the one it had come to replace. Rapid industrialization had not made Japanese society any more collectivist. "The fact is," Halstead maintained, "that the prevailing concept of competition is detested by the ordinary Japanese because everyone is involved in it in their real lives."[134] Imanishi's view of a nature that was not about the survival of the fittest provided a highly competitive and selectionist society with "a dreamscape of irrational sentimentality," so soothing that it had come to represent the average intelligent layman's understanding of evolution. The right welcomed the celebration of Japanese uniqueness; the left embraced the emphasis on cooperation and mutual aid. However, while Imanishi preached the primacy of the group for the rest of society, he and the entire Kyoto elite cultivated an extreme individualism, Halstead claimed: "Originality and innovation flourish in a secret enclave beyond the experience of the ordinary Japanese, condemned, as they are, to the rigid authoritarian feudal society that masquerades as one of the advanced nations of the world."[135]

Pamela Asquith warned against reducing Imanishi's view of nature to an expression of *nihonjinron* and its progenitors: "Those who fear right wing tendencies, see them in Imanishi's work; those who embrace these tendencies also claim to see them in Imanishi. What the Imanishi papers reveal is that he was, at least, a real and careful scholar who drew on all possible sources, tested them against his own observations, and thus lent more credence than most to his views of the human condition."[136] It is not just the inherent quality of Imanishi's research, but also its positive reception by primatologists like Frans de Waal or evolutionary theorists like Kiyohiko Ikeda and Atuhiro Sibatani that led Asquith to appreciate "how far ahead Imanishi was" in his exploration of humankind's place in the world of living things.[137]

From my perspective, what matters is not whether others proved Imanishi correct or whether Imanishi himself was a rightist, but that de Waal and Sibatani could build on his insights in radically different political contexts. De Waal adopted Imanishi's focus on cooperation to naturalize his own political commitment to the left.[138] Sibatani's cosmopolitanism enabled him to integrate the evolutionary theories of Imanishi and Darwin in a synthesis of his own.[139] In other words, even if cultural primatology had been born from the spirit of *nihonjinron*, it could easily shed this local political association and be recontextualized in a different time and place.

Two Mythologies

Had a polemic like the one over Imanishi's anti-Darwinism occurred between, say, American and French scientists, we would expect the attacked to reject the culturalization of their work and insist on the empirical evidence supporting their accounts. It is difficult to imagine articles about "French endocrinology" or "American ecology" written not by anthropologists or historians of science but by scientists presenting their own work. After all, science usually does understand itself as having no borders—a view shared by most Western and Japanese scientists.[140] But Imanishi had made a career of presenting his view of nature as uniquely Japanese rather than universal. When Sibatani presented Imanishiism and Darwinism as commensurable, Imanishi insisted on the irreconcilability of the principles of competition and coexistence and snapped: "Having lived for a long time in a foreign country, has [Sibatani] perhaps been influenced by the ways of thought there?"[141]

At that point, however, Imanishi no longer understood his work as science. From the start, he had distanced himself from a cultural activity that modern Europeans and Americans had developed out of early modern natural philosophy.[142] It had been no coincidence that he presented his view of the world of living things not as a scientific theory but as a self-portrait, and that he had first proposed the idea of monkey *kaluchua* in literary form.

In a commentary on Imanishi's work, Carpenter had already sensed that one of the questions raised by their cross-cultural dialogue was "What is good theory, and what is its effect on the quality and productivity of research, especially of field research?"[143] Although Carpenter recognized that an overemphasis on the testing of theories and hypotheses could prevent fieldworkers from noticing phenomena beyond their preestablished conceptual framework, he urged Imanishi to generalize from his observations and work toward a general theory that could accurately predict specific behaviors across species and genera. Imanishi referred Carpenter to German philosopher Wilhelm Windelband's distinction between two scientific approaches.[144] While the nomothetic perspective characteristic of the natural sciences seeks to identify general laws, an idiographic perspective informs the attention that the historical sciences paid to individuals and particulars. Imanishi maintained that the epistemology of his version of animal sociology and cultural primatology was decidedly idiographic.

In his "Proposal for *shizengaku*"—an intellectual testament of sorts—Imanishi officially parted "once and for all with the natural sciences."[145] Evolution was history, and as such it could not be verified nor could it be explained in terms of cause and effect, as Karl Popper had pointed out.[146] *Shizengaku* had nature as its object but "historical philosophy" as its epistemic form.[147] Defending Imanishi's view of nature against Halstead's dismissal that it was merely "a poetic vision, . . . Japanese in its unreality," bioenergy researcher Noboru Hokkyo suggested that when Imanishi wrote he was no scientist: he did not break with rigorous scholarship but sought "to free contemporary, intelligent laymen from their cultural fragmentation by making them more conscious of the way art, morality, religion and science have become specialized, censorial, constrictive to the unbroken wholeness of our cultural experience."[148]

Although international exchange has eroded the differences between American cultural anthropology and British social anthropology, or between continental and Anglo-American analytic philosophy, as knowledge cultures the humanities and social sciences remained more local than the natural sciences. Maybe it shouldn't come as a surprise that humanities-oriented animal researchers have also maintained a national identity. To this day, representatives of the Kyoto School publish articles on "Japanese primatology" and explain its distinctive commitment to holistic descriptions by "the general tendency that Westerners see a 'tree' while Asians see a 'forest.'"[149] Or, as Sugiyama put it: "At the bottom of our methodology, because of the influence of traditional oriental philosophy, the understanding and grasping of the whole structure of the materials is considered more serious than detailed description of each phenomenon, material, and behavior."[150]

In the 1980s, however, US-oriented cultural anthropologists like Japanese-born Ohnuki-Tierney began to chastise such essentialization of ethnic groups, in which both Japanese primatologists and their critics dealt, as an epistemic vice: "To talk about 'the Japanese' or 'Japanese culture' is to commit the anthropological sin of lumping the whole population under one umbrella."[151] By the early 1990s, Japanese anthropologists and historians had discovered Japan's ethnic diversity, sought to dismantle its national consciousness, and questioned the use of culture as an organizing framework.[152] Cultural hybridity became both content and form of the globalizing humanities.

But Imanishi had moved from appointments in zoology to faculty positions in social and cultural anthropology long before such qualms arose. He freely engaged in an auto-anthropologization that led him to promote cultural relativism instead of truth claims. Just like anthropologist of science Bruno Latour and primatologist Shirley Strum, Imanishi decided to look at natural histories as origin stories, which tell us where we have come from and where we might be going. Although particular audiences would perceive some stories as more satisfactory than others, Latour and Strum made no difference between scientific and mythic texts—they all had a pronounced fictive and speculative element, which provided orientation to their readers.[153] These narratives were not just social constructions; they also helped construct societies. In this spirit, Imanishi suggested that Darwinism was less a biological theory than "a kind of mythology" that appealed to "those Christian Westerners." Importantly, he also conceived of his own evolutionary theory as "founded upon a creation myth" and concluded: "There is no way to verify which is correct."[154]

Beyond Oriental and Occidental Primatology

When I visited Koshima in 2015, the biosociological approach of the Kyoto School no longer determined research on the island. Since the days when Imanishi and colleagues had challenged the anti-anthropomorphism of the behavioral sciences from this remote corner of Japan, the islet had sunk back into the quiet of Kyushu country life. The field station, located at the shore right behind a rice paddy and next to a closed roadside café called Happy Island, consisted of a main building and a few shacks. Inside, a heavy scent of mold hung in the air, and the sticky film covering the desks and bookcases might have preserved Imanishi's fingerprints. The postdoctoral researcher now in charge associated Imanishi's animal sociology with anthropomorphic misconceptions such as the idea that macaque societies practiced a division of labor. Trained as an ecologist, Akiko Takahashi still employed the Kyoto School's trademark methods. Based on food provisioning, she carried on long-term research and identified each of the by now over one hundred habituated individuals to document their social relations over time. But Imanishi's writings appeared obscure to her, their lack of testable hypotheses as antiquated as the office furniture of the research station.

Meanwhile, Takahashi explained, sociobiology had come to serve as the dominant interpretive framework—a field of study Imanishi had known only from hearsay but considered "some inferior imitation of Darwinism."[155] This intellectual reorientation from biosociology to sociobiology had gone along with the administrative transfer of the field station from Imanishi's heirs to the now dominant figure of Japanese primatology, cognitive scientist Tetsuro Matsuzawa.

The controversy over Japanese anti-Darwinism marked the fall of Imanishi-ism from representing Japanese primatology tout court to being one tradition among others. Despite Halstead's orientalist account of Japanese culture as a homogeneous and bounded totality that did not allow for dissent, resistance to the grand old man of the Kyoto School had also grown within primatology in Japan. The aggressive confrontations on the international stage had brought this domestic opposition into the open. For months, *Nature* received letters from readers siding with or against Imanishi and Halstead. Looking back at the debate, entomologist Yosiaki Itō dismissed the idea of an essentially Japanese anti-Darwinism: "That Imanishi's recent books have a large audience is by no means proof of a strong influence in Japanese biology, just as the publication of large number of Creationists' books in the United States does not indicate that Creationism permeates American science."[156] Apart from the primatologists Kawai, Kawamura, and Itani, few of Imanishi's coworkers and students supported his anti-Darwinism, Itō claimed.[157] Similarly, anthropologist Pamela Asquith wrote: "The majority of Japanese scientists simply disagree with [Imanishi's] recent popularization and find these views obscure and untestable."[158] If this majority had remained silent, it was not due to lack of freedom of debate, Itō wrote, "but because young scientists do not find it necessary to devote their time to do it."[159] Yet Halstead's defamation campaign stirred Osamua Sakura and three of his fellow graduate students into action at the Kyoto University Primate Research Institute. They also wrote a letter to the editor of *Nature*, in which they declared the end of an era: "Although Japanese primatologists, including us, are still hampered by the influence of Imanishiism, we criticize his theory and try to eliminate its negative influence."[160]

Before Sakura dropped out of primatology and became a professor of science studies at the University of Tokyo, he had studied the social life of chimpanzees in Bossou, Guinea.[161] Under the tutelage of Yukimaru Sugiyama, who had always kept his distance from Imanishi, Sakura interpreted his findings through the lens of Edward O. Wilson's *Sociobiology*.[162] Although sociobiology remained controversial within the Kyoto School, some of its members had received it readily since the 1970s.[163] The Japanese had heard about the political controversies surrounding Wilson's new synthesis of social and life sciences in the United States, where leftists had accused the biologist of naturalizing racism, sexism, and capitalism.[164] "They had seen E. O. Wilson criticized as if he were a Neo-Nazi," noted Sakura.[165] But, considering how Imanishi's theory was used to promote a "Japan-is-best ideology," Itō, a former communist also known as one of the "Three Musketeers" of Japanese sociobiology, felt that the "risk of bad use of Imanishi's theory is, in the present Japan, larger than that of bad use of sociobiology."[166] Among the translators of Wilson's controversial book was Matsuzawa, at the time an assistant professor at the Kyoto University Primate Research Institute and a collaborator of Sakura's in Bossou. With the import of sociobiology, this new generation of animal researchers reanimated their field's cosmopolitan spirit.[167] As Itō put it one year before Imanishi's death: "Western science, no; Eastern science, no; Universal science, yes."[168]

Political climate aside, primatologist-turned-science studies scholar Osamu Sakura claimed that sociobiology had met little resistance because it did not have to compete with other species of Darwinism. He rejected the often-repeated story that Buddhist belief in metempsychosis and familiarity with nonhuman primates living in Japan had saved Japanese primatologists from the illusions of human exceptionalism and had paved the way for Morse's introduction of Darwin's theory in the Meiji Era. What most Japanese accepted before Imanishi was Darwinism as a theory of *social*, not *biological*, evolution, Sakura contended.[169] Since biological Darwinism had never gained a foothold in Japan, a generation of young ecologists and evolutionary theorists recognizing the exhaustion of Imanishi's paradigm could eventually import sociobiology as the one and only Darwinism. Seeing Japan as an island previously isolated from Darwinian thought, Sakura interpreted this process of "conceptual evolution," in which a neo-Darwinian theory outcompeted Imanishi's, against the background of intellectual island ecology.[170] This variety of Japanese science studies not only left behind the opposition of oriental and occidental theories of evolution but also realized Wilson's dream of overcoming the division between natural sciences and humanities by interpreting the history of science in terms of population genetics.[171]

While Japanese leftists and liberals rejected Imanishiism and embraced sociobiology, the American reception went just the other way. Suspecting racist, sexist, and classist biases and agendas behind supposedly universal science,

Donna Haraway presented Japanese primatology as an alternative to the Western approach in a multicultural field: "both autonomous and autochthonous," it was proof of the cultural contingency of primatological knowledge.[172] In his foreword to the English edition of Imanishi's *A Japanese View of Nature,* one of its translators, Shusuke Yagi, a Japanese professor in South Carolina, situated this work from 1941 in the context of postmodernism, feminism, postcolonialism, and multiculturalism. He hoped it would enable his generation of scholars "to go beyond the Cartesian dichotomy, positivism, Eurocentrism, illusory objectivity and universalism."[173] Yagi promoted Imanishi's theory of evolution as an exemplar for the "indigenization of science."[174]

Considering Sakura's opposition to the Kyoto School, it cannot come as a surprise that he felt alienated by Yagi's attempt to make Imanishi a spokesperson for postcolonialism and multiculturalism. Sakura reminded readers of *A Japanese View of Nature* that its author had hardly been a subaltern finally allowed to speak. Instead Imanishi had himself served as an agent of imperialism in China, where he directed an institute in Japanese-occupied Manchuria.[175] In this new context of reception, which privileged cultural hybridization over purification, Imanishi's work became a projection screen for Japanese nationalism and antinationalism as well as American identity politics.

Imanishi's frequently echoed claim to represent a form of indigenous knowledge that had developed independently from animal behavior research in other parts of the world became an object of controversy. Many commentators continue to distinguish Japanese from Western primatology through the methodological triad of provisioning, individual identification, and long-term observation.[176] In *The Myth of Japanese Uniqueness,* Peter Dale objected: "The idea that in this there is something 'uniquely Japanese' stems from a curious amnesia, for it is clear that Imanishi borrowed the idea of both feeding monkeys and recognizing them by sight from the pioneering work of C. R. Carpenter."[177]

The autochthonous nature of Japanese primatology and evolutionary theory was also called into question on the theoretical level. Sakura did not buy into Imanishi's self-fashioning as a representative of indigenous thought: "[Imanishi] stressed the 'non-Western' origin and character of his theory. However, in my opinion, his idea was very close to that of the 'Chicago School' ecologists, including W. C. Allee (1885–1955) and A. E. Emerson (1896–1976), who highlighted the harmony of the entire ecosystem and discussed holistic organism[s], as Imanishi did. . . . The heart of his system-oriented holism is strikingly similar to that of the Western tradition of thought."[178]

Imanishi had grown up in the late Meiji Era. He was a child of Japan's opening to the West. Pamela Asquith's research on Imanishi's library revealed that Imanishi, as a young man who fluently read English and German, had followed the debate between Western ecologists over natural selection and had sided with the antiselectionist camp.[179] He adopted South African general Jan Smuts's rejection of the dichotomy between body and life in the name of a holist account of evolution and was familiar with Pyotr Kropotkin's view that cooperation and mutual aid had been key to the evolution of species—a theme Imanishi again encountered in the Chicago School of ecology.[180]

Imanishi had also read about the transmission of learned behaviors that did not necessarily advance survival between generations of animals, which English zoologist Charles Elton had called traditions and which Imanishi would dub *kaluchua*. Elton had not been the only Westerner thinking about animal cultures in the 1920s.[181] For example, American sociologists Hornell Hart and Adele Pantzer as well as anthropologist Alfred Kroeber had previously written about so-called subhuman cultures.[182] Primatologist Robert Yerkes declared "the anthropologist's dictum" that chimpanzees were cultureless "seriously misleading, if not demonstrably false," even though he considered the "cultural elements" in chimpanzee life "relatively unimpressive because unstable, fragmentary, variable, and seldom integrated into functionally important wholes."[183] The example he gave was of captive chimpanzees who had learned from observing each other how to use a push-button water fountain installed in their living cages. Imanishi's discovery of social learning in Japanese macaques in 1953 roughly coincided with James Fisher and Robert Hinde's discovery of social learning in birds. In other words, many features of Imanishi's highly original view of nature, including the idea that monkeys had culture, which he later presented as essentially Japanese and contrasted with Western thought, can be traced back to European and American sources or had Euro-American counterparts.[184]

If cultural primatology had been born from the spirit of Japanese uniqueness, from the 1970s onward it gained followers in Europe and North America. Subsequently, primatological knowledge also flowed the other way—to an extent that de Waal spoke of a "silent invasion" of Japanese conceptions such as the cultural transmission of behavior, undoing two millennia of Christian anti-anthropomorphism within decades: "Imanishi's approach to primate behavior amounts to a paradigmatic shift that today has been adopted by all of primatology and beyond. Thus, if we no longer perceive anthropomorphism as the problem it once was . . . , and if students of long-lived animals in the field—whether they watch dolphins or elephants—routinely identify individuals and follow them over their life span . . . , then we are employing techniques from the East initially mocked and resisted by the West."[185] That this resistance was about to give way to mutual recognition had already become evident in 1984, when the British Royal Anthropological Institute awarded Imanishi's student and coworker Junichiro Itani the Thomas Henry Huxley

Memorial Medal, sometimes described as the Nobel Prize of anthropology.[186] "There have been invasions from both sides," Itani's American student Michael Huffman, today a professor at Kyoto University Primate Research Institute, told me. "We have all grown into this broader identity of a hybrid primatology. It would be impossible for me to tell the difference between Western and Japanese primatology." In his eyes, the emphasis Asquith had placed on cultural difference in the 1980s "was more on target at that time than it would be now."

By the end of the twentieth century, a series of observations on a small Japanese islet began to shake animal behavior sciences across the globe. De Waal vastly exaggerated when he claimed that primatologists had unanimously accepted Imanishi's approach. But it did become an important point of reference in the ensuing debate over monkey and ape cultures. Since Europeans and Americans were hardly strangers to anthropomorphism, especially if their next of kin were concerned, some embraced the Japanese findings enthusiastically while others continued to resist or ridicule such "anthropomorphomania," as Daniel Povinelli called it in a 2015 lecture. The outcome was less a paradigm change than a controversy. Primatologist William McGrew dubbed it the chimpanzee culture wars.[187]

2

Multiculturalism beyond the Human

CULTURE IS A FIGHTING WORD. Today, it's so colloquially used that we forget that, time and again, people have been at each other's throats over how to understand culture, who has it and who doesn't, and what moral and political consequences this entails. The idea that humans were not the only cultured animals fell into a series of intellectual and social conflicts, which charged it with an animus exceeding the usual antagonism of scientific controversies.

In antiquity, the Roman *cultura* referred to the cultivation or tending of a particular entity such as the soil (*agricultura*) or the soul (*cultura animi*).[1] Although both the agricultural revolution and the birth of ancient philosophy as a work on the self provoked momentous changes in human life and European thought, I don't know of any disputes about their designation as *cultura*. But ever since the English word *culture* and its translations began to acquire their many modern meanings in the late eighteenth century, political controversies have abounded.

According to Raymond Williams, it all began with conservative critiques of the new ways of life that democratization and industrialization brought about and that supposedly lacked cultural values.[2] In telling the then-unfolding story, the British historian of ideas upped his own Marxist agenda: he promoted a conception of culture that no longer distinguished the cultured few, but included the supposedly uncultured masses.

At a time when *Kultur* still set Germans apart from French *civilisation*, English anthropologist Edward Burnett Tylor was already using the concept of culture in a more inclusive way that covered not only civilized nations but also primitive peoples.[3] Sometime between the 1870s and the 1930s—different historians of anthropology propose different dates—the democratization and pluralization of culture laid the ontological foundation for US cultural anthropology: now everybody got their own culture.[4] German immigrant Franz Boas and his American students, especially, managed to wrench the culture concept from the racial hierarchies of an evolutionary framework.[5] Culture no longer distinguished a privileged group—be they an intellectual aristocracy, Germans, or Westerners. It became a hallmark of humanity and distinguished humans from animals.

Or at least that's what more and more American anthropologists thought from the 1920s onward. Since the British and French did not follow suit and continued to harbor deep-seated reservations about the culture concept, Alfred Kroeber and Clyde Kluckhohn speculated that their colleagues' resistance to the significance of cultural differences was due to English and French imperialism, while the heterogeneous origins of Americans had made them especially receptive to the momentous idea of cultural pluralism.[6] They opened their review of the literature that had amassed by the mid-twentieth century with a swagger: "In explanatory importance and in generality of application [the culture concept] is comparable to such categories as gravity in physics, disease in medicine, evolution in biology."[7] Others were less sure whether culture would remain at the heart of the discipline or whether it would dissipate like so many intangible epistemic objects before: "Culture, like the 'ether' of the nineteenth-century physicists, plays an essential role today and will do so for a considerable time to come," concluded Fred Eggan in his presidential address at the annual meeting of the American Anthropological Association in 1953.[8] "The distant future is more difficult to predict—I think it was Whitehead who remarked that the last thing to be discovered in any science is what the science is really about."

At the end of the twentieth century, anthropology still did not know what it was about, and culture again became engulfed in controversy. Sociologist James Hunter coined the expression "culture wars" to describe the escalating battles among American citizens over so-called conservative and progressive values.[9] In this heated atmosphere, many cultural anthropologists decided to abandon the concept of culture because they worried that culturalism was still a form of racism.

Simultaneously, European and American animal behavior researchers followed in the footsteps of Japanese primatologists and suggested that not only the uneducated classes and non-Western people possessed culture, but some nonhuman animals did, too.[10] Ethologists started to promote a progressivism beyond the human. Unconcerned by the moral and political qualms of cultural anthropologists, American cultural primatologist William McGrew would later describe their project in an idiom of political inclusion, which the left had popularized in the 1980s: "We humans may need to rethink the boundaries of multiculturalism. We may need to be more inclusive in extending our appreciation of cultural diversity beyond anthropocentrism to admit our cousins the great apes."[11] In the ensuing "chimpanzee culture wars"—a term that McGrew coined in hindsight—he had thrown one of the first stones.[12]

This chapter provides a historical account of the debate over chimpanzee cultures as it spilled from Japan to Europe and North America. It covers many important bones of contention in the scientific controversy: competing definitions of culture, different categorizations of social learning, the methodological limitations of field and laboratory research, the method of exclusion, incongruities between anthropomorphic and parsimonious interpretations of animal behavior, human-animal continuity and discontinuity, the role of language, and the question of how to return from chimpanzee cultural diversity to chimpanzee universals. Other facets of the debate will be addressed only in later chapters. My goal is not to provide a systematic historical overview but to answer two specific questions: How exactly did the chimpanzee culture controversy intersect with the politics of the culture wars? And how did the culture wars foil the convergence of cultural anthropology and the nascent field of cultural primatology?

Chimpanzee Cultural Diversity

In 1975, after fifteen months at Jane van Lawick-Goodall's Tanzanian chimpanzee field site Gombe, the young William McGrew visited the nearby Japanese research station in the Mahale Mountains. At a conference, he had met Kinji Imanishi's successor Junichiro Itani, who had invited him to visit their site. In 1958, two years before Goodall began following the Gombe community, the Kyoto School had extended the geographical and taxonomic scope of its endeavor from Japanese macaques to the African great apes. Yet unlike Goodall's British Kenyan mentor Louis Leakey, they lacked colonial connections and knowledge of the continent. Consequently, it took them until 1965 to find and establish their own field site in East Africa.

Taking her cues from American sociology, Swiss ethology, and Japanese primatology, Goodall had speculated about chimpanzee culture.[13] But, being familiar with only the Gombe group, she lacked evidence for behavioral variation between communities. For the same reason, Itani, although he had been privy to the Japanese discovery of a nonhuman primate culture and had visited Gombe in 1961, did not realize that some of the behaviors they observed were particular to Mahale.[14] "Didn't all chimpanzees do this?" Itani shrugged, as McGrew expressed his surprise when, on his very first day in Mahale, he saw a behavior he had never noticed in Goodall's group.[15]

At the Japanese field site, the chimpanzees groomed each other's armpits by extending their arms overhead and clasping their hands. Considering that both Mahale and Gombe chimpanzees belonged to the same eastern subspecies, McGrew excluded a genetic explanation of this so-called grooming handclasp. Ecological causes also appeared unlikely because McGrew could not identify any differences in habitat that appeared relevant: "It was a behavior that couldn't be linked to anything physical: it wasn't about food or shelter, but plainly social," he remembered in an interview. It appeared like the kind of *kaluchua* that facilitated bonding between individuals and allowed them to form an orderly group. Despite their own paradigmatic discovery of sweet-potato washing, Imanishi and his followers favored this focus on social culture over the one on food and material cultures, which would soon dominate the chimpanzee culture controversy.[16] Since McGrew felt that the natural sciences could not account for such nonutilitarian variation, he turned to anthropology and began to explore what seemed to be the only alternative explanation: the two communities had formed different cultures.

At the time, many cultural and evolutionary anthropologists considered culture a uniquely human phenomenon. As Boas's first American PhD student, Alfred Kroeber had helped to define culture as a nature-free realm, but he had at least considered the possibility of ape culture when reading German psychologist Wolfgang Köhler's laboratory observation of chimpanzee dancing: to count as cultural it would have to be newly invented, adopted by others in a standardized manner, and survive the influence of the inventor as it was passed on to future generations or other communities.[17] McGrew and his coworker Caroline Tutin argued that the behaviors Goodall had documented for the Gombe chimpanzees met all these criteria.[18] Consequently, they no longer spoke of preculture, protoculture, or subculture, but simply of culture. Being about to make an inflammatory intervention in anthropology, McGrew preferred being more papal than the pope, though, and added two more criteria to Kroeber's.

First, cultural behaviors could not serve the animals' subsistence. He hoped that presenting chimpanzee culture "as social and transcendent" would make Tutin and him "look a bit more Durkheimian," McGrew told me.[19] Culturally learned behaviors should not be confused with biological characteristics of the species or psychological quirks of individuals. Unlike eating washed potatoes on the beach of Koshima and fishing for termites with sticks in Gombe, the grooming handclasp fulfilled the nonsubsistence criterion: the behavior was independent of the animals' intake of nutrients.

Controversy was still sure to follow because this onerous criterion excluded some of the very paradigmatic cases of human cultural activities such as agriculture. In *Homo sapiens* subsistence-related activities counted as cultural; in *Pan troglodytes* they did not. For the sake of scientific rigor, the subsistence criterion applied double standards to humans and animals despite McGrew's intention to demonstrate human-animal continuity.

Second, cultural behaviors had to be naturally adaptive. The requirement to be adaptive gave culture "a sociobiological slant" somewhat at odds with the nonsubsistence criterion.[20] It suggested that culture did not transcend nature. Culture was a product of and continued to be subjected to natural selection: it was part and parcel of hominoid biology.

But the emphasis was equally on culture being *naturally* adaptive—that is, not the result of human intervention. This preempted critiques of the Japanese discovery of macaque *kaluchua*. In 1953, the group around Imanishi, which included his student Itani, had observed how a macaque on Koshima Island invented a new behavior: in a small stream flowing across the beach into the ocean, Imo washed the sand off sweet potatoes before eating them. Subsequently, the researchers documented how other members of the troop learned from her how to wash potatoes. When the research group first presented this observation of what they interpreted as cultural transmission in animals to an international audience, some suspected that the Japanese had taught the monkeys this new behavior. Indeed, the Koshima macaques had not washed any potatoes before the arrival of the scientists because this product of human agriculture had not been part of their diet. Imanishi's group provisioned them with potatoes to habituate them to human presence.[21] In one of the first presentations of Japanese primatology to an anglophone audience, John Frisch asked: "Are not some items of social behavior modified or suppressed, or perhaps intensified, by the new situation created by provisioning?"[22] McGrew considered the dismissal of macaque *kaluchua* nonsense. Even if potato washing was an artifact of the research process, it made no difference to the animals' capacity for innovation and social learning where the potatoes had come from. Nevertheless, he did not want to leave this flank open.

McGrew still had a problem, though. In Tanzania, Goodall and Itani had also facilitated the habituation of their chimpanzees by providing prized foods. Thus, neither the Gombe nor the Mahale chimpanzees—the two groups McGrew and Tutin had compared—met the criterion of natural adaptiveness. As long as no data from unprovisioned communities became available, the "ethnography of the chimpanzee" would remain an unfinished project.[23]

Ethnography of the Eternal Now

This insistence on a conceptual and methodological decontamination of chimpanzee culture from human influence throws into relief the ontological background assumptions of cultural primatology, at least in its most prominent Euro-American varieties. If we were to lump rather than split ontologies, we could place cultural primatology in a very broad family of intellectual projects—ranging from sociobiology to multispecies ethnography and from ethnoprimatology to science studies—that have abandoned the opposition of nature and culture (or society), which had organized European thought since the nineteenth century.[24] Many of these initiatives, in both the natural and the human sciences, can be traced back to the 1970s. The fact that such diverse fields challenged the nature/culture dichotomy seems indicative of a momentous cosmological transformation.

Upon closer examination, however, chimpanzee ethnography did not collapse nature and culture the way science studies, multispecies ethnography, or even ethnoprimatology did. All these fields focus on what science studies scholar Donna Haraway and ethnoprimatologist Agustín Fuentes called "naturecultures": intersections of the human qua culture and the nonhuman qua nature.[25] By contrast, the object of chimpanzee ethnography has been what Boesch called "wild cultures": cultures that nonhuman primates developed on their own.[26]

At a point in natural history when human activities affected virtually all other forms of life, this project had to confront a tricky cosmological problem. Being naturalists, McGrew and Tutin recognized *Homo sapiens* as another African ape that had "a legitimate place in the biosphere as at least a hunter-gatherer since prehistoric times."[27] Moderns, however, seemed to have lost that legitimacy. Did they act contra naturam, violating a natural order that early modern deists and their present-day conservationist heirs had endowed with moral authority?[28] Who or what else decided the legitimacy of a species' existence?

The controversy over chimpanzee cultures began at about the same time as a new generation of cultural anthropologists vehemently renounced the idea of the primitive, of people living in the state of nature, as anthropology's epistemic object. Instead of looking at their interlocutors as representatives of an earlier stage of human or social evolution, they recognized them as contemporaries. They weren't Stone Age people but historical actors adjusting to colonization or different forms of culture contact with Europeans and other moderns.[29] Analogously, the omnipresence of modern humans in what had originally been imagined as pristine nature deprived chimpanzee ethnographers of unspoiled apes.

McGrew responded to this conundrum by "designating arbitrarily 1950 as the endpoint to the 'ethnographic present.'"[30] Data from before this time could be taken as unaffected by human presence. At least since the 1960s, anthropologists had used the concept of the ethnographic present to characterize a particular form of writing about other cultures, which George Stocking traced back to one of the founding documents of the discipline in its twentieth-century guise, Malinowski's *Argonauts of the Western Pacific*.[31] The use of present instead of past tense suggested that in the described forms of life, nothing ever changed. It endowed them with a perennial quality. By blocking out observations that would have muddied the image of an eternal now, the ethnographer constructed the present as it supposedly existed in the traditional past and not at the historical moment when he conducted his fieldwork.[32] After all, his very presence already indicated that he had come too late to record a way of life uncontaminated by the encounter with European or American moderns. As cultural anthropologists began to question the validity of these partial representations, archaeologists used the epistemological vacuum to improve their own standing in American anthropology, suggesting that only they could contribute accounts of unacculturated cultures to the ethnological project of documenting cultural diversity.[33]

But it was not only for epistemological but also for political reasons that the ahistorical usage of ethnographic accounts was pilloried just as chimpanzee ethnography became part of the tool kit of evolutionary anthropology. The narrative device constructing other people's timelessness obliterated the colonial situation under which most anthropological subjects suffered. And it denied their contemporaneity with the anthropologist and his audience, argued Johannes Fabian.[34] Since the 1980s, moral indignation about this political epistemology has provoked a widespread dismissal of the ethnographic present under the banner of a decolonization of anthropological thought—so widespread that João de Pina-Cabral felt the need to remind his colleagues that many anthropologists who had written in the ethnographic present had themselves openly challenged colonial domination.[35]

Should we conclude that politics and epistemology are less intimately entwined than a radicalized postpositivism has led us to believe? Or, should evolutionary anthropologists catch up and acquire the kind of self-reflexivity that cultural anthropology has developed since the 1980s, as Jeremy MacClancy and Agustín Fuentes demanded?[36] Wouldn't a chimpanzee ethnography that took into account the presence of humans, including that of the scientists, have to call into question the heuristic fiction of an ethnographic present, in which McGrew's chimpanzees had never actually lived?

Biological Humanism Defended

At the time of McGrew and Tutin's discovery of chimpanzee cultural diversity, it received a very different response, less concerned with the ahistorical representation of animals than with the historical association of human-animal continuity and racism.[37] Their critics still remembered nineteenth-century illustrations of a pre-Darwinian ladder of nature, ascending from so-called lower life-forms to apes, Africans, and Europeans.[38] "The humanization of the orang-utan went hand in hand with the animalization of the 'savage,'" remarked historian Silvia Sebastiani.[39]

Meanwhile, anthropologists had replaced this great chain of being with a belief in the fundamental equality of all human beings. Now many worried that attributing culture to chimpanzees might undo the political ontology on which the discipline had come to rest.

In a scathing letter to the editor of *Man*, the American doyen of a "new physical anthropology," Sherwood Washburn, and cultural anthropologist Burton Benedict worried that McGrew's postulation of nonhuman primate cultures undermined the ontological and epistemological premises of twentieth-century anthropology. Washburn had shifted the focus of physical anthropology from measuring skulls of different human races to understanding the functional integration of anatomy and behavior into adaptive wholes.[40] As a naturalist, he proposed to treat humans as primates and insisted that the social should be studied with reference to the biological.[41] He also knew that chimpanzees learned from other chimpanzees how to use tools.[42] Nevertheless, Washburn and Benedict defended the dogma of US anthropology that cultural phenomena were not reducible to biology and that "the traditional study of culture (human customs) assumes that the biology of contemporary human populations is fundamentally the same, and that the differences in behavior between populations are to be understood in terms of learning and history."[43] They defended such "anthropological definitions of culture" as part of the "necessary intellectual foundation of social science" against McGrew's chimpanzee ethnography, which apparently ignored that "freeing the study of social behaviours from nineteenth-century biology was a major intellectual achievement."[44] This conceptual framework suggested a clear-cut division of labor between physical and cultural anthropologists: while the former focused on the evolution of the human species, the latter studied socially learned differences in customs.

This position, which Washburn and Benedict upheld—that human populations were biologically alike and differed only culturally—had become an article of faith in biological humanism, which the United Nations had promoted since World War II. "UNESCO Man," Donna Haraway christened this figure of the human in her sometimes cheeky, sometimes moralizing history of primatology.[45] Even before the defeat of Germany and Japan, the young Washburn had reviewed the available anthropological knowledge about human races to counter German and Japanese claims to racial supremacy.[46] But he also worried that after the war, Nazi biologists would be "laughing with delight" when "nearly as absurd claims about the nonexistence of varieties of *Homo sapiens*" would be translated into German, as if dark pigmentation was of no functional importance to people living in the tropics.[47] Racial differences were real but shallow because "every living race has had at least one hundred times as much of its human ancestry in common with all the other races as it has had alone," and "none of them is close to ancient man, let alone the apes."[48]

UNESCO's statement on race—a document drafted and signed in 1949 by prominent social anthropologists and evolutionary biologists such as Claude Lévi-Strauss, Ashley Montagu, and Theodosius Dobzhansky—also considered race a meaningful biological concept to categorize inherited physical differences between human populations but emphasized that these could not account for differences in mental characteristics. "The long-standing confusion of race and culture has produced fertile soil for the development of racism," UNESCO noted and took it upon itself to prevent future misunderstandings by releasing a statement that was both scientific and political.[49] Some physical anthropologists expressed concern, not about the general spirit or the main conclusions but about insufficient evidence. Not even half a decade after the Holocaust, however, an authoritative rebuke of racism could hardly wait for the accumulation of more data. UNESCO jumped the scientific gun and declared all human beings to be endowed with the same innate capacity for intellectual and emotional development. Differences in appearance should not distract from the fundamental biological unity of humankind.

When Washburn became president of the American Anthropological Association (AAA) in 1962, at the time of the civil rights movement, he reluctantly returned to the subject matter.[50] Scientifically, it seemed of little interest to him. To the contrary, Washburn repeatedly emphasized that "race isn't very important biologically" and that anthropologists should rather focus on the evolution of the entire species.[51] Politically, however, he felt the need to speak out against "a ghetto of hatred" where the denial of education, medical care, and economic progress "kills more surely than a concentration camp."[52] Apart from a few glum-looking physical anthropologists, the AAA members greeted Washburn's spirited presidential address with standing ovations.[53]

Primatologist Alison Jolly noted that when, in the wake of the Nazi horrors of World War II, "anthropologists found it politically correct to stress the unity of the human species," they took up again the quest for "a Rubicon: a single adaptive shift which transformed our lineage into humans—perhaps toolmaking, or hunting, or language."[54] In his book on the current anthropogenic mass extinction event that also threatened the great apes, Richard Leakey (son of Jane Goodall's mentor Louis Leakey) also observed how antiracist politics had induced a momentous ontological shift: "Viewed as more primitive than white Caucasians, the 'inferior races' formed something of a bridge between the ultimate expression of *Homo sapiens* and the rest of the animal world. When all races were regarded as being equal, that bridge disappeared, and a gap opened up, making modern humans appear even more separate from the world of nature."[55]

In the 1970s, a decade before Haraway revealed the historical contingency of biological humanism, its anthropological presuppositions had come under attack from two directions. On the one hand, sociobiologists proposed genetic explanations of human social behavior, which Washburn dismissed as a resurgence of biological determinism. On the other hand, chimpanzee ethnographers proposed cultural explanations of animal social behavior, which also called into question a human exceptionalism that Washburn vigorously defended: "The argument that the human species is just a species that is distinct from others the way any species is distinct seems to me to be nonsense."[56] In a society torn by racial inequality, any plea for human-animal continuity touched a raw nerve. Would it not reintroduce a ladder of nature with whites on the top rungs, apes at the bottom, and people of color in between?

Only a categorical difference between humans and animals could prevent such a relapse into nineteenth-century conceptions of orthogenetic evolution. In Washburn's eyes, what fundamentally distinguished humans from apes was language. The linguistic capacity of *Homo sapiens* had itself evolved as a biological attribute of the species. But it had emancipated culture from biology. Without language there could be no innovation, dissemination, or standardization. Without language there could be no culture. The only reason Kroeber had not included language in his list of criteria that chimpanzees had to meet to qualify for culture, Washburn and Benedict claimed, was that "anthropological definitions of cultures have assumed language as a given, since there are no groups of non-linguistic humans."[57] Language was so essential to culture that it had not seemed worth mentioning. Not until McGrew came along.

Including Chimpanzees, Excluding Humans?

When McGrew began to observe wild chimpanzees in Tanzania he had been trained in the natural sciences and already held a doctoral degree in psychology. However, this experience led him onto a different track. "I was seeing things, especially behavioral variations, for which I could not find a credible explanation in the natural sciences," he explained to me in an interview. "So I started reading anthropology as an outsider." Conceiving of the difference between the Gombe and the Mahale community as cultural, he wanted to enter into a conversation with the discipline that had almost come to own the culture concept—at least in the United States, where German immigrant Franz Boas and his students had established the subfield of cultural anthropology around the idea of *Kultur*, or culture as collective and plural.[58] It was no coincidence that McGrew and Tutin did not publish their first findings on social customs in chimpanzees in an ethology journal but in *Man*, today known as the *Journal of the Royal Anthropological Institute*.[59] In the mid-1980s, McGrew taught psychology as senior lecturer at the University of Stirling. At the time, British social anthropology had begun to open to the German American lineage of cultural anthropology, and after hours McGrew embarked on a second doctorate in social anthropology.

In 1986, at the Conference on Hunting and Gathering Societies in London, McGrew sought to demonstrate that no unbridgeable gap segregated human and animal cultures. He compared the material culture of supposedly extinct Tasmanian foragers, the Parlevar, with that of extant Tanzanian chimpanzees in Gombe and Mahale. The former had been documented by the British chief protector of aborigines in the 1830s, the latter by European, American, and Japanese primatologists in the 1960s and 1970s. The Tanzanian chimpanzee population was known for its relatively complex tool kit, the Tasmanian hunter-gatherers for the simplest tool kit to be found in the ethnographic archive.[60] In his construction of an ethnographic present, McGrew deliberately excluded all tools and techniques introduced by Europeans from his analysis.[61] He modeled hominization by taking the apes as the closest living analogue of the last common ancestor of both humans and chimpanzees. The tropical hunter-gatherers served as the closest analogue of the early humans who had evolved from this hominoid. A quantitative analysis of the complexity of these human and chimpanzee cultures revealed that one of the most sophisticated chimpanzee material cultures was almost as complex as the most primitive human material culture on record. McGrew concluded that "the differences shown here could easily be cultural, not phylogenetic."[62]

Politically, the comparison of Parlevar culture with that of another species was inconsiderate. In the nineteenth century, British settlers committed to a rather un-Darwinian hierarchy of species had compared native Tasmanians to orangutans, considering it "a disgrace to call them Man."[63] Although McGrew's study was not informed by the metaphysics of such a great chain of being, aboriginal students at the University of Tasmania took offense. Had they controlled the data, they explained, they would not have allowed him to compare it with Goodall's and Itani's data sets.[64] They objected to the political legitimacy of who spoke, not to the scientific validity of what the British primatologist said about nineteenth-century Tasmanian material culture. The primitive nature of Parlevar technology, mostly stone and bone tools, remained uncontested and had also been noted by anthropologists highly critical of colonial science.[65] But whether Parlevar instruments for food acquisition resembled human artifacts from the Pleistocene more closely than they resembled the multiplex instruments and machines that populated British kitchens in the 1980s was not a question McGrew's critics cared about.

Explicitly conceiving of cultural primatology as "positivist," McGrew must have separated what logical positivist Hans Reichenbach had called the historical context of discovery from the corresponding context of justification.[66] It did not occur to McGrew that the social conditions under which Tasmanian material culture had been documented could affect the validity of that documentation. Positivism had long served as a term of abuse, which interpretive cultural anthropologists leveled against scientifically minded colleagues who still believed in observable facts free of theoretical presuppositions and political biases. By the 1980s, it had become "a swear-word, by which nobody is swearing," as Raymond Williams put it.[67] Nobody but a defiant McGrew. McGrew did not care about the philosophical nuances of the term any more than most humanities scholars and interpretive social scientists who dismiss quantitative approaches as positivism, even if their practitioners vehemently reject the label. "I just mean that having been trained in science, I favoured evidence-based arguments over opinionated-impressionistic ones," he explained to me in an email. "I am essentially a naturalist, a chimpanzee chaser, who spent time over 40 years in the African bush. I am theoretically unsophisticated. . . . I just went out and watched, then came back and reported what I saw." Of course, the idea that one could simply observe and report without theoretical, moral, and political presuppositions had been one of the characteristic and much contested topoi of the philosophical lineage spanning from Hume's empiricism to logical positivism and beyond. McGrew's display of epistemological naiveté and his political inconsiderateness appeared to represent an ill-defined yet principled stance in line with his makeshift philosophy of science.

Self-declared antipositivist Tim Ingold served as external examiner of McGrew's dissertation in social anthropology and witnessed the uproar McGrew's thesis provoked at the London conference.[68] "Was he really trying to tell us that Tasmanian hunter-gatherers had scarcely advanced beyond the apes, that they were stuck in an evolutionary time warp?" he would later write. "McGrew comes close to returning to the overt racism of an earlier era of anthropology."[69] Ironically, Ingold also followed the logic of orthogenesis when he relegated chimpanzees to the lower developmental stage from which he sought to free the Parlevar, as if late twentieth-century apes were any less McGrew's and his evolutionary contemporaries than nineteenth-century foragers. As the diffusion of cultural relativism into all branches of social science had made it embarrassing to employ "primitives" to represent the protohuman past, primates filled this Western cosmological slot for continued use in the era of decolonization, as historian Susan Sperling noted. This was the place of chimpanzees both in McGrew's model and in Ingold's polemic against it.[70]

But the commonalities ended there. Ingold maintained that the differences between the ways of life of human Tasmanians and chimpanzee Tanzanians were not cultural but phylogenetic: "What does seem incontrovertible is that a Tasmanian Aborigine, transported to the twentieth century and raised in an affluent part of the world, would have no particular difficulty in becoming, say, an airline pilot or a software engineer. But I would not, for my money, take a plane piloted by a chimpanzee!"[71] Behind the apparent similarity of chimpanzee and hunter-gatherer tool kits, the anthropologist suspected a fundamental difference of capacity, which no acculturation would ever be able to bridge.

Ingold seemed equally as irritated by the zoomorphization of *Homo sapiens* as by the anthropomorphization of *Pan troglodytes*. McGrew's symmetrical primatology, which applied the same quantitative measures to *Pan* and *Homo* cultures, threatened to include nonhuman apes in an extended multiculturalism while excluding human foragers from coevality with their modern conspecifics. McGrew's plea for extending our appreciation of cultural diversity beyond anthropocentrism had been suffused with human politics from the start. Despite its originator's professed positivism, it was never judged on epistemological grounds alone.

Ditto for the critics: "Wanting to compare human and animal societies is perfectly legitimate," French philosopher Dominique Lestel wrote in defense of McGrew.[72] "Its denial is not an expression of methodological prudence but of theological closure." In Lestel's eyes, neither anthropologists nor anyone else knew what a culture was, and comparisons across human and animal cultures were a way of posing the question anew, without deciding in advance what the answer would be.

The Cultural Left against Sociobiology

Looking back at the ensuing scientific controversy over chimpanzee cultures, McGrew compared it to the culture wars that have ravaged American society and especially American universities since the 1980s: "The 'culture wars' in 'cultural studies' are about essentials and jurisdiction, and, ultimately, about identity. The same issues affect what is called here 'the chimpanzee culture wars.'"[73] The analogy was hardly more developed than McGrew's profession of positivism. I found it evocative enough to make it the title of this book because it highlights that the disagreements over nonhuman primate cultures had a political dimension from the start. Moreover, the first accounts of chimpanzee cultures happened to coincide with the beginning of the "culture wars" in the United States. Speaking of "cultural diversity" and "multiculturalism," McGrew applied the political vocabulary of the cultural left to chimpanzee life.[74] So how did the culture wars play into the chimpanzee culture controversy?

American sociologist James Davison Hunter coined the term *culture wars* to describe a family of intensely polarizing conflicts over race and gender relations, abortion, gun rights, recreational drug use, free speech, separation of church and state, education, immigration, and multiculturalism.[75] The term translated into the German *Kulturkampf*, with which Rudolf Virchow, a Berlin liberal parliamentarian, physician, and head of the German Anthropological Society, had designated an anti-Catholic campaign in the 1870s. In the unifying Second Reich, Franz Boas's future mentor redefined *Kultur* as a distinctively modern form of life, which was incompatible with backward Catholicism. In the academic realm, this *Kulturkampf* posited anthropology against humanist scholarship, which established continuity between moderns and premoderns by reviving the past in the present. Anthropology offered a different kind of political education: instead of studying ancient Greece and Rome for the purpose of European autobiography, it used natural scientific methods to study people without culture, so-called *Naturvölker*. Thus anthropologists endowed the German *Kulturvolk* with a new modernist notion of culture, which set them apart from non-European peoples—and Catholics.[76]

In the late nineteenth and early twentieth centuries, American social conflicts also followed denominational lines. But, Hunter argued, in recent decades it was no longer different brands of Christianity, Judaism, and secular humanism that defined cultural conflicts over the moral character of the United States.[77] Increasingly, the orthodox of these denominations joined ranks with each other against the progressives of their own creeds. The culture wars differed from the earlier *Kulturkämpfe* in that the frontlines had ceased to divide religions but pitted progressives and conservatives against each other.

As these tensions informed radical politics in the academy, they led a new generation of critical scholars and scientists to unmask the racist, sexist, and capitalist biases of bodies of knowledge supposedly affirming the social order. In the behavioral sciences, it was predominantly sociobiology that attracted the wrath of activist groups. Since the 1960s, British and American life scientists had used the tools of population ecology and population genetics to study the biological basis of social behavior and the organization of societies in all kinds of organisms. When Edward O. Wilson synthesized these theoretical developments and empirical studies in his book *Sociobiology*, which gave the field its name, he immediately came under heavy fire by the International Committee Against Racism as well as by Science for the People.[78] The latter, a collective of Marxist life scientists, denounced the Harvard biologist from the Deep South as justifying the status quo of class, race, and gender inequalities on genetic grounds: "For Wilson, what exists is adaptive, what is adaptive is good, therefore what exists is good."[79]

Anthropologist Marshall Sahlins decried the sociobiological explanation of human behavior in terms of competitive self-maximization of individual genotypes as projecting the culture of capitalism into nature, which in turn naturalized the prevailing economic order.[80] Against the moral authority of nature, Sahlins mobilized the autonomy of culture as the symbolically organized domain of freedom, which hinged on the arbitrary relationship between signifier and signified. Since the contemporaneous ape language projects had failed to persuade him that animals could use symbols, he dismissed anthropomorphic talk about the "culture of animals" as a politically dangerous obfuscation of our uniquely human freedom to act otherwise.[81] In Sahlins's eyes, Wilson's attempt to tear down the ontological division between nature and culture amounted to an abuse of biology, which left humans bereft of all hope for the realization of political alternatives, unencumbered by genes and environmental pressures.

Biographically, however, Wilson was hardly the rightist that his leftist opponents sought to portray him as. He self-identified as a Roosevelt liberal turned centrist. He indeed welcomed free markets, but regarding racial inequality, he presented himself as a biological humanist.[82] Wilson even went beyond the UNESCO statement on race and insisted that any attempt to define discrete human races was futile because "such entities do not in fact exist."[83] Geographical variations between human traits such as skin color or the ability to digest milk, which did exist, should not carry with them any value judgments. In his eyes, sociobiology validated the unity of humankind. Wilson regarded an "irrationally exaggerated allegiance of individuals to their kind and fellow tribesmen" as the "force behind most warlike policies."[84] Firmly committed to Enlightenment universalism, he preferred assimilation to an identity-conscious multiculturalism.

Historians Ullica Segerstråle and Neil Jumonville argued that it was not right and left, racists and antiracists who clashed in the sociobiology debate, but those who believed that identity should be centered on the species instead of the race and a new ethnos-centered left privileging the recognition of difference over Martin Luther King's dream of a color-blind society.[85] American philosopher Richard Rorty would speak of "the cultural left."[86] These proponents of a burgeoning identity politics contended that talk about our common humanity had all too often served as a cover-up of racist domination.[87] The 1970s were also the time when concerns about official racism such as segregation or the denial of voting rights gave way to concerns about institutional racism. As institutions of higher education and scientific research, universities quickly moved center stage in this battle. Segerstråle suggested that it was no coincidence that the sociobiology controversy concurred with legal challenges by proponents of affirmative action to the Civil Rights Act of 1964, which had prohibited discrimination in hiring practices.[88] In 1978, the US Supreme Court upheld the right of the University of California to discriminate against white applicants in order to overcome discrimination against ethnic minorities. The postwar emphasis on the moral value of common membership in a species, which had found its most prominent expression in the UN Declaration of Human Rights in 1948, clashed with the emergence of an ethnos-centered pluralism.[89]

In 1975, McGrew's first observation of what he interpreted as a cultural difference between two chimpanzee communities coincided with this conflict between old and new forms of progressive politics in the sociobiology controversy. As chimpanzee behavior came to be seen as culturally variable and human behavior as genetically determined, the ontological gulf between *Homo sapiens* and *Pan troglodytes* seemed to vanish. Treading in Imanishi's footsteps, McGrew would even come to consider the possibility that chimpanzees had their own identity politics: "It may be that the ultimate function of culture for community-living apes is social identity. Just as human languages proliferated in areas where there were many distinct human groups, so may it be for our nearest living relations, where cultural identifiers tell who you are and where you come from."[90] In McGrew's eyes, the "us versus them" logic of such self-perceived common identity increased the reproductive success of groups over less cohesive aggregations of individuals.[91] His cultural primatology offered nothing less than a natural history of rampant identity politics, but an identity politics no longer confined to human societies.

Thus the chimpanzee culture wars were as much about cultural as about species identity. In his book about the philosophical ramifications of the discovery of animal cultures, Lestel claimed that "man is confronted with the greatest identity crisis of his history."[92] Did cultural primatologists, as they moved from the universalist commitments underlying Wilson's sociobiology to McGrew's chimpanzee multiculturalism, extend the ethnopolitics that would become so prominent among cultural anthropologists into a realm beyond the human? What exactly would an appreciation of cultural diversity in nonhuman species entail politically?

Writing for and against Animal Culture

McGrew evoked the culture wars in cultural studies, which in turn reflected and sought to intervene in the larger political conflicts going by that name.[93] But the bulk of his essay revolved around the debates within primatology and comparative psychology, as they related to cultural anthropology. The idea of animal cultures in general and chimpanzee cultures in particular had gained traction in the United States and Europe after a full-day symposium on protoculture at the Fourth International Congress of Primatology in 1972 brought together American, European, and Japanese students of primate cultures.[94] Against Sherwood Washburn, who had reminded the conference participants that a theory of culture was a "pure creation of human theorists," primatologist

Emil Menzel insisted that "a certain degree of 'naive realism' is not only warranted but essential for a practicing science of culture."[95] He refused to get bogged down in a debate over whether to call the object of their field of study culture, protoculture, preculture, infraculture, or biological bases of culture: "Whether or not any of these terms can be defined in a fashion that is acceptable to all is another problem. This problem might disturb deductively-oriented theorists; but it should not disturb the inductively-oriented researcher who wants to know first of all, what do animals and men *do*, and why?"[96] This tension between the empirical and the conceptual, between a positivist emphasis on observations and a constructionist emphasis on definitions, would haunt the chimpanzee culture controversy in the decades to come. One of its protagonists, Andrew Whiten, told me that there were two main causes of the controversy: disagreements over definitions and uses of different methods.

From the start, the warring factions clashed over the question of whether culture presupposed language: no, claimed the proponents of animal cultures; most certainly, replied the skeptics.[97] In part, this was a matter of what different researchers took the term *culture* to mean. In his interview with me, Whiten emphasized that there was "no empirical way of establishing those definitions. Sometimes scientists write that a definition is right. But that simply cannot be the case. Definitions are just things that we have to converse on, so that we know that we are talking about the same thing."

Preferences for different definitions, however, also reflected different intuitions about empirical questions such as the relative weights of observational learning and verbal instruction in human development. For example, chimpanzee researcher Toshisada Nishida downplayed the centrality of language by citing a Japanese proverb: "Children do not as parents tell them to do, but do as parents do."[98]

Yet language soon lost its pride of place and gave way to a more foundational debate over social learning, which enabled not only language but many other behaviors shared within groups. The question of whether apes could ape one another emerged as a linchpin of this discussion. This brought comparative psychologists to the scene. "The most virulent critiques of the idea of animal cultures came from psychologists, and not from ethnologists who largely ignored these new developments," noted philosopher Lestel.[99]

The opponents of animal culture had to unravel the paradigm case, which Japanese primatologists had established in the 1950s. "Imo was not so creative a 'genius' as the secondary literature suggests," asserted American psychologist Bennett Galef.[100] Elisabetta Visalberghi and Dorothy Fragaszy had reported that tufted capuchin monkeys and crab-eating macaques had easily learned how to wash their foods.[101] Galef inferred that each Japanese macaque on Koshima might have invented potato washing independently.

Consequently, Galef also questioned whether the monkeys had indeed traditionalized sweet-potato and wheat washing by imitating Imo, as Masao Kawai had suggested.[102] If they had really learned these behaviors from each other, Galef reasoned, the supposed custom should have spread faster and faster as more and more practitioners passed it on to their conspecifics at an accelerating rate, until all individuals had acquired the new behavior. Reanalyzing the data presented by Kawai, Galef inferred that instead the propagation of sweet-potato washing had been "painfully slow" and had not conformed with what mathematical models of social learning predicted.[103]

Moreover, Galef called into doubt whether the monkeys had maintained the behavior on their own account. From Steven Green, an American primatologist who had done fieldwork on Koshima in the 1970s, Galef had learned that the woman who provisioned the macaques had given sweet potatoes only to those individuals who washed them.[104] What appeared to be cultural transmission might have been conditioning, plain and simple, Green had surmised. In conversation with Galef, he had even speculated that Satsue Mito might have deliberately maintained the potato washing because it attracted scientists and tourists who spent money at the local inn and gave tips to caretakers.[105] Suddenly, the paradigmatic case of cultural primatology appeared less like a scientific revolution and more like a hoax, which sly locals had played on naive Kyoto professors and nature-loving day-trippers.

Having begun his career studying how Norway rats learned from each other how to avoid poison baits, Galef had no intention of categorically ruling out the possibility that Japanese macaques were capable of social learning. Instead he sought to demonstrate that the Japanese field observations were inconclusive and that alternatives to imitation, including "more humble types of social learning" as well as individual learning, could just as well account for what Imanishi and his students had taken for a case of nonhuman culture.[106]

This controversy over whether animals other than *Homo sapiens* imitated can be traced back to Darwin, who noted that "apes are much given to imitation, as are the lowest savages," and who speculated that imitating others, especially their inventions of new weapons and means of defense, would gradually give the more innovative and imitative tribe a competitive advantage over other tribes.[107] Around 1900, however, comparative psychologists like C. Lloyd Morgan, Edward Thorndike, John Watson, and Leonard Hobhouse argued that their experiments could not corroborate imitative faculties in other animals, at least not if interpreted as parsimoniously as possible.[108] In the emerging field of animal psychology, anti-anthropomorphism became a "badge of professionalism."[109]

Almost a century later, the debate over imitation flared up again. It emerged as the most loaded concept in the psychologists' typology of social learning because it came to be seen as the foundation of cumulative culture. So-called stimulus enhancement was another kind of social learning based on the observation of others: seeing them use a particular object motivated the observer to try it out herself, triggering a process of individual trial-and-error learning. An experiment conducted by psychologist Michael Tomasello and others at the Yerkes National Primate Research Center suggested that captive chimpanzees learned the use of a new tool through a variant of stimulus enhancement: it was not just the tool that appeared more salient to them when they saw another member of their group use it, but the tool in its function as a tool.[110] Tomasello called this third type of social learning *emulation*. Emulation sought only to reproduce the completed goal (e.g., an open nut), while the means remained unspecified and subject to trial-and-error learning.[111] By contrast, "true imitation" (not to be confounded with those unclear learning processes that Darwin and Imanishi had called *imitation*) required copying the demonstrator's actual behavior, not just its outcome. "Only imitation may be considered cultural transmission," Tomasello maintained.[112] The minuscule difference between imitation and emulation would eventually give rise to the gulf between the primitiveness of animal traditions and the vast complexity of human cultures, Tomasello speculated.

Tomasello's account of human uniqueness was soon challenged from within comparative psychology. First Andrew Whiten and others criticized that his colleague had tried to answer the question of whether the entire chimpanzee species could imitate by enticing subjects whose "levels of culture and imitation . . . may have atrophied" in captivity "to reveal their maximum competence in copying arbitrary, nonfunctional acts."[113] Whiten sought to avoid this mistake by asking the more "naturalistic question" of whether two particular groups of chimpanzees from Yerkes Primate Center and the Madrid zoo imitated "in a situation designed to be an analogue of foraging, a prime candidate for functional imitation in the wild." For this purpose, he designed an "artificial fruit," a Plexiglas box that bore no resemblance to real fruit other than that it contained food and could be opened in different ways, which human models demonstrated to both human children and adult chimpanzees. Whiten found that even these relatively unenculturated captive chimpanzees imitated the specific technique they had been shown by another primate species, although less faithfully than the children (but of course, no imitation was ever completely exact and fidelity was always a matter of degree, he noted). What Tomasello had presented as a qualitative difference between *Pan troglodytes* and *Homo sapiens,* Whiten turned into a quantitative difference. And Tomasello restored the qualitative difference by insisting on a conceptual opposition between emulation (Whiten's less faithful imitation) and true imitation, which could be observed only in humans.[114]

Frans de Waal pursued a conceptual line of attack reminiscent of how ordinary language philosophers have challenged scientists whose technical terminologies sowed confusion by distorting the ordinary meanings of words: if primatologists adopted the definition of culture propagated by experimental psychologists, which centered on imitative problem solving, then clothing, ornamentation, religion, food preferences, music, art, dance, and social styles would no longer be considered cultural. "Needless to say, this is not how we usually employ the term," de Waal added.[115] In other words, while there is no truth as to how to define culture, definitions that diverge so far from ordinary language make nonsense of it.

Masako Myowa-Yamakoshi and Tetsuro Matsuzawa took a reconciliatory stance.[116] They had also tested the ability of captive chimpanzees to imitate how a human model opened a box by using different tools (although in the face-to-face setting to be discussed in chapter 5). Like Whiten and his colleagues, they found that the animals did not reproduce the demonstrator's technique faithfully because they were less sensitive to his body movements. Their observation that the apes paid closer attention to where the tool was applied to the box was also in line with Tomasello's claim that chimpanzees emulated the completion of a goal instead of imitating the exact way that had led to this end point. Matsuzawa and his colleagues concluded: "Monkeys never ape, but apes ape in a few cases, even though their imitative abilities are a long way from those of humans."[117] He never committed to Tomasello's terminology and sidestepped the debate over how to define imitation, which he considered the origin of the controversy.[118] In contrast to his American colleague, Matsuzawa attributed culture to chimpanzees and defined it as "a set of behaviors that are shared by members of a community and are transmitted from one generation to the next through nongenetic channels," leaving open through what psychological mechanism such social transmission occurred.[119] He tried to shift the conversation from dissensus over words to consensus over experimental observations.

Since its beginnings in the late nineteenth century, the animal imitation debate had been perpetuated by the continual splitting of imitative phenomena and other forms of social learning into more and more categories: true imitation was opposed to instinctive imitation, stimulus enhancement, local enhancement, response facilitation, social facilitation, matched-dependent learning, observational learning, observational conditioning, contagion, copying, goal emulation, and so on. As participants and observers of this controversy, Hoppitt and Laland lamented that "the field of social learning suffers from such a bewildering array of often overlapping and inconsistent terms"—and yet "new classification schemes continue to be proposed."[120]

Amid this proliferation of conceptual distinctions, comparative psychologists managed to keep imitation at the center of the culture controversy. At first glance, this emphasis on doing as others do gave the concept of culture a conservative bent. Imitation enabled the most faithful transmission of all forms of social learning, accurately preserving useful behavioral adaptations across generations. But it also spared youngsters the need to reinvent the wheel, or at least the details of its manufacture, time and again. Hence, imitation enabled them to acquire skills at a younger age and to spend their adulthood improving them. "This so-called ratchet effect is what constitutes cultural progress in human societies, and it simply does not seem to be part of chimpanzee 'culture,'" Tomasello claimed.[121] "And a good argument could be made that it is precisely this progressive quality of human culture that underlies many of the unique aspects of human evolution."

The idea that human culture differed from primate protoculture in its ability to go beyond what had been learned from previous generations had already been entertained by mid-twentieth-century anthropologists and biologists.[122] But it seems as if cumulative culture had come to replace language, tool use, and so forth as the most prominent *differentia specifica* between humans and other animals only since the publication of Tomasello's book *The Cultural Origins of Human Cognition*.[123] Several recent publications have presented the ability of *Homo sapiens* to build on the shoulders of our predecessors as "the secret of our success."[124] At the beginning of the twenty-first century, cumulative culture became the new human nature. Or maybe not that new: in 1755, Rousseau had already claimed that "it is well demonstrated that the Monkey is not a variety of man; not only because it is deprived of the faculty of speech, but especially because it is certain that this species lacks the faculty of perfecting itself which is the specific characteristic of the human species."[125]

The resurgence of this understanding of human nature made cumulative culture one of the most hotly contested positions in the chimpanzee culture wars. Based on observations of how chimpanzees in Taï, Loango, and the Congolese Goualougo Triangle used complex tool sets to extract honey from beehives, Christophe Boesch inferred that a cumulative cultural evolution had brought about more and more complex and efficient technologies.[126] Considering that cumulative culture had been found even in homing pigeons, McGrew argued that ratcheting was not uniquely human.[127] Nor was it universally human because the evidence for ratcheting in the ethnographic literature was slim: "Most ethnographers of traditional societies report stasis, not dynamic change," McGrew claimed.[128] For him, the insistence on the accumulation of cultural complexity culminated in what he called the space shuttle fallacy: "What chimpanzee ever designed a space craft? Or composed a symphony? Or even cooked a soufflé? None, obviously. But in fact, most individual humans have done none of these (have you?), nor have whole human societies (foragers). Using such a criterion from modern industrial society would exclude entire traditional populations of *Homo sapiens* from humanity."[129]

Against an ethnocentrism that supposedly denied culture to fellow human beings (which neither Galef nor Tomasello nor any other participant in the chimpanzee culture controversy ever did), McGrew recommended realistic self-knowledge, humility, and a more inclusive appreciation of multicultural diversity.[130] Tomasello, on the other hand, complained in an interview with a popular philosophy journal about such politicization of animal research: "A political motive, which is praiseworthy and which I fully support, is the desire to conserve animals in their natural habitat and to protect them in zoos and other places where they are kept. Some people believe that, to be able to do so, they constantly have to proclaim how much they are like us. That's all well and good, but it's not science."[131] In an interview with me, American field primatologist Kevin Langergraber shared Tomasello's annoyance with his colleagues' moralism: "It's funny to hear this sneering in people's voices when they talk about the psychology department in Leipzig, like 'oh, aren't we morally superior, they don't even think that chimps can do that.' It's the same thing as with political correctness and racism, just with animals." As the front line shifted from cultural anthropology to comparative psychology, the struggle for the moral high ground continued.

Galef considered the controversy over animal cultures as one instance of a "wider debate regarding the relationship between human and animal behavior," which had begun in the 1980s with claims about the deceits of chickens and Frans de Waal's inclusion of chimpanzees in the once exclusively human zoo of political animals.[132] The ensuing conflicts pitted lumpers against splitters, claimed Galef. In this scheme, Japanese primatologists ended up in the lumper camp with such strange bedfellows as sociobiologist Edward O. Wilson, who had argued that "culture, aside from its involvement with language, which is truly unique, differs from animal tradition only in degree."[133] On the other side, comparative psychologists like Galef himself and Michael Tomasello joined ranks with fellow splitters such as cultural anthropologist Marshall Sahlins to defend human exceptionalism against a lumper zeitgeist.[134]

The controversy over monkey and ape culture revolved as much around primate behavior as it did around the behavior of primatologists. At stake was how the latter were to conduct good science. "Some feel (and I confess that this is my own bias) that it is probably best for scientists to be conservative," wrote Galef, "to adopt the simplest descriptions and explanations of behavioral phenomena consistent with the available evidence."[135] Following

Morgan's Canon, which advised comparative psychologists to refrain from interpreting an animal activity "in terms of higher psychological processes if it can be fairly interpreted in terms of processes which stand lower in the scale of psychological evolution and development," Galef recommended that his colleagues replace anthropomorphic interpretations with parsimonious ones as they sought to understand the development of locale-specific behaviors in animals.[136] Good scientists did not proclaim scientific revolutions but exercised "a healthy skepticism" and a "commitment to empiricism."[137]

Although Galef preached empiricism to the choir, not all choir boys agreed. His colleague Frans de Waal, for example, wondered why a committed skeptic and empiricist had "himself made so little effort to verify his own assumptions, for example, by actually visiting the island in person," where Satsue Mito could have explained to him that in a strictly hierarchical monkey troop, the presence of dominant males would have prohibited her alleged operant conditioning of sweet-potato washing by preferentially handing out food to a low-ranking individual such as Imo.[138] Of course, Mito's observations had been largely anecdotal, but de Waal preferred such "firsthand accounts by experienced animal watchers" to Galef's "overly skeptical [story] about human influence," which he dismissed as mere "armchair speculation."[139] Thus the methodological question of what form primatological empiricism should take opened up yet another front in the controversy over animal cultures.

Experimental Psychologists versus Field Biologists

Taking his cues from primatological fieldworkers like Goodall, McGrew, Boesch, and Nishida, comparative psychologist Michael Tomasello felt that "one of the most exciting findings to emerge from recent observations of free-ranging chimpanzees in equatorial Africa is that different populations behave differently."[140] Behavioral variation within primate species had emerged as a major concern since the 1960s.[141] Its interpretation had already been at the heart of another primatological debate. In the infanticide controversy, field primatologists fought over the question of whether killing infants was normal primate behavior, or whether it reflected abnormal ecological conditions at particular field sites.[142] In this case, variation had been used to call into question the naturalness of the contested behavior. In the chimpanzee culture controversy, by contrast, the proponents of chimpanzee culture pointed to behavioral differences between groups to *corroborate* their claim that cultural learning was part of chimpanzee nature: it was perfectly normal (and not the effect of pathogenic environments) for all communities of *Pan troglodytes* to behave differently because all members of the species acquired new behaviors through social learning. While the infanticide controversy raged between fieldworkers who took the particularities of their research site as representative of the entire species, the main front line of the chimpanzee culture controversy divided fieldworkers and laboratory researchers. Neither side asked whether it was normal for chimpanzees to crack nuts with hammer and anvil or to prepare tools for honey fishing. But experimentalists like Tomasello doubted that uncontrolled field observations could validate claims about the learning mechanisms underlying the behavioral differences their colleagues reported from their field sites.

About to make a career of comparing the cognitive capacities of human children with those of captive apes, Tomasello sided with Galef in claiming that "only imitation may be considered cultural transmission."[143] He denied the fieldworkers' methodological ability to discern how a behavioral difference between chimpanzee communities had come about. Consequently, they were in no position to declare the observed behavioral difference a cultural difference.

Field observations could not "distinguish sensitively among different types of learning that may be involved," Tomasello maintained.[144] They could not tell whether members of a group had acquired a distinctive behavior by individually learning to deal with a particular environment, or whether they had learned it from each other—and, if they had indeed learned it socially, whether they had imitated their fellows. Tomasello agreed with Galef that only controlled experiments could discern what kind of social or individual learning the animals employed.[145] Thus, what had started out as the conceptual question of whether culture required imitation and as the empirical question of whether Japanese macaques or chimpanzees were indeed capable of imitation gave rise to a controversy over the relative value of laboratory and field research.

Considering that naturalistic observation tended to be the preserve of field biologists while most psychologists preferred to stay in their laboratories, Tomasello's thrust also pitted two academic disciplines against each other—unless, of course, biologists accepted the psychologists' proposed division of labor, which ultimately meant that they could have the first but not the last word in the debate over animal cultures.

In his own account of the chimpanzee culture wars, McGrew proposed a different scheme for cooperation between the disciplines: anthropologists asked *what* constituted culture (artifacts, rituals, etc.), psychologists asked *how* these elements constituted a culture (especially how they were invented and spread), and zoologists asked *why* culture had evolved in the first place, answering this question in terms of neo-Darwinian evolutionary theory (adaptation, fitness, survival value).[146] Cultural primatology synthesized these three perspectives, and comparative psychology was reduced to a mere feeder discipline.

Such proposals for how to structure interdisciplinary research on animal cultures might have been less contentious if the different factions represented only different methodologies. But the simultaneously cooperating and competing disciplines were also identified with conflicting ontological positions as they disagreed over whether to attribute culture to chimpanzees and other nonhuman species: "Ethologists (Boesch, Goodall, Nishida, Wrangham), who study chimpanzees observationally in nature, tend to say yes," noted McGrew.[147] "Comparative psychologists (Galef, Premack, Tomasello), who study apes or rodents experimentally in captivity, tend to say no." Exceptions such as chimpanzee culture–denying field primatologist David Watts and chimpanzee culture–affirming psychologists Frans de Waal and Andrew Whiten proved the rule.

Mutual methodological skepticism and strife over ontology hardened the antagonism. For example, psychologist Daniel Povinelli, a renegade who, "like a drunken man slowly sobering," had sworn off the "myth about an animal who, behind a thicker coat of hair, a more prognathic face, and an eerie silence, carefully guarded a mind nearly identical to our own," denounced a "cult-like atmosphere surrounding studies of chimpanzees."[148] He and his father wrote a venomous review of Christophe and Hedwige Boesch's monograph on the Taï community: "The authors can revel in the intelligence of wild chimpanzees—the pure, uncontaminated minds of animals whose brains tick in harmony with the problems against which their neural systems were honed across hundreds of thousands of generations."[149] But their assumption that chimpanzees were smarter than the lab worker thought was supported largely by "the prodigal friend of the primatologist: the Anecdote," contended the Povinellis.[150] The Boesches, on the other hand, reminded their readers that "in everyday life as well as in clinical experience most of our knowledge of human beings is based on anecdotes," which could compensate for the lack of statistical analysis if multiple observations turned out to be consistent with each other.[151] The Povinellis noted downright hostility between researchers who studied chimpanzees in the forests of Africa and those who studied them in captive settings, and they accused the Boesches of "not seeking so much to resolve the tension between the field- and lab-worker, as much as to define, dig, and man the trenches in what they clearly see as a very long battle in a war whose most likely end is stalemate."[152]

The opposition of field and laboratory scientists had already constituted one of the front lines in the sociobiology debate. Segerstråle noted that the leading figures of sociobiology had been trained as fieldworkers, whereas many of their most prominent critics were experimentalists.[153] Each side accused the other of doing bad science. In the eyes of the bench workers, the adaptationist just-so stories and plausibility arguments of sociobiologists fell short of the standards of controlled experiments, whereas sociobiologists felt that their hard-nosed critics got bogged down in devising the right control conditions and lacked the naturalist's desire and enthusiasm for understanding the world of living things and the meaning of human existence. Although cultural primatologists emphasized social rather than genetic transmission of behavior, McGrew, Boesch, and Whiten welcomed sociobiology, and sociobiologists like Wilson and Dawkins embraced the idea of animal cultures.[154] The association between their shared commitment to human-animal continuity and their penchant for fieldwork appears contingent, but, following Segerstråle, it might well have provided the basis for reading each other's work sympathetically in the right naturalist spirit while laboratory scientists adopted a critical distance toward the epistemic cultures of both sociobiologists and cultural primatologists.

Chimpanzee Ethnology: Primate, All Too Primate

In the course of the 1990s, cultural primatology developed from a seemingly exotic quirk of the Japanese and a pet project of a few radical Americans and Europeans to a fully formed scientific field. As a growing number of chimpanzee ethnographers documented the cultures of their ape communities, McGrew's dream of moving beyond chimpanzee ethnography to an "ethnology of the species" that systematically compared those cultures was about to come true.[155] Representatives of the seven most long-term study populations, including Gombe, Mahale, Bossou, and Taï, teamed up to form the Collaborative Chimpanzee Cultures Project (CCCP).

The resulting landmark paper "Cultures in Chimpanzees" by Andrew Whiten and colleagues came out in the journal *Nature* and put the geographical variation of chimpanzee behavior on the map far beyond primatology.[156] It quickly became the most widely cited of all European animal behavior research papers between 1999 and 2010.[157] Soon students of other large-brained mammals employed the CCCP approach to demonstrate culture in orangutans, capuchin monkeys, whales, and dolphins.[158] Making front-page news in the *New York Times,* the chimpanzee study also introduced cultural primatology to the public, which led Frans de Waal to exclaim: "Perhaps this paper will permanently alter our own culture."[159]

In the face of its success, however, the CCCP also revealed some of the obstacles that hindered the kind of collaboration necessary for cross-cultural comparisons. For one, it speaks volumes that the first author of the most widely cited publication on the cultures of wild chimpanzees had made a name for himself with fieldwork on baboons. On chimpanzees Andrew Whiten had conducted primarily captive as well as some field experiments, but he was no chimpanzee ethnographer. Although the first phase of the project, compiling a comprehensive list of candidate cultural variables, had been conducted by him and Christophe Boesch, it was Whiten who served as the "honest broker," bringing on board Jane Goodall, William McGrew, Caroline Tutin, Toshisada Nishida, Vernon Reynolds, Yukimaru Sugiyama, and Richard Wrangham. In an interview with me, Whiten explained that the reason these notoriously competitive researchers had all agreed to collaborate was probably that he (not Boesch, that is to say) had issued the invitations: "Someone, on the one hand, they could trust as an ecologist, who studied baboons. And—that was the other side—I wasn't a chimpanzee scientist." Rumor had it that especially the Japanese would not have accepted Boesch as lead author. As one prominent student of chimpanzees later said to me: "I have no reason to pant-grunt Christophe." In other words, the researchers whom Galef had lumped together as the lumpers turned out to be split among themselves.

The glaring absence of some important field sites from the *Nature* article provided another hint that field primatologists did not stand united against comparative psychologists, as McGrew's and Povinelli's accounts of the controversy suggested. Representing the Ngogo Chimpanzee Project in Kibale National Park, David Watts wrote to me after a talk I had given in the Anthropology Department at Yale University: "I am skeptical about how far some of my colleagues extend 'culture,' which they threaten to make synonymous with 'social learning.'" Apart from the controversy between Boesch and Tomasello about what specific kinds of social learning chimpanzees were capable of, "there's also another debate," Watts pointed out, "or maybe a non-debate, between Andy Whiten, Bill McGrew, Christophe, et al. and those who dismiss such considerations, sometimes out of hand. To me, what chimps do is wonderful and fascinating, but we have cognitive abilities that they simply lack and that allow us to generate the human version of 'culture.'" Asked why this "non-debate" remained confined to corridor talk and email exchanges, McGrew offered a sociological explanation: "David, probably out of consideration for his fellow high-ranking males, isn't gonna go out in public and trumpet his criticism of those of us who do cultural primatology. I think another reason why there are people like David is that they work in departments where there are frictions with sociocultural anthropologists. They would much rather not get involved in that debate."

Scientists are always competitive, of course. Since the inception of science studies in the 1980s, anthropologists and sociologists have examined this human, all too human—or should we say "primate, all too primate"?—underbelly of one of the most highly prized domains of modern life. But the rivalries obstructing collaborations that would have facilitated the transition from chimpanzee ethnography to chimpanzee ethnology also reflected the elevated status of the field. The charisma of their research subjects brought chimpologists an undue share of public interest, one of them confided. An absolutely ingenious student of lorises still wouldn't get half the attention that a miserable chimpanzee researcher could count on. Organgutan researcher Carel van Schaik told me: "Within primatology, there is a hierarchy: the chimp people are at the top, followed by the other great ape people, then the monkey people, and at the bottom are the lemur researchers. The citations are always asymmetrical: chimp people only cite chimp people, but if you're a lemur person, you have to cite the monkey and ape literature." This chimpocentrism, as American primatologist Benjamin Beck had called it, amounted to bad science, van Schaik claimed.[160] The limelight also poisoned human relations between chimpanzee researchers. "If you study chimpanzees, it's all politics," a doctoral student in Taï lamented. "No matter what you find out, it will always make you powerful enemies. By contrast, if you study one of the many little known primate species, everybody will be happy when you present your results. Nobody feels challenged."

The Method of Exclusion

The controversy around "Cultures in Chimpanzees" also highlighted a fundamental methodological problem that had haunted chimpanzee ethnography and ethnology since their inception. Describing geographical variation of behavior as cultural diversity presupposed that social learning rather than individual adaptation to different shared environments had brought about the recorded differences between chimpanzee populations. The ethnographic method underlying the *Nature* article by Whiten and colleagues sought to demonstrate the cultural nature of the described behaviors by excluding alternative genetic and ecological explanations.[161] Since Boesch and Whiten had compiled a list of behaviors that varied between *Pan troglodytes* populations, they considered genetic differences an unlikely explanation, just as cultural anthropologists had come to consider racial differences an unlikely explanation of behavioral diversity in humans. The primatologists pointed to one particular case to make this assumption more plausible: no nut cracking had been observed east of the N'zo-Sassandra River, which divided Côte d'Ivoire, even though the same chimpanzee subspecies, *Pan troglodytes verus*, lived on both shores.[162] It seemed as if the technique had been invented only once in far western Africa but not a second time on the other side of the uncrossable barrier. This account seemed to rule out a genetic explanation of nut cracking in Taï and Bossou.

As Whiten and Boesch pruned the list of candidate cultural variants, the CCCP members also considered whether local ecological conditions could explain the absence of a behavior at some field sites, in which case they struck the variant off the list. For example, it seemed reasonable to assume that only the Bossou chimpanzees had been observed fishing for algae because other populations encountered hardly any algae.[163] Thus, the authors dismissed environmental causes. In the debate over animal cultures, this so-called method of exclusion was often used as a synonym of "the ethnographic method."[164] But this way of establishing culture by eliminating genetic and ecological explanations of geographical differences in behavior had come under attack.

Early on, Tomasello had already rebutted that naturalistic observations could rule out environmental factors rather than social learning in determining how chimpanzee communities lived at different field sites: "It is impossible *in principle* for ecological analyses by themselves to answer questions about learning processes. On the one hand, failure to find ecological differences between groups does not mean there are or were none. . . . On the other hand, the converse of this is also possible; that is, it is also possible that a population difference in behavior may be accompanied by an ecological difference that seems important to us, but in reality is irrelevant."[165] Although Whiten and colleagues assured the readers of *Nature* that they had considered all relevant ecological factors, they provided no information on which factors they had actually taken into account, as Kevin Laland, Jeremy Kendal, and Rachel Kendal complained.[166] The possibility remained, for example, that chimpanzees learned different methods of ant dipping individually rather than socially as each ape figured out for itself how to best eat the local species of ants without being bitten.

During my fieldwork at the Max Planck Institute for Evolutionary Anthropology in Leipzig, Langergraber, who visited the Department of Primatology from the Ngogo Chimpanzee Project, raised an objection to the ethnographic method as Whiten and colleagues had applied it: failure to find a behavior in one population that had previously been observed in another population did not necessarily mean that there really was a behavioral difference.[167] For instance, he called into question whether chimpanzees living east of the N'zo-Sassandra River indeed did not crack nuts. No primatologist had habituated any communities in this region. During excursions Boesch and his colleagues had not noticed the characteristic sounds of nut cracking, nor did local people they had interviewed report such noises.[168] "But that's as if I went for a stroll in Leipzig and afterwards I would make claims about what people here did *not* do," Langergraber told me. And sure enough, Bethan Morgan and Ekwoge Abwe found that chimpanzees did use stone hammers to crack open *Coula* nuts in Ebo Forest, Cameroon, about 1,700 kilometers east of the riverine information barrier.[169] This suggested either that western chimpanzees had invented nut cracking on multiple occasions, or that they had invented it a very long time ago and the nut-cracking traditions of communities between the N'zo-Sassandra River and Ebo Forest had all gone extinct. "Understanding the extinction of chimpanzee traditions holds promise for explaining why ape culture has never blossomed as it did, critically, for humans," noted Richard Wrangham.[170] The third explanation was that they simply hadn't been studied well enough, as Langergraber suggested.[171]

Where Whiten and colleagues had identified real behavioral differences, Langergraber objected, they had failed to rule out genetic causes. Evolutionary biologists Kevin Laland and Vincent Janik had already criticized that only one-third of the supposed cultural variants had been observed in the same chimpanzee subspecies, while the rest could just as well be attributed to differences in DNA between *Pan troglodytes troglodytes, P. t. schweinfurthii, P. t. verus,* and *P. t. vellerosus.*[172] And indeed Langergraber found that behavioral dissimilarity between chimpanzee groups correlated with their genetic dissimilarity. This was not to suggest that genes had caused any of the supposedly cultural behaviors—migrants could have carried both cultural variants *and* genes between neighboring populations—but to demonstrate the insufficiency of the method of exclusion as the dominant approach used to identify animal culture in the wild. "A strict application of the method of exclusion may lead to an underestimation of the true number of cultural variants that exist in the wild," Langergraber and colleagues pointed out in a reduction ad absurdum: "Indeed, if the method of exclusion were applied to humans, the strong correlations between behavioural and both genetic and ecological similarity would indicate that a considerable amount of human between-group diversity is not necessarily cultural in nature."[173]

Laland, Kendal, and Kendal put it even more boldly: "We suspect that were the ethnographic approach to be rigorously applied, it would reject most genuine cases of culture."[174] Their talk about "genuine" cases of culture revealed that, unlike the splitters Galef and Tomasello, this group of critics did not want to negate animal cultures. Quite the opposite—as arch lumpers, Laland and his colleagues advocated a conception of culture qua social learning that was broad enough to include even guppies.[175] In McGrew's eyes, such a "dumbed-down version" made a "mockery of culture as the complex phenomenon it was originally defined to be by anthropologists."[176] Laland and his coworkers conceived of cultures as adaptations to local environments and predicted a co-variation of social and ecological factors, which the CCCP had not tested statistically. Consequently, they rejected the purification of behaviors as genetic, ecological, or cultural: "This categorical thinking evokes memories of the nature-nurture debate."[177]

Yet primatologists and cetologists who were committed to the method of exclusion, which they distinguished from ethnography, knew that the dichotomy of innate and learned behaviors couldn't vet candidate cultural variants. In their response to Laland and Janik, Michael Krützen, Carel van Schaik, and Andrew Whiten clarified that, as ethologists in the tradition of Robert Hinde and Hans Kummer, they did not apply the genetic/learned dichotomy to a particular category of behavior.[178] Instead they followed Kummer's proposal to resolve the nature/nurture controversy by considering that "only a difference between traits, not a trait as such, can be called 'innate' or 'acquired.'"[179] Kummer had provided the following example: speaking French was both acquired and had an inherited genetic basis (the capacity to speak), and the same was true of speaking Italian, but the difference between speaking French and Italian was purely acquired. In this sense, Krützen and colleagues argued, the method of exclusion could demonstrate that "a difference in behaviour between two populations could be caused only by social learning and not by either genetics or individual learning."[180]

At stake in the debate over the method of exclusion was how to constitute culture as an object of natural scientific research. The attempt to isolate it from genetics and ecology sought to make a stronger case for animal cultures. It served to demonstrate an ontological autonomy reminiscent of the freedom of human culture from nature, which continues to inform the philosophical tradition—except that cultural primatologists considered social learning as much a part of nature as genes and ecosystems. But just like McGrew's nonsubsistence criterion, the method of exclusion invited allegations of applying double standards to *Homo sapiens* and all other animals. Both the practitioners of the method of exclusion and their critics understood themselves as naturalists in the metaphysical sense. Nobody assumed any nonphysical stuff. But they disagreed about the place of culture in the wider biological world.

Symbolic Cultures in the Forest

Did the method of exclusion endow the concept of chimpanzee culture with the autonomy that anthropologist Marshall Sahlins had demanded for human culture?[181]

Sahlins's attempt at purging culture from biology rested on a narrow definition of culture as symbolic culture. He loved to quote Leslie White's bon mot that "no ape could appreciate the difference between holy water and distilled water—because there isn't any chemically."[182] Maybe it is no coincidence that Kummer had exemplified what would come to be known as the method of exclusion with a paradigm case of symbolic culture. By definition, symbols refer to the world in arbitrary fashion. No inherent quality of apes makes Frenchmen call them *singes* while Italians use *scimmie*. By referring to the same creatures differently, speakers of these languages adapt to nothing other than social conventions, and at least collectively, they would be free to adopt different terms. This capacity for symbolic thought distinguished *Homo sapiens*, anthropologists like White and Sahlins believed. It wasn't that "the lower animals can do these things but to a lesser degree than ourselves," White insisted; "they cannot perform these acts of appreciation and distinction *at all*."[183] This dictum presupposed more than a purely referential understanding of the symbol as something that stands for something else. "The meaning of a symbol can be grasped only by non-sensory, symbolic means," White had argued, and "the keenest senses cannot capture the value of holy water."[184] But then Christophe Boesch confided to me that as an atheist, he couldn't tell the difference, either.

Kummer's former doctoral student Christophe Boesch took on Sahlins's challenge by describing and comparing what he considered symbolic cultures in chimpanzees. In 1991, he described a system of symbolic communication in the Taï chimpanzees who drummed on buttresses to coordinate their travels.[185] But he had to move from chimpanzee ethnography to ethnology to demonstrate that their gestures were indeed linguistic signs. Every sign had two sides, a perceivable shape and a mental concept, which were connected by convention alone. Structuralist linguists spoke of signifier and signified. In his book *Wild Cultures*, Boesch summarized a growing number of field studies of gestural communication in great apes and suggested that not only enculturated but also wild chimpanzees (and possibly other species such as bonobos, orangutans, dolphins, and whales) could use different signs to convey the same meaning and conveyed different meanings with the same sign. For example, at Taï, a young male who wanted to signal to a female his desire for sex without attracting the attention of higher-ranking males would knock with his knuckles on a tree trunk, whereas at Mahale he would clip a leaf. For Taï chimpanzees, however, such leaf clipping would have announced that he was about to display, and at Bossou it meant "I want to play with you."[186] Considering the geographic proximity of the Taï and Bossou populations and the abundance of leaves and tree trunks at all three locations, Boesch excluded ecological and genetic explanations of these geographic differences in gestural communication and took them as evidence that chimpanzee communities had different material as well as symbolic cultures.

Symbolic culture in chimpanzees possesses one of the key attributes of culture in humans, namely a totally arbitrary nature disconnected from ecological constraints. The meanings of the sign codes are based purely on "social shared conventions" that are understood by all group members. This could be viewed as the first step toward the development of belief, myth, art, and discourse.[187]

The use of the method of exclusion to purify culture as an epistemic object should not be confused with an ontological commitment to a dualism of nature and culture. Boesch's brand of naturalism was monist through and through. He maintained that culture had emerged to cope with quickly changing environmental conditions. Especially long-lived species such as hominoids and cetaceans profited from social learning because it happened at a much faster pace than genetic mutations and their natural selection. The partial decoupling of cultural evolution from ecological determinants was secondary: "Once you have acquired the ability to learn socially, why not do it to respond to purely social challenges as well?" Boesch asked rhetorically in an interview with me.

Boesch used the case of symbolic culture to satisfy critics who insisted on the exclusion of alternative explanations and who refused to acknowledge hybrids of, say, social learning and ecological adaptation as cases of genuine culture. But, unlike symbolic and interpretive anthropologists, Boesch refused to make the ontological autonomy of the symbolic realm paradigmatic of culture per se. Even if the relationship between a group's particular signifiers and the entities that these gestures or drummed codes signified was arbitrary, the use of symbolic communication as such served an adaptive purpose. For example, the Gombe chimpanzees might not communicate their travel directions by drumming on buttresses, Boesch speculated, because they did not live in a dense forest full of leopards who would be attracted by frequent contact calls between individuals unable to see each other.[188] If culture had eventually gained some autonomy from genetic and environmental conditions, that still did not endow chimpanzees—or humans for that matter—with the kind of "freedom from biology" that anthropologists of Marshall Sahlins's ilk had postulated.[189]

Onward to Chimpanzee Nature

At the time of my fieldwork between 2013 and 2016, the controversy over chimpanzee culture had taken yet another turn. Since McGrew and Tutin's first article on the subject, the emphasis had always been on cultural difference.[190] In the following three decades, the literature documented more and more behavioral variation between communities of great apes, whales, dolphins, and other animals.[191] McGrew should have been pleased. Over time, however, he grew concerned about the tendency of cultural primatologists to get lost in minutiae. Although not motivated by the politics of difference, they were about to take a road similar to the one cultural anthropologists had traveled during the culture wars, which had led many to become so suspicious of human universals that they had given up anthropology's original epistemic object, *anthropos*. Instead they reoriented the discipline from being concerned with the question of what makes us human to problems of racial, cultural, and ontological otherness.[192] At a congress of the Primate Society of Japan in 2015 on the campus of Kyoto University, where Imanishi and his students had first postulated that nonhuman primates had culture, too, McGrew warned: "In chimpanzee studies, we live for contrasts. We look for differences, for new things that go beyond what has been reported at other sites. In the process, we may lose track of the importance of the opposite of variation, by which I mean chimpanzee universals."

McGrew was especially concerned about what had happened after the publication of the landmark paper in *Nature*.[193] As chimpanzee ethnographers offered more detail, they split up the behavioral categories that they had compared in "Cultures in Chimpanzees." To detect cultural variation they had to demarcate one cultural trait against another. But where to draw the line if behaviors varied by degree, not kind? Primatologist Claire Watson and colleagues noted that the method of exclusion, as applied by Whiten and his coworkers, disregarded such quantitative variation and focused entirely on determining the presence or absence of qualitative intergroup differences.[194] Sociocultural anthropologists with a penchant for positivist analysis faced the same problem, Boesch explained in an interview with me: "When I look at studies of pottery, I wonder what is considered a trait. They measure how long the necks of different bottles are. Each time it's a little bit longer, it counts as a different trait. If they find a bottle with a long neck, they take it as the product of cumulative cultural evolution. Does it really make any difference whether the neck is one or two centimeters longer?" In principle, of course, any quantitative difference could be translated into qualitative differences such as the use of "long" (ca. 70 cm) and "short" sticks (ca. 37 cm) for driver ant dipping at Taï and Gombe, respectively.[195] Yet nothing prevented researchers from populating the interval between long and short sticks by more and more intermediaries, thereby multiplying cultural differences ad infinitum.

This trend toward finer distinctions proved to be a serious obstacle to a second round of the Collaborative Chimpanzee Cultures Project. One and a half decades after successful completion of the first, CCCP-2 had still not come to fruition. This even more ambitious undertaking included twice as many field sites, and the number of behavioral variants skyrocketed from 39 to 571—not because cultural primatologists had discovered more than five hundred completely new behaviors but because they had noticed more and more nuanced variation in the already described behavioral patterns. A table comparing all twelve communities, trait by trait, now comprised almost seven thousand cells. To complete phase 2 of the project the researchers would have to assign each of these cells one of six codes. Ultimately, this massive tableau of chimpanzee cultures would comprise more than forty thousand coding options.[196] And each time behaviors were split or lumped, the categorization scheme and thus the whole matrix had to be revised. "We ended up with a vast catalogue, which is why we still have not published it," Whiten told me in an interview. "Not surprisingly," McGrew pointed out, "energy has waned." Should the results of this grueling effort still see the light of the day, it would paint a pointillistic picture of staggering cultural diversity among chimpanzees.

More importantly in McGrew's eyes, lumping together the nuanced variations in a preliminary analysis had allowed identification of about thirty-five universal behavioral complexes such as bed making, drumming for communication, and branch dragging in agonistic displays, which young and old, male and female, dominant and subordinate individuals showed all over Africa. This return from the diversity of chimpanzee cultures to the unity of chimpanzee nature differed from claims about *the* chimpanzee that had preceded cultural primatology. For CCCP-2 no longer universalized the behavior of one group, most often Goodall's somewhat atypical woodland-living Gombe chimpanzees, unaware of any geographic variation. Rash talk about whole species assuming that chimpanzees were chimpanzees no matter where one found them was not what McGrew had in mind when he noted "the total absence of publications on the subject of universals in nonhuman primates."[197] Establishing true universality in a species was hard work, and it could be achieved only by way of collective observation: "One needs to have a decent (statistically representative) sample of populations that have been observed enough, and then to find a behavioral pattern present in all. . . . The ethnographic record for apes was not substantial enough for such an exercise until now."[198]

Culture in Crisis

The controversy over chimpanzee cultures got going in Europe and the United States just as the culture wars began to repoliticize the culture concept—a development that profoundly transformed the discipline of anthropology. In the 1950s, Alfred Kroeber, Clyde Kluckhohn, and others had tried to establish an objective science of culture, and in the following generation, Clifford Geertz had made a detached hermeneutics of culture the paradigm for interpretive cultural anthropology.[199] In the 1970s and 1980s, anti-imperialist, antisegregationist, and feminist issues repoliticized the discipline. As debates over multiculturalism polarized the North American and European public, the concept of culture again became an object of intense problematization—so intense that many cultural anthropologists decided to discard the culture concept just as cultural primatologists adopted it. What had happened?

At first glance, the discourse of multiculturalism simply restored the original political impetus of the culture concept. Johann Gottfried von Herder and his fellow Romantics had already celebrated the unity and plurality of cultures.[200] *Kultur* became the rallying cry for the inhabitants of thirty-nine independent German states that sought to govern themselves as one nation no longer subject to the rule of noble families and royal houses. It also became the rallying cry for those who defended the unique character of each *Volk* against the universal progress of Enlightenment rationalism and western European "civilization." Just as early nineteenth-century Germans had mobilized the culture concept to save their country from "the long road west," which it would only really take after 1945, it also became an important discursive resource for non-European peoples who wanted to create and defend their collective identities against Westernization, modernization, and imperialist exploitation and domination.[201] The Kyoto School's insistence on a genuinely Japanese primatology that resisted the Anglo-American hegemony of disciplinary norms and forms could serve as a case in point. Despite their reliance on colonial infrastructures, cultural anthropologists often understood themselves as advocates of the people they studied and used the culture concept not just for analytic purposes but also to defend their research subjects' alterity and autonomy. In the United States, for example, Franz Boas, Alfred Kroeber, and others had demanded that the US government recognize the cultural, moral, and political equality of their Native American research subjects. From the start, cultural anthropology had been as much a philosophical and scientific as a political project. The mobilization of the culture concept by an ethnos-centered or cultural left only picked up where Franz Boas and his students had left off.

In the 1980s, however, the debate over multiculturalism brought out the tension between the unifying and differentiating impulses of the culture concept. The proponents of multiculturalism challenged the idea of the United States as a "melting pot," in which different cultures and races blended into a homogeneous society, and proposed to think of this ideal republic as a "salad bowl," in which people got to preserve their individual identity and cultural distinction while being transformed into something new.[202] Just as the biological humanism of the species-centered left had come under suspicion of serving as a cover-up for the racist discrimination that it had set out to fight, the ethnos-centered left rejected calls for cultural assimilation and unity as a conservative ploy to impose Anglo-American dominance on ethnic minorities in the name of national culture.

Soon this struggle for the self-determination of all peoples turned against the ontological presupposition of multiculturalism itself, namely that every human being belonged to a particular culture. British anthropologist Marilyn Strathern proposed that Melanesians had neither nature nor culture.[203] Eduardo Viveiros de Castro later argued that Amerindians had reversed the Western cosmology of mononaturalism and multiculturalism: their animist ontology posited one culture that humans shared with all other animals, and many natures that set the species apart as each inhabited its own kind of body.[204] Of course, anthropologists committed to the biological unity and cultural diversity of humankind, originally established as a bulwark against the polygenist doctrine that different races had descended from independent pairs of ancestors, could always maintain that Melanesians and Amerindians had culture but did not know that they did—just as Thibaud Gruber and colleagues had argued for apes.[205] Yet Strathern, Viveiros de Castro, and more recent proponents of the so-called ontological turn challenged their colleagues to bracket their own ethnocentric cosmology, including their "culture cult." Instead they conceived what Roy Wagner had called a "reverse anthropology," which did not take the European and American conceptions of nature and culture for granted but made a serious effort to look at its own organizing categories through the eyes of people—and species!—who imagined and inhabited very different worlds.[206]

Other anthropologists observed a cultural diffusion of the culture concept itself. In the second half of the twentieth century, more and more people came to understand and assert themselves in terms of their cultures. For example, Terrence Turner reported that in the 1980s, the Amazonian Kayapo adopted the Portuguese word *cultura* to defend their way of life against assimilation into the larger Brazilian society, and as we saw in the last chapter, the Japanese bypassed competition with the West and restored their national pride as they came to understand themselves and their monkeys in terms of *bunka*.[207] Right after the end of the Cold War, conservative intellectuals such as American political scientist Samuel Huntington agreed with their multiculturalist opponents about culture as a locus of resistance but looked at it with hawk eyes as the spark that could ignite a coming clash of civilizations: "The great divisions among humankind and the dominating source of conflict will be cultural."[208] It seemed as if left and right, the West and the rest had finally found some common ground. Marshall Sahlins noted that anthropologists were now "faced with a world Culture of cultures," which enabled humanity "to realize itself culturally as a species being." What an irony of intellectual history: via the detour of Romantic particularism, the culture concept would have realized the dream of Enlightenment universalism. We are all united in our insistence on being different from each other!

At the same time, the growth of international flows of people, goods, money, and ideas like culture undercut the analytic purchase of the culture concept at the height of its success. In 1755, Rousseau had already noticed that "travels and conquests bring different Peoples closer together, and their ways of life grow constantly more alike as a result of frequent communication."[209] Two dizzying centuries of spiraling globalization later, historian of anthropology James Clifford wrote: "In a world with too many voices speaking all at once, a world where syncretism and parodic invention are becoming the rule, not the exception, an urban, multinational world of institutional transience—where American clothes made in Korea are worn by young people in Russia, where everyone's 'roots' are in some degree cut—in such a world it becomes increasingly difficult to attach human identity and meaning to a coherent 'culture' or 'language.'"[210] This "'postcultural' situation" further complicated the job of the anthropologist as a researcher who was "in culture while looking at culture" because the unprecedented overlay of traditions, beautifully exemplified by cultural primatology, muddled the question from which culture he was looking at which other culture: the boundedness, homogeneity, and authenticity of either now appeared less fact than fiction.[211]

And yet Clifford could not yet do without the deeply compromised idea of culture, however strategic and selective its uses, because the construction of cultural identities benefited indigenous and other marginalized people in their political and legal battles for land claims and against assimilation. Instead he was "straining for a concept that can preserve culture's differentiating functions while conceiving of collective identity as a hybrid, often discontinuous inventive process."[212] That it equally served identitarian movements in Europe and more recently in the United States certainly added to the predicament of culture, but it did not divert Clifford from a politics of cultural difference.[213]

Others took more radical positions and abandoned the dubious concept of culture altogether. There was always an ontological side to these critiques: the conviction that culture, if understood as a quality shared by all members of a distinct social group, did not exist. But the fury with which it was dismissed did not derive from a scholastic investment in metaphysical questions but from deeply political sentiments. Arab American anthropologist Lila Abu-Lughod demanded that her colleagues stop "writing culture" because "'culture' operates in anthropological discourse to enforce separations that inevitably carry a sense of hierarchy."[214] Extending the practice of participant observation from the epistemological to the political realm, engaged cultural anthropologists did not conceive of hierarchy as the dominance rank order of particular communities of *Homo sapiens* to be documented ethnographically, but as a social ill to be fought in the name of egalitarianism.

Like many of her colleagues, Abu-Lughod rejected the ethos and the "language of power" of positivist social science: "Cultural anthropologists have never been fully convinced of the ideology of science and have long questioned the value, possibility, and definition of objectivity."[215] Since the anthropologist's knowledge was always situated, the cultivation of interpretive and normative restraint appeared not only in vain but a moral failure to live up to the demands of critical scholarship, which simultaneously represented and intervened in the social world.

Although representing others in terms of their culture was better than race because culture is learned, can change, and "removes difference from the realm of the natural," the culture concept still operated much like race in that it essentialized and froze differences, Abu-Lughod worried.[216] She concluded: "If 'culture,' shadowed by coherence, timelessness, and discreteness, is the prime anthropological tool for making 'other,' and difference, as feminists and halfies [people of mixed national and cultural identity] reveal, tends to be a relationship of power, then perhaps anthropologists should consider strategies for writing against culture."[217]

The solution was not to return to Enlightenment universalism and biological humanism. From Edward Said, Abu-Lughod had learned that the elimination of the cultural difference between "the Orient" and "the Occident" did not mean the erasure of all differences but the recognition of even more differences: postcultural anthropologists would abstain from all generalizations and write "narrative ethnographies of the particular" about individuals in their time and place.[218] She was a splitter, if ever there was one.

This plea for "writing against culture" fully unfolded a self-contradiction that had lain dormant in the political heart of the concept of culture all along. Any plea for uniting people in the name of culture could also be resisted in the name of a culture. The inherent pluralism of the modern culture concept allowed not only the demarcation of one culture from other cultures, but also the undermining of any such demarcation. Cultural difference could always be claimed at a more local level to resist integration into a larger unit, all the way down to the point where every individual claimed his or her own—still infinitesimal "cultural"?—identity.

The identity politics of America's cultural left provoked a backlash not only by conservatives concerned about the coherence of American national culture but also by leftists and liberals who worried that the multiculturalist insistence on the right to be different undermined the fight for the right to equal treatment. South African anthropologist Adam Kuper stated this bias up front in his history of the culture concept in anthropology. As a "liberal in the European rather than American sense," he had "limited sympathy for social movements based on nationalism, ethnic identity, or religion, precisely the movements that are most likely to invoke culture in order to motivate political action."[219] He noted that the Boasian argument that culture made people what they were and that respect for cultural difference should be the basis for a just society might have been a benign argument in the United States, but in South Africa it had served to justify apartheid. Its commitment to a "separate development" that would produce higher Bantu culture rather than black

Europeans had its intellectual underpinnings in the ethnology of Werner Eiselen, whose school of thought cited American cultural anthropology approvingly.[220] Consequently, Kuper fully embraced the reservations that British social anthropologists had long harbored against the culture concept and expressed his "moral objection" to culture theory: "It tends to draw attention away from what we have in common instead of encouraging us to communicate across national, ethnic, and religious boundaries, and to venture beyond them."[221] As the culture wars spread from American universities to institutions of higher education in Europe, Australia, Japan, and elsewhere, a sense of moral indignation took hold of all parties involved.

Although the American William McGrew had studied social anthropology in the United Kingdom, as a cultural primatologist he felt let down by his new colleagues: "Some proportion of sociocultural anthropologists find the concept of culture to be outmoded, and even obstructive," he noted with reference to Kuper's book.[222] "This is hard for nonspecialists to understand, almost as if musical chairs could somehow be played in silence. How strange to think that finally when cultural primatology realizes how much it needs cultural anthropology, the latter may drop its central tenet." In 2000, cultural anthropologists and primatologists (including Christophe Boesch) participating in a Wenner-Gren symposium titled "Culture and the Cultural" conveyed the spirit of their discussions in the form of a mock newspaper headline: "Apes Have Culture; Humans Don't."[223] Just as primatologists had documented enough chimpanzee cultures to move from chimpanzee ethnography to ethnology and even to a general "cultural panthropology," which used cross-species comparisons to shed light on the evolution of hominoid culture at large, culture seemed to vanish as cultural anthropology's epistemic object.[224] Frans de Waal expressed disappointment that cultural anthropologists did not put up a fight even when primatologists snatched what used to be the defining concept of their discipline: "This lack of territoriality is due to [the anthropologists'] own ferocious internal battles combined with postmodern nihilism."[225] Between cultural primatology and cultural anthropology it was truly a *rencontre manquée*.

Simian Orientalism and Amoral Primatology

Primatologists not only lost interlocutors as more and more cultural anthropologists, especially at the more prestigious American departments, abandoned the project of a science of culture, eventually repurposing their interpretive approaches to activist ends. At the height of the culture wars, primatologists also found themselves the objects of such critical scholarship when Donna Haraway published *Primate Visions*, a history of Western primatology as "simian orientalism."[226] Taking her cues from Said's book

Orientalism, she read the discourse of primatology as an imaginative projection and fabrication of changing political orders. Just as European and American students of the Orient had a long colonial history of defining themselves in opposition to exoticizing portrayals of the peoples of the Near and Far East, so primatologists marked off their humanity in contradistinction to equally fictitious representations of monkeys and apes. But Haraway replaced Said's concern with how orientalists constructed cultural differences with her own focus on how primatologists constructed species differences between humans, apes, and monkeys: "Here, the scene of origins is not the cradle of civilization, but the cradle of culture, of human being distinct from animal existence."[227] Haraway's observations were decidedly second order: she observed how primatologists observed primates without making any effort to observe the animals herself part of her approach. In Haraway's readings of primatological texts as a peculiar form of "science fiction," the lives of other primates became the projection screen for "White Capitalist Patriarchy (how may we name this scandalous Thing?)."[228]

In *Primate Visions,* the decisive turn of events occurred in the 1970s when a new generation of female primatologists—Jeanne Altmann, Linda Fedigan, Adrienne Zihlmann, and Sarah Blaffer Hrdy—reinvented their discipline as "a genre of feminist theory" just as the incipient conversation with Japanese and Indian colleagues decolonized and transformed primatology into a "multicultural field."[229] Drawing from Pamela Asquith's work on the Kyoto School, Haraway rethought the boundaries of multiculturalism in a different way than McGrew and his fellow cultural primatologists: we may need to be more inclusive in extending our appreciation of cultural diversity to admit the sciences, these supposed cultures of no culture, she could have said.

Haraway's deconstruction of the opposition between science and society presented primatology as a field wide open to other cultural domains such as politics, economics, gender relations, art, literature, or movies like *King Kong.* In the early twentieth century, philosophers and sociologists of science had been united in their quest for a demarcation criterion distinguishing science from pseudoscience and other products of intellectual activity such as metaphysics, religion, fiction writing, and so forth. *Primate Visions* established a genre of cultural history of science that deliberately muddled science and nonscience.

Among primatologists, this challenge to the autonomy of their scientific field did not receive a particularly warm welcome. In fact, it represented one of the transition points at which the culture wars of the 1980s morphed into the science wars of the 1990s, pitching natural scientists against their social constructionist critics. While many of the scientists involved in this late twentieth-century contest of faculties were physicists, the sociobiology debate represented another battlefield that drew animal behavior researchers into the science wars.[230] Three distinct but related issues provoked their ire: Haraway's style, her insistence that primatology was "politics by other means," and her unabashed moralization of science.[231]

French philosopher Michel Foucault once remarked that "in the West the combat of forms has been just as hard fought, if not more so, than that of ideas or values."[232] When *Primate Visions* first came out in 1989, it left many of its readers, especially the primatologists among them, flabbergasted. "I have never come across anything like this," wrote Oxford chimpanzee researcher Vernon Reynolds. From Henrietta Moore's *Feminism and Anthropology* he knew and admired what he called "the deconstructionist tradition," but that had been "a normal book," whereas Haraway's history of Reynolds's own discipline appeared "way 'over the top.'" He might have responded in kind when he praised it as "a masterpiece in the sense that a great work by Constable or Rembrandt is a masterpiece" while noting that "its more purple passages are the purest verbiage, flowing over the reader like tidal waves, threatening to demolish something: Science? Primatology?"[233] Others responded less wryly to "Haraway's openly subjective and often contemptuous style," full of "vaporous French prose" and "gossipy anecdotes": "It is this style that has infuriated many primatologists (and reviewers), some of whom are the historical figures treated rudely by the author," wrote American biological anthropologist Craig Stanford.[234] But even the most infuriated of all, Matt Cartmill, agreed with Reynolds that, as a contribution to "postmodernism," a style of thinking and talking "increasingly prevalent in the humanities and social sciences," *Primate Visions* was indeed a masterpiece, for "all its deficiencies are deliberate products of art."[235]

That Haraway did not even pretend to write a "disinterested, objective study," that she made no effort to hide or restrain her political biases, her commitment to "the broad left, anti-racist, anti-colonial, and women's movements" as well as to animals, had epistemological reasons, which her primatologist readers did not accept.[236] She belonged to the founders of science studies, a field that had grown out of a radicalization and, according to American philosopher of science Philip Kitcher, an overinterpretation of postpositivism, including its signature doctrine that all observations were already laden with theory and therefore could not settle scientific disputes.[237] Willard van Orman Quine, Paul Feyerabend, and Thomas Kuhn had brought up this theory-laden nature of observation against the logical positivists' strict separation of theory and observation.[238] If such observation without theoretical presuppositions was indeed possible, as the positivists of the Vienna Circle had contended, hypotheses could be tested against the raw data of perception and recorded in a neutral language, which would serve as the bedrock of scientific progress as theories were refined and replaced. The most famous result of the postpositivist challenge to the positivist faith in science as an exemplar of cumulative culture was Kuhn's thesis that scientific knowledge did not accumulate in a steady fashion but underwent revolutionary paradigm shifts, which led subsequent observers to see the same thing in a radically different and even incommensurable fashion. Haraway took this challenge two steps further, from a mere underdetermination of theories by evidence to the equation of science with "science fiction" and from the resulting global skepticism to a situated moralism: "Facts are always theory-laden; theories are value-laden; therefore facts are value-laden."[239]

Cartmill objected that Haraway's syllogism was "like saying, 'This bus is full of skeptics; skeptics are full of doubt; therefore this bus is full of doubt.' Buses do not contain skeptics in the same way that skeptics contain doubt, and facts do not contain theories in the same way that theories contain values."[240] If facts were really value laden, he argued, we could infer something about a person's values from any declarative statement. Since this wasn't always possible, the association of facts and values had to be contingent, not logically necessary: "Facts must therefore be, at least in principle, independent from values."[241]

Nobody doubted a weaker reading of Haraway's constructionist philosophy of science as postulating contingent associations of facts and values. Stanford recognized a "seed of truth" in her general insight that "science is a product of a given social and political context" and concluded that scientists should acknowledge their biases.[242] On a more substantive level, Reynolds agreed with Haraway's story that the influx of women primatologists had replaced the "androcentric paradigm" of the 1960s by a "gynocentric" one that shed light on the behavior of female primates and thereby aided understanding of the social strategies of both males and females.[243] But he also reminded her that "part of the social process called science (or science fiction, if you will) is not just story-telling; it interfaces with the physical world as well."[244] After all, such paradigm shifts come from discovering new or previously disregarded facts, not just from the way women see the world. While Reynolds found Haraway's attack on the authority of the sciences "quite harmless" because it did not touch on "the real scientific enterprise," Cartmill responded in a more thin-skinned manner.[245] He compared her alleged "denial of external reality" to the brainwashing of the hero of George Orwell's *1984*, whom the inquisitor tells that reality is inside the skull and that he needs to let go of nineteenth-century ideas about the laws of nature to accept whatever assumption the Party currently favors: "The postmodernist sensibility displayed in this book is strangely reminiscent of the official philosophy of Orwell's posttotalitarian state."[246]

That primatologists responded to Haraway's *Primate Visions* with a vitriol that Kuhn's *The Structure of Scientific Revolutions* had been spared reflected her reinterpretation of the theory-laden nature of observation as value laden. It subjected the natural sciences to legitimate "criticism on the level of 'values,' not just 'facts.'"[247] What irritated the scientists was not simply the suggestion that science had values. Historians Lorraine Daston and Peter Galison showed that the moral economy of science confronted researchers with difficult choices over whether to privilege objective or ideal-typical representations, mechanical measurements or a trained eye, increased precision or commensurable experimental results, and so forth.[248] These epistemic values are in the service of the true, not the good, and therefore are usually indifferent toward moral values. They had often originated in the ambient culture of science. Originally, they might have been of moral significance but they had been reworked to form a scientific way of life.[249] The primatologists reviewing Haraway's book defended a subset of scientific values such as objectivity and replicability against her contention that objectivity was an "ideological fiction" veiling hidden political and economic interests, or against her emphasis on anecdotes, which made primatology appear as a mere "story-telling practice."[250] What made feelings run high was that Haraway's narrative foregrounded values with a decidedly moral slant.

If scientific facts were contingent on moral values, they had to be judged good or bad. Many primatologists saw Haraway as dividing their discipline into "patriarchal villains" and "feminist heroines"—whom she depicted as "sardonic deconstructionists like herself," leaving even many of the praised feeling alienated.[251] "The men (Stuart Altmann, C. Ray Carpenter, and Robert Yerkes, to name only a few) are portrayed as unconscious flag-bearers of capitalism, colonialism, and chauvinism," wrote Stanford.[252] "The women are portrayed as either unwitting protagonists of a colonial theme in primatology (Jane Goodall as the penetrator of the darkest of all Africas, the world of African apes) or as modern feminists (Sarah Hrdy and Adrienne Zihlmann) who have challenged the past to breathe some fresh air into their science."

Allegations that the characters populating *Primate Visions* lacked nuance might themselves have profited from more nuance. For example, Haraway refrained from painting Kinji Imanishi and his students in black and white: Japanese primatologists emphasized the spiritual kinship of humans and animals but justified the latter's abuse as a form of legitimate domestic violence.[253] They appeared as men who dominated the even smaller number of women in their field (no Japanese Jane Goodall or Dian Fossey to this day). Not feminism but the important role of female shamans, female deities, and the sun goddess Amaterasu in Shintoism had allowed them as early as the 1950s to recognize the matrilineal organization of macaques and to identify the female Imo as the great innovator of the Koshima troop. But this complexity of character was still cast in moral terms: Japanese primatology was "not innocent,"

Haraway warned her readers—nothing in *Primate Visions* ever is.[254] Even Hrdy's liberal feminism, which made her sociobiology "good science" in comparison with the "sexism" of Edward O. Wilson's science, propagated a capitalist market logic, which a self-identified feminist socialist like Haraway considered the source of moral ills.[255] But Haraway's ambivalence toward even the supposedly feminist primatologists hardly took her beyond good and evil.

Although the protagonists of *Primate Visions* developed into richer characters than stock heroines and villains, their complicity with or resistance against what primatologist Peter Rodman described in his book review as the three "dark forces" of colonialism, racism, and sexism decided their place in this history of science.[256] Did their research support or subvert the occupation and exploitation of other people's home countries, discrimination against fellow human beings on the basis of their ancestry, and the oppression of women? The stark asymmetry of these distinctions—it was never in doubt what the good and the bad sides were—led the reviewers cited above to understand Haraway's interpretations as expressions of approval or disapproval of whole persons. Niklas Luhmann noted that such moralization had a high potential for escalating polemics.[257] In Haraway's case, it sure did.

This mode of cultural critique was at odds with the ethnographic and historiographic cultivation of interpretive charity. As a professor of anthropology familiar with his colleague's desire to build rapport and understand the so-called native's point of view, however bizarre it might at first seem, Cartmill noted that Haraway's "deconstruction" of primatology was "not a friendly act" and made "no effort to understand or to sympathize with the intentions of scientists" or "to deal with the past on its own terms, to give an account of people's actions in terms of their own ideas" because she sought only to reveal "the scandalous Thing lurking within."[258] Usually, primatologists thought about their research findings as either true of false, but not as good or bad. The way in which Haraway pulled scientific facts into the moral register appeared to them inappropriate and even "rude," "hostile," and "contemptuous."[259] And yet, for a new generation of anthropologists Haraway's approach would come to serve as a model for how to critically engage with science.

An intellectual historian might read *Primate Visions* as a document of the culture wars of the 1980s. Its attempt at deconstructing the dichotomy of nature and culture could have paved the way for a fruitful exchange between the emergent fields of science studies and cultural primatology at a time when cultural anthropology largely fell out as an interlocutor. Simultaneously, the old division of labor between disciplines studying human phenomena such as culture, society, and history and disciplines studying natural phenomena such as animals and evolution had broken down. But McGrew's and Haraway's pleas for extended notions of multiculturalism never made common ground.

The collaboration between Californian primatologist Shirley Strum and French anthropologist of science Bruno Latour probably represented the only flourishing encounter between primatology and science studies during these years.[260]

The science wars have usually been interpreted as a conflict between scientific realists and social constructivists who vehemently disagreed about the role of human and nonhuman agency in the creation of scientific knowledge.[261] My reading of the book reviews of *Primate Visions* suggests that it was not just the radicalization of a postpositivist philosophy of science that led many primatologists to reconfirm their own professional identity as positivistically minded scientists by habitually bashing so-called postmoderns. It was also—maybe even more so—the moralism with which Haraway and her generation of humanists and posthumanists approached the sciences that stirred up the primatologists' bile and ignited the science wars as a secondary theater of the culture wars.

Second-order observations don't have to be hostile acts: they simply reveal that other ways of knowing and doing are possible. Examining the contingency of facts on epistemic, political, and even moral values can open a spectrum of genuine alternative perspectives on a given problem. Realizing that their accounts are less tightly coupled to reality than they had previously imagined could also be of help to practicing scientists.[262] It is the moral coding of the alternatives that not only endows second-order observations with a polemic quality but immediately closes the newly opened space of possibilities. When Haraway labeled a research program sexist or racist, she revealed the empirical indeterminacy of scientific knowledge only to eliminate it in the name of moral determinacy. This brand of social constructionism presented the establishment of scientific facts as contingent on human agency, but here agency appeared closer to the freedom to sin with which the God of Christianity had endowed human beings. The choice between good and bad is hardly a real choice.

This chapter reconstructed the debate surrounding cultural primatology in a manner that paid more attention to the moral and political stakes than cultural primatologists do in their own publications. But I did not intend to heat up this ongoing controversy. The primary goal of this chapter has been to clarify the stakes: What exasperated the participants involved, and where did they draw the front lines in the chimpanzee culture wars? Observing how different parties politicized and moralized the culture concept eventually adopted by primatologists brings down the temperature—although I do realize that this might in fact infuriate those invested in the project of critique. I do not claim that their moralism is mistaken.

With almost thirty years of hindsight, but at a time when, at least in the United States, culture war talk again resounds throughout an even more deeply divided land, the moral critiques of sociobiology and cultural primatology discussed above remind me of Luhmann's saying that "perhaps the most pressing task of ethics is to warn against morality."[263] That is not to say that the sciences should never become matters of moral concern. Like all human behavior they can be evaluated morally. For example, McGrew's comparison of human Tasmanians and chimpanzee Tanzanians to challenge human exceptionalism might have been bad because it offended contemporary Tasmanians. But that did not render its results false. The true, the good, and the beautiful do not necessarily coincide. There are always multiple ways to look at things, and depending on the observer's interests, it is not always wise to talk about a given phenomenon in terms of good and bad. Luhmann's point was that there are even ethical reasons to refrain from moralization because it provokes an overengagement of participants, harbors a potential for violence, and drives wedges between people who could otherwise find common causes—cultural primatologists and cultural anthropologists, for example.

3

Chimpanzee Ethnography

ROUSSEAU, SARTRE, AND DARWIN cracked nuts in the Ivorian rainforest and bore names that reminded readers of Christophe Boesch's ethnographies of the chimpanzees' much fought over place in European philosophy. Boesch's book *Wild Cultures* is swarming with chapter epigraphs from Herodotus and Epicure to Voltaire and Durkheim.[1] These mute evocations of the classics do not serve as direct commentary but put the reader's hermeneutic skills to the test.[2] They show us a glimpse of the author's mental furniture and make the experiential knowledge garnered in the field talk back to ancients and moderns. The cited philosophers often led the primatologist to examine and call into question their speculative claims about human nature. Of course, he did it in a more empirical fashion than the cynic Diogenes, who responded to Plato's definition of man as a "two-footed featherless animal" by bringing a plucked cock to his school: "This is Plato's man."[3] Do behavioral observations of chimpanzees support Voltaire's contention that only humans knew that they would have to die?[4] Was Rousseau (the philosopher, not the ape) right when, sitting in his armchair, he maintained that humans and possibly apes but not monkeys were able to improve upon their nature?[5]

In 1755, Rousseau had already been critical of the very practice that allowed him to make such claims. "Philosophy does not travel," he noted, and consequently it was full of "ridiculous prejudices" because "the only men we know are the Europeans" and yet "philosophasters" assumed that men were everywhere the same.[6] He doubted that the sailors, merchants, soldiers, and missionaries from whose reports he had learned about the Pongoes and Engecoes in the kingdom of Loango were good observers. Rousseau speculated that these apes (today known as gorillas and chimpanzees) might have been savage men, and he wished that a Montesquieu, a Diderot, or a Condillac (but not himself) would travel to such remote locations to determine whether a given animal was a human or a beast.[7] Chimpanzee ethnography would appear almost like a late modern realization of Rousseau's dream, if only its practitioners were polymathic savants rather than highly specialized scientists.

Boesch had spent more time observing a West African chimpanzee he had named Rousseau than reading the oeuvre of the animal's French namesake. The French Swiss ethologist had studied the ways of African apes for four decades, initially only with his wife, Hedwige, in Côte d'Ivoire, then as director of the Max Planck Institute for Evolutionary Anthropology in Leipzig, Germany, with a large team of researchers and field assistants. Although they studied all aspects of chimpanzee life, from reproduction to feeding ecology, what had gradually taken center stage was the question of culture. By the early twenty-first century, there was no longer any doubt that chimpanzees were not human, but Rousseau's question of whether they shared the specifically human faculty of perfecting themselves had been rearticulated in the controversy over whether chimpanzees had culture, maybe even cumulative culture.

In the last chapter we saw that disagreements between field biologists and experimental psychologists over how to relate naturalistic observations and controlled experiments to answer this question constituted one of the front lines in the chimpanzee culture wars. This chapter takes a close ethnographic look at how primatologists conduct and write chimpanzee ethnography on an originally philosophical battleground.

The Taï Chimpanzee Project

In the 1970s, Christophe Boesch, not even a doctoral student, sought a life in nature but hit upon another culture—or at least, this is how he would gradually come to understand his experience among the chimpanzees of Taï Forest in Ivory Coast. Born in the Swiss German town of St. Gallen and raised in Paris, where, as the son of cultural psychologist Ernst Boesch, he and his friends played hide and seek in the corridors of the Sorbonne, Boesch described himself as a city boy.[8] Soon enough, however, he became a Boy Scout, spending weekends in the forests surrounding the French capital. *King Solomon's Ring*, a book by ethologist Konrad Lorenz, which his father had handed him at age fourteen, awakened his desire to study animal behavior.[9] At Collège Calvin in Geneva, where he studied after the student revolts of 1968 had shaken the French educational system, he read George Schaller's *The Year of the Gorilla*.[10] In 1973, he had already helped Dian Fossey with a survey of the very mountain gorillas that Schaller had observed in what was then called Zaire.

Physician and ecologist François Bourlière, who had introduced Boesch to Fossey, also told him about reports that West African chimpanzees used hammers to crack nuts. Reports about nut cracking could be traced back to the sixteenth century, when Portuguese missionaries first mentioned tool use by chimpanzees in Sierra Leone.[11] Darwin already knew from Thomas Savage and Jeffries Wyman that "the chimpanzee in a state of nature cracks a native fruit, somewhat like a walnut, with a stone"—exactly as humans did.[12] In the mid-twentieth century, American physician Harry Beatty had observed how chimpanzees cracked oil palm nuts during a trip to Liberia.[13] And yet European anthropologists had remained so committed to the idea of "man the toolmaker," first proposed by Galen in the second century AD, that Louis Leakey responded to Jane Goodall's 1964 discovery of the Gombe chimpanzees' use of sticks for termite fishing with the famous telegram: "Now we must redefine 'tool,' redefine 'man,' or accept chimpanzees as humans."[14]

Soon doubts arose over whether the uses of these different tools were indeed characteristic of "the" chimpanzee in the state of nature. American primatologist Thomas Struhsaker and his French colleague Pierre Hunkeler collected evidence for termite fishing in the rainforests of Cameroon, but in Taï Forest they found no such thing.[15] Then they discovered sites in Taï, but not in Cameroon, where the chimpanzees seemed to have smashed nuts—although the two researchers never actually saw the behavior.

Intrigued by this geographical variation, which Struhsaker and Hunkeler had not interpreted any further, Boesch first traveled to Taï in 1976. He documented the nut-cracking workshops and the tools left behind.[16] During three months in the rainforest, Boesch saw one chimpanzee holding a stone in her hand. He didn't see her opening any nuts with it, but he heard her hammering. Based on this one observation, he set out on a PhD thesis, supervised by Hans Kummer at the University of Zurich. He sought "intimacy with a really pristine place" and hoped to observe animals that were "as undisturbed as possible."[17] In 1979, he returned to Ivory Coast, where he conducted his doctoral research in Taï Forest until 1983.

The fieldwork advanced at an excruciatingly slow pace.[18] Boesch saw the approximately eighty chimpanzees mostly from behind: during the first year he got to observe them only 1 percent of the time; in the second year, it was 5 percent. It took five long years before the community later known as the North Group got used to his presence. Since he submitted his thesis before all chimpanzees had been habituated, it amounted to what, in an interview, he called a "bottom study." Lorenz's writings had taught him that to endure the enormous amount of observation time necessary to gain even a most basic understanding, one had to love animals.[19]

But the brief and distant contacts with the Taï chimpanzees sufficed to establish a number of facts about how they cracked nuts. The pounding could be heard from a kilometer away, so, although afraid of humans, the animals involuntarily gave away their whereabouts and could be followed with relative ease during the nut-cracking season. Even though Boesch couldn't see them, he heard how many times they had to strike the hard shells of *Coula edulis* and *Panda oleosa* fruits and could infer how many nuts they ate. When they ran

FIGURE 4. Tracking chimpanzees: Christophe Boesch examines feces of an unknown animal that recently passed by. Photo by author.

away from him, he could tell from their buttocks whether the animals were male or female and whether they were adults, juveniles, or infants. And then he could examine the wooden and stone hammers they had left behind. After Struhsaker and Hunkeler had already provided circumstantial evidence for this behavior, Boesch managed to rule out all doubts that the Taï chimpanzees cracked nuts in this way.[20] Yet the dispute over the philosophical and anthropological interpretation of these observations had only just begun.

In the humanities, even calling the apes' rocks and chunks of wood "tools" remained controversial. American philosopher Barry Allen, for example, denounced the "myth of the chimpanzee's tool" because it obscured the unique contribution of technical culture to human ecology.[21] If the apes were deprived of their so-called tools, he maintained, they could easily switch to other food sources that did not require any instruments, whereas humans could not possibly survive without their tools. Moreover, real tools were not simply objects used to achieve an end, as primatologists stipulated, but they helped coordinate cooperative endeavors.[22] Faced with the choices from Leakey's cable, Allen refused to redefine humans or grant chimpanzees humanity and instead redefined the concept of a tool. Exorcized from primatology for some decades, the humanist ghost of a *Homo faber* who alone could control the world and his own fate with tools continued to haunt select quarters of the humanities until the very end of the millennium.

By contrast, as Boesch observed Salomé and her son Sartre opening nuts with granite hammers and preparing sticks to fish for grubs hidden under tree bark, he remembered Friedrich Engels's contention that labor distinguished humans from animals.[23] The philosopher's claim that all other species just used their environment without mastering it while humans prepared the means to make it serve their ends led Boesch to wonder: "Were Salomé and Sartre humans or animals?"[24]

After completing his dissertation, Boesch managed to escape for another eight years the brouhaha of academic clashes between anthropomorphists and anti-anthropomorphists, which were about to give rise to the chimpanzee culture wars. Together with his family, he returned to a secluded life in Taï National Park, which he established as the third major chimpanzee field site. After Goodall's Gombe Stream National Park and Itani's Mahale Mountains National Park, both in East Africa and home to *Pan troglodytes schweinfurthii*, the ample resources of the Max Planck Institute would eventually allow Boesch to develop Taï National Park into the largest field site in West Africa where *Pan troglodytes verus* lived.

The taxonomic designation *Pan troglodytes verus* could be translated as the "the real chimpanzee." Boesch's book of the same half-ironic, half-serious title, however, emphasized that there was no such thing as a typical chimpanzee. He dismissed the widespread tendency of fieldworkers to consider their own study population to be the one truly representative of the species.[25] In cultural anthropology, such privileging not of one's own people but of the people one studied over others was known as allo-ethnocentrism. It had played no small part in the infanticide controversy between field primatologists.[26] Fully conscious of this pitfall, Boesch did not claim that the Taï chimpanzees were more real than any other wild chimpanzee communities (although we will see in the next chapter that Boesch did not extend this relativism to the captive communities studied by comparative psychologists). This also meant that no other population was more representative than the one inhabiting Taï Forest. Nevertheless, Boesch criticized, Jane Goodall's savanna-dwelling chimpanzees were frequently taken as typical of the whole species.[27]

It was an anthropological theory that had originally inspired the special interest in the Gombe community. The savanna hypothesis, first postulated by Raymond Dart, that our species had evolved from ancestors who had left the forest for open grasslands, had led mid-twentieth-century anthropologists to look to baboons and savanna- and woodland-dwelling chimpanzees as models of early *Homo*.[28] But the majority of chimpanzees lived in rainforests between Congo and Sierra Leone. In Boesch's eyes, Louis Leakey's decision to send Jane Goodall to Gombe reflected her mentor's anthropocentrism: "Leakey intentionally selected a population of chimpanzees living in a marginal habitat, a habitat that best paralleled the conditions of our early ancestors, and not a habitat that is representative of that of most living chimpanzees in Africa."[29] If anything, the Gombe chimpanzees lived under rather atypical ecological conditions. Yet because of Goodall's charisma and successful public relations work as well as the large volume of publications produced by her field site, this community gained paradigmatic status and came to represent "the" chimpanzee. Despite its divergent theoretical orientation, the Kyoto School's work on the same subspecies inhabiting a similar environment in neighboring Mahale only reinforced the bias.

By focusing on the apes of Taï Forest, Boesch set out to correct this distorted image. He believed that chimpanzees were originally inhabitants of the forest, where they had acquired most of their abilities.[30] Only later had they adapted to drier and more open habitats. Shortly after Yukimaru Sugiyama established an intermittent Japanese research presence in 1976 in nearby Bossou, which was also home to western chimpanzees but had little intact primary forest, Boesch set out to conduct the "first long-term project to study chimpanzees in a dense tropical rainforest with the aim of learning how ecological conditions affected the behaviour of this species."[31]

Although Boesch assumed that the Taï chimpanzees lived under conditions closer to those under which the species had evolved, his primary concern was not a monolithic state of nature, the way of life of primeval chimpanzees, but an understanding of the variability of the species. In the 1950s and 1960s, some of the first modern primatological field studies had discovered behavioral variability in langurs, baboons, and other monkeys. At the time, it was usually explained in ecological terms: different environments led primate groups of one and the same species to behave differently.[32] In European and American primatology, culture did not play a significant role yet. In 1971, Boesch's thesis supervisor Hans Kummer had published *Primate Societies* in a book series on human cultural ecology. Kummer discussed the behavior of the Koshima macaques as "the closest primate parallel of human culture studied in detail to date" but emphasized that it did not amount to symbolic culture.[33] He was an ethologist in the tradition of Lorenz, who saw in Imo's sweet-potato washing the modification of an innate behavioral potential: the brushing and rolling of food items to remove dirt, which Kummer had also observed in baboons, was now performed in water. Nevertheless, this transition gave ethology a new face. It complemented the field's original focus on adaptations of evolving genotypes by studies of how populations with the same genotype modified their behavior either in response to different local environments or through social learning. While Lorenz had concentrated on instinctual behavior, Kummer acknowledged the importance of traditions and thus put pressure on "a frame of thought that has virtually no place for culture as this term is usually understood."[34]

FIGURE 5. Chimpanzee infant learning from his mother how to crack nuts with a stone hammer and anvil, Bossou, Guinea. Photo by author.

In one of their first coauthored articles, Christophe and Hedwige Boesch proposed that tool use for the purpose of nut cracking could be looked at as "a precultural adaptation of chimpanzees towards eating hard nuts."[35] At the time, Boesch still focused on ecological explanations of the behavior. Doubts first arose when he visited Gombe in 1982 and 1983 and experienced a potentially cultural difference firsthand. He noticed that Goodall's community ate the flesh around oil palm fruits without cracking the nuts inside. There weren't many oil palms in Taï Forest, but Yukimaru Sugiyama and Jeremy Koman had just reported that the chimpanzees of nearby Bossou, where the local Manon people cultivated these trees, opened their nuts with stone hammers and anvils: Why didn't the Gombe chimpanzees avail themselves of this highly nutritious source of food?[36] By the end of the decade, enough evidence had accumulated for the Boesches to suggest that some tool uses, culinary preferences, and different ways of butchering monkeys seemed independent of environmental conditions.[37] As more and more behavioral diversity came to light—in some cases so poorly adaptive that "the rule would be 'do what others do' and not 'search for the

FIGURE 6. Learning chimpanzee culture: the author trying to crack *Coula* nuts. Photo by author.

best solution'"—they too raised "the question of attributing culture to chimpanzees."[38] Boesch concluded that "it is time to follow in the footsteps of Margaret Mead and Claude Lévi-Strauss and make a detailed ethnography of chimpanzee culture if we want to obtain a better understanding of what is unique about and what is shared by human and chimpanzee cultures."[39]

Healthy Distance

On a Friday afternoon in late March 2014, I arrived in Taï in a Land Rover packed with field assistants who had been picked up in the villages surrounding the national forest. Even before I could follow the primatologists to the chimpanzees, it became abundantly clear that the ethnography of African apes wasn't what it used to be. In the 1960s, *National Geographic* magazine could still photograph Dian Fossey cuddling with gorilla infants. Jane Goodall's later husband, Hugo van Lawick, filmed her grooming the Gombe chimpanzees. When Boesch and his wife entered the field two decades later they already maintained a respectful distance toward the animals, but they still took liberties. Apart from monkey meat, they sampled all the foods consumed by the chimpanzees. When the Boesches needed something from a shop, they or their assistant, who doubled as nanny, could make the one-hour drive to town and they could still resume their tasks in the forest the same day. The boundaries between humans and animals, town and forest, nature and human culture, if you will, were relatively permeable. Paradoxically, by the early twenty-first century, when these ontological divisions had also become porous in anthropological discourse, the wilderness began to resemble a high-security laboratory to be entered and exited through a number of locks.

Everybody who had newly arrived in the forest that day in March was first quarantined in a camp set up around the house where Christophe and Hedwige Boesch had raised their two children. German primatologist Roman Wittig, to whom Boesch had passed on the direction of the Taï Chimpanzee Project (TCP), greeted me on his last night before returning to Leipzig. French camp manager Sylvain Lémoine, a dreadlocked doctoral student who had learned in Nigeria and Borneo how to organize the demanding logistics of a research station, took mouth swabs from the newcomers, which he froze in a liquid nitrogen tank. Until a year ago, the samples had immediately been tested for viral infections in the ramshackle field laboratory. But the DNA primers for the polymerase chain reaction machine had proved too expensive for this precaution. The frozen specimens would allow at least a retrospective analysis, should a human disease be transmitted to the apes despite the other safety measures. To prevent this from happening, the camp management trusted that new arrivals would develop symptoms within the five days during which they isolated us from everybody interacting with the chimpanzees. If no infections caught in human settlements or on airplanes manifested during our confinement at the quarantine station, we could move on to the research camps.[40]

This system made the lives of researchers and field assistants psychologically and socially taxing. Before the introduction of the quarantine in 2008 in response to a respiratory disease outbreak among the chimpanzees, the assistants had been able to regularly see their families in the village or even to live with them in one of the research camps. The scientists could spend a day off watching a soccer game in a bar or take the UN helicopter to Abidjan, if they needed to recover from the at times oppressive darkness of the rainforest at the beach of Grand-Bassam. This was several years before the small resort town and former colonial capital became the scene of a jihadist shooting rampage in 2016. Now every time someone left the camp, even if only for a day, it took a week to resume work, tearing a hole in the dense fabric of behavioral observations. Consequently, even though the field assistants' villages were only a one-hour drive away, they worked for three weeks in a row before returning to their families for one week. The doctoral students often stayed in the forest for months on end, without ever seeing the horizon. Email or telephone communication with friends or loved ones required climbing on top of an inselberg, an elevated rock formation, where a small clearing usually allowed smartphones to pick up a signal. This sylvan seclusion might explain why interpersonal relations occasionally got tense. "People can get a little crazy," one forest-dwelling scientist confided. Since Boesch had decided to implement the quarantine, I was told, he had not subjected himself to such a week of idleness, which would have allowed him to see again the chimpanzees of Taï Forest. The busy schedule of a Max Planck Institute director might also have spared him the saddening experience of seeing his beloved North Group dwindle away.

Leaving camp in the early morning hours to pick up the chimpanzees wherever they had built their nests the previous evening, we had to pass through *la barrière*. At this makeshift wooden construction marking the entry of each of the four field stations, we all had to clean our boots with bleach, change all our clothes, and disinfect our hands before going into or returning from the forest. To pee you had to move out of sight of the chimpanzees and cover the puddle of urine with dead foliage. Our excrement we collected in plastic bags, which the *homme de camp* would later burn at the field station.[41]

When Boesch introduced this measure to protect the chimpanzees against infections with human pathogens, the Ivorian field assistants refused to honor it for fear that, if misappropriated, their feces might allow a sorcerer to kill them. Two claimed to have already lost relatives that way. "Communication between cultures is very difficult," Boesch explained to me. "Even in the sciences we don't always manage to communicate across disciplines. Although I'm interested in cultural differences, I did not tolerate what they did because I knew the consequences." Since their employer stuck to his own understanding of disease transmission but failed to convince the field assistants that they should worry about germs, not curses, the Taï Chimpanzee Project eventually resolved this clash of ontologies pragmatically. The assistants changed their eating habits so that they no longer had to relieve themselves during the time they worked in the forest.[42]

In the presence of the chimpanzees, all researchers and their assistants were required to maintain a minimum distance of seven meters and put on surgical masks to minimize the risk of infection—a measure introduced after twelve animals had died in 1999 of a flu-like disease presumably transmitted by one of their human observers. Many chimpanzee communities in regular contact with researchers, tourists, or local people suffered such losses, just as human populations in the Americas had often succumbed to infectious diseases in the wake of their first contact with European colonizers. *Pan* is close enough to *Homo* to be susceptible to most bacteria and viruses that afflict our kind without having had enough exposure to develop sufficient immune responses. This punishing apparatus of interspecies borders and checkpoints served to reduce

FIGURE 7. *La barrière*: an outdoor changing room with buckets of bleach and hand sanitizer erects a hygienic barrier between camp and forest, Taï National Park, Ivory Coast. Photo by author.

the risk and burden that the apes' habituation to human presence had imposed on them.

The various distancing devices also protected human researchers and field assistants against some of the perils lurking in the jungle. Anthrax and Ebola were endemic in Taï Forest. When I visited at the end of March 2014, five chimpanzees had already died of anthrax infections since the beginning of the new year. Only days before my arrival the news of an Ebola outbreak among the human populations of neighboring Guinea and Liberia had hit. With public health officials and NGOs spinning to save Côte d'Ivoire from the devastating medical and economic consequences of the epidemic, the sudden death of forest animals became a matter of serious concern.

Every time the scientists, or usually the chimpanzees they followed, found a dead duiker, anteater, monkey, or fellow ape, the researchers informed the veterinarian on their team. After twelve hours of dissertation research on the play behavior of malaria-plagued chimpanzee infants, she then had to go back into the forest at night to dissect the cadaver on the spot, occasionally under the hungry eyes of a leopard. During one such autopsy in 1994, in the middle of an Ebola outbreak among the Taï chimpanzees, a Swiss ethologist had been infected with a new strain of the virus, which would come to be known as Ebola Côte d'Ivoire. Taken to her home country for medical treatment as the first known person carrying an Ebola virus off the African continent, she was most fortunate to survive the hemorrhagic fever.[43] With this cautionary tale in mind, the postmortems on the forest floor were conducted wearing a white full-body plastic suit, an apron, three pairs of disposable gloves, protective goggles, and a necropsy mask, through which one had to gasp for warm steamy air. The specimens taken from the chimpanzees deceased in early 2014 tested positive for anthrax at the Robert Koch Institute in Berlin.

The deadly bacterium appeared to be spreading as global climate change manifested in stronger harmattan winds blowing from the Sahara over West Africa, where they slowly desiccated the rainforests. Whereas Boesch had still sampled every fruit and leaf the chimpanzees ate, a strict sanitary regime barred the current generation of students from such an experience-near approach for fear that they might ingest anthrax spores. The hygienic measures at the *barrières* also served to keep this deadly bacterium out of the camps.

While the veterinarian of the Taï Chimpanzee Project collected tissue samples from dead animals, the activists of Boesch's Wild Chimpanzee Foundation used the Ebola outbreak as an opportunity to remind the local population that the consumption of bushmeat, especially of great apes, might have been responsible for the unfolding disaster, which would eventually cost the lives of more than eleven thousand West Africans and devastate the economies of the affected countries.

Côte d'Ivoire was largely spared. But Ivorians were no strangers to zoonotic diseases. While HIV-1 had originated in Central African chimpanzees and gorillas, HIV-2 had begun its career as a simian immunodeficiency virus in West African sooty mangabeys, which primatologists also studied in Taï Forest. Virologists had just discovered a new lineage of HIV-2 in the villages surrounding the national park, which turned out to be identical with an SIV strain found in its mangabey population.[44] A monkey researcher I met at Taï speculated that it had crossed over during the civil war in 2002 when many people sought refuge in the national park and had to rely even more heavily on poaching, wiping out an entire mangabey study group, which had lost its fear of humans in the process of habituation. The Ebola epidemic provided an opportunity to again warn against hunting wild primates.

In the last two decades, a growing interest in interspecies relationships gave rise to the formation of new fields of research. When Leslie Sponsel introduced the term *ethnoprimatology* to describe the interface between human and primate ecology, it included investigations of monkey populations as reservoirs for human diseases.[45] Yet ethnoprimatologists have rarely attended to

FIGURE 8. Autopsy of a mangabey that the Taï chimpanzees found dead shortly after the outbreak of an Ebola epidemic in West Africa. Photo by author.

symbioses of primates and primatologists. This is not surprising considering that scientific work with habituated apes and monkeys represents a rare and most unusual kind of human-animal relationship. Typically, it is butchering, not observing, animals from the wild that transmits zoonoses. But the stern hygienic measures implemented by the Täi Chimpanzee Project reflected a growing awareness of the risks posed by human-chimpanzee contact.

The practices of detachment distinguishing contemporary chimpanzee ethnography from the fieldwork of, say, Jane Goodall call into question the usually positive connotations attributed to practices of attachment in multispecies studies. While ethnoprimatologists have maintained a largely natural scientific orientation, a broad coalition of animal studies, environmental humanities, and posthumanist scholars as well as anthropologists thinking beyond the human approached the borderland between *Homo sapiens* and other species from a humanities angle. As Donna Haraway moved from the critical second-order observations that had characterized *Primate Visions* to a first-order philosophy of so-called multispecies entanglements, she became a figurehead of this field of study. With reference to the constitutive role of bacteria, fungi, and protists in our existence, she claimed that we have never been human because we are always "becoming with" many other "critters," so-called companion species such as pets and livestock but also microorganisms and feral animals living in our midst.[46] They constitute us and we constitute them, breaking down the barriers between humankind and nature in a materialist *unio mystica*. Yet chimpanzee ethnographers did not conceive of their research subjects as a companion species.

Unlike baboons and macaques, which could muddle through in anthropogenic environments and even forage in suburbs and cities, chimpanzees and the other great apes did not thrive in the presence of their next of kin. Unless they had been habituated by scientists or were forced to share their habitat with humans as in Bossou, they stayed away from our kind. And, as we will see in chapter 6, the sustainability of such cohabitation was in question. It was precisely in response to the close genetic proximity of humans and chimpanzees and their shared susceptibility to infectious diseases that primatologists kept their distance from the animals. Thus, paradoxically, a philosophical anthropology emphasizing relatedness and continuity between *Pan* and *Homo* manifested in an exacting research practice that erected barriers and cloaked itself in protective layers of disposable plastic and bleached rubber. In Taï Forest, disentanglement had become a precondition for close-up observations of chimpanzee culture.

Habituation, Not Participant Observation

Since Bronislaw Malinowski's and Margaret Mead's immersion in unfamiliar forms of human life had become paradigmatic, social and cultural anthropologists came to equate ethnographic fieldwork with participant observation.[47] The rationale was that joining one's research subjects' everyday activities would make it easier to understand other cultures. Anthropologists usually stopped short of "going native," but the Malinowskian rule of thumb was that when on the Trobriand Islands, do as the Trobriand Islanders do. Some primatologists, especially in the tradition of the Kyoto School, aspired to this ideal, as far as their human bodies allowed them to blend into nonhuman ways of life. Itani had taught his American student Michael Huffman: "To learn about monkeys, you have to become one first. Experiencing their lives to know them is more important in the beginning than any textbook."[48] For the chimpanzee ethnographers of the Taï Chimpanzee Project, however, it would have been anathema to do as the primates themselves did. And, as I observed their interactions with the apes, I adjusted the maxim of participant observation for my own purposes: when with primates, do as the primatologists do.

"I always turn around, close my eyes, so that I don't see what he is doing, and then I hope it won't hurt so much," the doctoral student whispered as Jacobo got worked up. Rocking from side to side, he began to pant-hoot, at first almost imperceptibly, then more and more vigorously, until he charged toward me with a scream, dragging a whooshing branch behind him. He hurled it at me while darting past. Proclaiming his dominance, the leader of the South Group flung dead wood and big green *Strychnos* fruit at researchers and field assistants alike—and at me. So much for second-order observation. Fortunately, the anatomy of the chimpanzee shoulder doesn't lend itself to targeted throwing, and only once did such a projectile hit my hand. I never managed to close my eyes as advised, but I did force myself to stand absolutely still, not giving even a twitch when this ferocious animal, hair on end, rushed toward me. In the meantime, the German PhD student from the Max Planck Institute for Evolutionary Anthropology whom I followed that day took a note on the alpha male's behavior on her handheld device.

Screaming in terror, the other apes, not just the females but even Kuba, Jacobo's predecessor in the alpha position, fled up the trees when Jacobo began his dominance displays. Eventually, they acknowledged their subordination in a strictly hierarchical society by submissively pant-grunting at their superior. Feigning indifference toward Jacobo's pretension to power, we hoped, would protect us from what had happened to Jane Goodall.[49] In the 1960s, she had experienced the magic moment of grooming her favorite chimpanzee. David Graybeard served as the primatological equivalent of a key "informant" who introduced Goodall to the social customs of his society.[50] She paid the price for this intimacy with the Gombe community when David chased her through the jungle. Looking back at her field experience, she later noted that she would have kept a greater distance had she known that she would conduct a long-term study.[51] I was told that at one of the East African field sites, chimpanzees had dragged researchers around after they had displayed back at the animals. As a chimpanzee ethnographer, Boesch still insisted that "culture should personally be experienced or else, at first, it can be very confusing and hard to understand."[52] Yet like most field primatologists today, members of the Taï Chimpanzee Project made every effort *not* to actively participate in the social lives of their subjects. Not only because it might alter observed behavior but also because becoming involved in often violent chimpanzee politics could easily threaten a researcher's life and limb.

The primatologists' approach was not participant observation but habituation. The term *habituation* originated in physiology, where it designated the waning of a behavioral response as a result of repeated stimulation. In field biology, it came to refer to a procedure first introduced in the 1930s by primatologist Ray Carpenter. He gradually reduced the "tolerance distances" from which he could observe howler monkeys without changing their behavior, until they had lost their fear and behaved as if the at times strenuously disengaged human observer did not exist.[53] "A primary aim of habituation is for the observer to be a neutral element in the habitat," primatologists Elizabeth

Williamson and Anna Feistner noted.[54] More poetically, Japanese primatologist Tetsuro Matsuzawa described habituation to me as the scientist becoming to the chimpanzees "like a breeze, like the air, like a rock."

Since the inception of habituation, its techniques and processes had only rarely been described, remarked Williamson and Feistner, supposedly because researchers saw habituation as a means to an end.[55] Since all methods are mere means and yet many have received intense scholarly attention, other motives might have played a role, too. Noticing my interest in the topic, one junior researcher warned me to conduct my inquiry with tact. After all, the stories I had collected suggested that the scientists' presence affected the animals' behavior, possibly calling into question its naturalness. Moreover, my account could become grist for the mill of conservationists opposed to habituation because they feared that any human presence stressed the animals and was detrimental to their health.

A monkey researcher not involved in the study of great apes told me that, regarding habituation, he sensed "a lot of guilt in chimp people—whether it is about giving them respiratory diseases, keeping their study group from attracting unhabituated females, or enabling them to win intercommunity encounters against groups not used to humans." Another fieldworker studying chimpanzees at a different field site acknowledged the transformative effect of human presence: "We impact everything from the very obvious (disrupting a hunt, providing 'security' during patrols near unhabituated neighbors, impacting the rank of males born to peripheral females who take longer to habituate . . .) to the not so obvious (it matters who we sit next to, where we sit, if we sit or stand, and so on . . .)."

One could hardly expect a species with such a highly evolved social cognition as *Pan troglodytes* to actually consider its closest relatives in the animal kingdom to be part of the landscape—although that was how meerkat researchers conceived of habituation.[56] "If habituation is the process through which the presence of human observers becomes neutral, it doesn't happen," explained Liran Samuni, an Israeli doctoral researcher at Taï. "The word *habituation* is not suitable. They are getting used to these new creatures around them, but they never ignore us."[57]

But maybe that's exactly what becoming like a rock means to a species using rocks to crack nuts. Even familiar features of an ape's environment remain salient, anthropologist Lys Alcayna-Stevens pointed out, especially if there is a possibility that—unlike rocks—they will do something unexpected.[58] For example, not only the most timid females but also the bold Jacobo moved away when I pointed my telephoto lens at him. Is it conceivable, Alcayna-Stevens asked, that *Pan* does not habituate to humans tout court but only to specific aspects of human behavior?

FIGURE 9. Field assistants observing and being observed by a habituated chimpanzee. Photo by author.

If human behavior mattered, the process of habituation might also have required behavioral modifications on the part of the apes' observers, especially of volunteers, beginning doctoral students, and field assistants as well as visiting anthropologists of science not used to moving around in rainforests and chimpanzee groups. Novices like me frequently stumbled over vines, fell down, and were too preoccupied with not stepping on imagined and real snakes to notice that they had just cut off a particularly timid individual. But even highly experienced researchers such as Boesch whose ability to blend in was legendary encountered their limits. He related how he once followed a patrol into their neighbors' territory. During this potentially life-threatening enterprise, the chimpanzees formed a single line and silently marched in unison—until their human observer stepped on a branch. Upon the loud crack, the group came to a halt and they all turned their heads, giving the large maladroit primate clomping behind them a look.[59] When Boesch began to study how the chimpanzees hunted colobus monkeys, he practiced how to follow them as quietly as possible as they sneaked up on their prey. If his clumsiness gave them away, he would not only deprive them of a meal but also reduce their hunting success rates and thus spoil his own data. Alcayna-Stevens described this human adaptation to the unfamiliar environment as the habituation of field scientists.[60] Although this is not how the practice is described in the primatological literature, habituation could be understood as a two-way process, in which chimpanzees adapt to the presence of humans while humans also adapt to the chimpanzees' world.

Primatologists had noticed an element of social learning in the habituation of chimpanzees. Before entering the Leipzig doctoral program, Samuni had helped to implement her mentor Catherine Hobaiter's new habituation strategy in Uganda's Budongo Forest. By letting the chimpanzees set the pace, they managed to habituate a new group in record time. But they also attributed the swiftness of the Waibira community's habituation to the chimpanzees themselves, especially to two female immigrants from the neighboring Sonso group. The latter had grown up with human observers, since the Sonso chimpanzees had already been habituated in the early 1990s. Samuni and colleagues argued that these two individuals came to serve as models from whom the nonhabituated Waibira chimpanzees learned to tolerate the researchers' presence.[61] "They see that the habituated individuals are not harmed by us," Samuni explained to me. The authors presented this as a case of "social learning" but stayed clear of the heated controversy over primate "culture," which in this case appeared to adapt to the culture of primatology.

Differences in their attitudes toward humans also set apart the three study populations in Taï Forest. In the South Group, Jacobo was not the only one to test us. Shogun, a twelve-year-old male, occasionally walked right past me, almost grazing my legs. He clearly did not respect the seven meters that study protocols required researchers to keep between themselves and the chimpanzees. On my first day, Oscar, the by now six-year-old child star of the Disney movie *Chimpanzee*, which Alastair Fothergill and Mark Linfield had shot in Taï, swaggered toward me with a stick, lashing out at some shrubs right in front of my feet.[62] He also poked female researchers in the belly or grabbed their hair from a branch above their heads. Following protocol, they did not step away but froze, letting Oscar play with their hair for fear of encouraging his behavior by showing any response. I couldn't help wondering whether freezing was not a response, too.

How very differently the East Group related to humankind. Its alpha male, Arthos, never charged any of us. He hardly challenged his fellow chimpanzees. Most of his displays remained brief and matter of fact. He did not single out particular individuals and appeared a calm and self-confident leader who casually accepted his inferiors' pant-grunts without chasing them up the next tree. Samuni pointed out that he regularly displayed by hitting the ground with his flat hand. In response to her imitation of his gesture, he moved a few meters away from us. "That's the difference between East and South Group," she commented. Under Kuba's reign, however, the South Group had also enjoyed a more laid-back life, the research team's veterinarian remembered. The character of a community could change a lot, depending on the leader's character, she explained. "Each alpha male has his own style," wrote McGrew.[63] "But individual differences are usually seen as something for personality psychologists to study, not as matters of culture."

In the East, many of the females continued to be fearful, especially if researchers pointed technical equipment like cameras or microphones at them. When the group had enough of its human observers, they occasionally shook them off and disappeared into the forest. Consequently, the researchers treaded more carefully. Where, in the South, they would have passed by an individual at a short distance, in the East, they occasionally circumvented the entire group to avoid disturbing anyone. Samuni wondered, "if we want them to be as habituated as the South Group, whether we shouldn't treat them the same way."

Why the two groups' attitudes toward humans differed nobody knew for sure, but everybody had an opinion. There was widespread consensus that the South Group had been "overhabituated," and several researchers blamed this on Disney.[64] But whereas the film team's presence had supposedly made this community of chimpanzees lose their respect vis-à-vis our species, it was said to have made the East Group wary because their habituation had been completed only one year before the camera-wielding humans arrived in 2006. In the two communities, one and the same stimulus seemed to have provoked almost opposite responses.

Williamson and Feistner had already noted that too much familiarity with observers could change their status among their research subjects "from being a 'piece of the furniture' to a social tool, available for inclusion in [the primates'] social relations."[65] They wrote, "Being 'used' by your study animals is rarely mentioned in the literature, although it probably happens regularly." Among the "unpublished secrets" of primatological fieldwork, which one researcher at Taï readily shared, was that lower-ranking individuals often sought the proximity of the scientists because they served as shields against attacks from superiors.[66]

Other explanations of the obtrusive behavior of certain South Group individuals included the story of how, one day, a field assistant had been hit by a big fruit falling from a tree. Before dropping unconscious in front of the chimpanzees, he had screamed out. Ever since, Shogun and other males had supposedly grown curious about how to elicit such cries from their all too composed observers. One assistant believed that the South Group's aggressions against our kind had been triggered because they had experienced more poaching than the East Group (but why wouldn't they run away like their neighbors, then?). A female German researcher claimed that Jacobo's charges were aimed mostly at white men and had never been so pronounced before I arrived. An Ivorian assistant, on the other hand, dismissed this allegation of interspecies racial discrimination. He assured me that Jacobo had always been like this and that he himself had frequently been hit by pieces of wood. Transgressive behavior toward humans, shown mostly by adolescent males and young adults, was a matter of age and would sort itself out in time, some speculated, while others were concerned that it might get even worse as naughty children turned into fully grown chimpanzees. In other words, nobody really knew why these neighboring ape communities related so differently to humans.

From the researchers' point of view, both of these human-animal relationships posed problems: the East Group's incomplete habituation made them difficult to follow; the South Group's overhabituation made them potentially dangerous to be with. Usually left out of these discussions, Boesch's original, now-dwindling North Group seemed to conform best to what primatologists expected of habituated chimpanzees. Ideally, the apes simply ignored their observers, acting naturally in every other respect than having overcome their fear of the most lethal predator in the jungle.

A different interpretation of habituation prevailed in multispecies studies. Donna Haraway and Vinciane Despret followed baboon researcher Barbara Smuts's heterodox understanding of habituation as aiming not at indifference but at an intersubjective relationship, in which the observer also modified her own behavior as she learned to act in accordance with what she took to be the monkeys' social conventions.[67] From Smuts's point of view, it was impossible for primatologists to neutralize their presence among primates, and any attempt to do so only irritated them. She offered an example from her brief experience with the Gombe chimpanzees where every effort to ignore the physical transgressions of an adolescent male called Goblin aggravated the situation further—until she punched him in the face. Smuts compared her behavior to that of a female resisting submission in the dominance rank hierarchy.[68] It was reminiscent of how Japanese primatologist Naoki Koyama had gained the respect of the macaque troop of Arashiyama by kicking the alpha male in the face.[69] His American colleague did not reveal how and especially for how much longer her relationship with Goblin developed subsequently. But chimpanzee researchers with more long-term experience expected all females to eventually subject themselves to male domination. That is to say, if Smuts had not soon thereafter left the Gombe community for other reasons, her retaliatory action might have set her up for even more violent confrontations with a much stronger animal, leaving her the choice between battered submission and terminating her participant observation of the chimpanzee group.

When I discussed this approach with a seasoned chimpanzee researcher, she explained: "I have a lot of sympathy with the position that we are not and can never be a neutral presence as observers." But she also felt deeply troubled by the potential fallout of Smuts's approach: "Perhaps most worryingly (given researchers know the risks they take) what happens when a female who is used to treating humans as 'part of the group' immigrates to a group that regularly crop-raids or interacts with other people (increasingly true now that we're officially in the 'anthropocene'). It's unlikely to end well for anyone." Even habituated chimpanzees did not figure as the primatologists' companions.

Observing Chimpanzees

Alfred Radcliffe-Brown explained the opposition of British social anthropology to the culture concept by its lack of empiricism. Ethnographers could observe the behavior of individuals, including their speech acts and the material products of past actions, which revealed a complex network of social relations, but nobody had ever seen culture: "We do not observe a 'culture,' since that word denotes, not any concrete reality, but an abstraction."[70] Similarly, evolutionary biologists William Hoppitt and Kevin Laland remarked that "social learning and imitation cannot reliably be deduced from the casual observation."[71] If cultural primatology was "positivist," as McGrew claimed, and positivists took sensory perceptions as the best foundation of knowledge, as philosopher of science Ian Hacking noted in an ideal-typical portrayal of their epistemology, then what and how did chimpanzee ethnographers actually observe?[72]

Boesch offered a rather disenchanted glimpse of chimpanzee ethnography when we visited the ecotourist sites of his conservationist NGO Wild Chimpanzee Foundation near Taï. Instead of live primates we found only stones and pieces of wood next to roots with cuplike depressions. The shells of *Coula edulis* nuts indicated that these places had served as nut-cracking workshops. After documenting the traces, Boesch noted wryly: "That's how you study culture in the wild." He mostly wanted me to understand that his dissertation research had hardly amounted to immersion in an exotic form of nonhuman life: before the habituation of the North Group was complete, he had to rely mostly on indirect evidence. But his comment also suggested that chimpanzee ethnography was not a particular method of data collection. Whether direct or indirect, observations would count as ethnographic only if their object turned out to be chimpanzee culture.

As Itani's failure to recognize the grooming handclasp of the Mahale chimpanzees as cultural showed, culture could be hiding in plain sight. Only in retrospect did the founders of cultural primatology realize that they had worked ethnographically. Goodall, McGrew, Boesch—they had all initially looked at chimpanzee cultures without knowing it. Only after the fact did they arrive at this conceptualization, looking back and comparing what they had seen at different field sites. Whereas many cultural anthropologists take culture to pervade human life to an extent that all their fieldwork amounts to ethnography, the classification of great ape behavior as cultural remained confined to particular activities (only Japanese primatologist Michio Nakamura argued for "the ubiquity of culture in every domain of behavior in chimpanzees and perhaps other nonhuman animals").[73] Since demonstrating that the cultural nature of these activities could happen only in the course of data analysis, usually after the researchers had left the field for an office desk, the practice of chimpanzee ethnography, as conducted in the forest, was largely that of chimpanzee fieldwork in general.

At Taï, following the apes could be demanding. Depending on where they had slept the night before, we got up at 4:00 or 5:00 a.m. to arrive at their nests before they left. These hikes through the dark usually took about an hour but could also take twice as long since the South Group had begun to take over a more remote territory after most of its original inhabitants had been poached. If it didn't rain hard, marching down the nocturnal trails cut through the forest had a meditative, sometimes dreamlike quality, especially when the huge eyes of a little galago gleamed in the beam of a headlamp or a dazzled armadillo slowly climbed up a tree at the wayside. Occasionally, a wondrous mushroom robed in a white net raised its foul-smelling cap from the ground. One time the field assistant going ahead made a gigantic leap backward and landed in my arms when a cobra shot up right in front of his feet. Eventually, we would leave the trail and turn off our lights as we approached the GPS-recorded location of where the chimpanzees had built their nests the night before. Thick vegetation could make this last part of the commute arduous. In the dense secondary forest that had overtaken abandoned plantations at the edges of the national park, we sometimes crawled on all fours through impenetrable undergrowth. The reward was to see the chimpanzees' descent from the treetops in a dawning rainforest, sometimes slow and sleepy, sometimes loud and theatrical when the males breezed down on bending young trees, starting screeching displays as soon as they hit the ground running.

The everyday life of chimpanzees oscillates between moments of high drama with fast-paced action and long hours of feeding, grooming, and staring into space. Researchers' observations of these periods of idling could require much patience, although primatologists who studied chimpanzee nutrition or how picking parasites from each other's fur facilitated social bonding got particularly busy when their subjects were at rest. Some students and I killed time reading e-books or scientific papers on hand-held devices used to record behavioral data and field notes. But the best observers never averted their eyes. Things could change very quickly. By the time a turn of events became noticeable to the distracted among us, important interactions had already taken place: the silent origins of a noisy conflict or the old female who had signaled that it was time to move on would remain unknown.

Once the chimpanzees decided to travel to the next place, they often darted off at a pace that was difficult to keep up with—especially for bipedal humans who tripped over spiky vines (known among Leipzig doctoral students as "asshole plants"). This hateful obstinacy of a nature not to our design quickly dissipated the nature mysticism I otherwise entertained on well-maintained footpaths. But if the researchers and field assistants lost the chimpanzees during such a chase, it could take days to relocate them. Thus, heavy backpacks, muggy air, sweat-soaked facial masks, fire lianas, and fear of snakes notwithstanding, we had to stumble after the apes as fast as we could.

Usually the chimpanzees headed for another food tree. They seemed to know when its fruit would be ripe, and they managed to find it with admirable precision in a boundless sea of trees. The dense forest offered so few landmarks that urbanites like me and the other Europeans, and even most Ivorian field assistants who had grown up between villages, roads, and plantations would have been lost without GPS and compass. If it had poured all night, the chimpanzees often steered toward the log of a fallen tree that had torn a hole in the canopy and allowed some sunlight into the forest and onto their wet coats. Nightly rainstorms usually announced lazy days since the unsheltered animals had not had much sleep up in their trees.

But the humdrum everyday life of chimpanzees could give way to moments of intense focus if the group decided to enter the territory of a neighboring community, silently marching in single file. Occasionally, the group stopped and listened. Only a loud fart escaping a chimpanzee gut would interrupt their intent silence. Often these patrols amounted to nothing more than the hushed raiding of a contested fig tree or a hunt for colobus monkeys that flourished in the borderland between two chimpanzee communities, like the wildlife that used to thrive in the death strip of the Berlin Wall.

Collective Empiricism

Once, however, the furious neighbors did attack. While the South Group munched on *Sacoglottis* fruits, I withdrew with one of the primatologists behind a big tree trunk where we could enjoy our own lunch out of sight. Never seeing us eat, shit, or piss, the apes must have considered *Homo* a rather ethereal kind of animal.[74] At the first spoonful of cold rice and avocado, a pandemonium of shrill screaming broke loose. In the ensuing chaos, I could not find the researcher I was shadowing and ran with my lunch partner after the animal he was observing that day.

Unfortunately, Woodstock turned out to be an aptly named peace-loving chimpanzee who preferred to defect. While the other males fought and, as we learned via walkie-talkie, even briefly captured a female from the enemy group, he escaped to a safe distance behind the front line, where he supported his party by loudly pant-hooting and drumming against a tree. As primatologists have frequently noted, the presence of human observers put the habituated group at an advantage over their opponents, who had to face not only their own kind but also a most frightening predator.[75]

In such a dizzying turmoil and low visibility, it was impossible for a single researcher to capture everything that was happening. Even the team of three doctoral students that I followed that day was bound to miss many important events. Undoubtedly, however, their collective observations offered a more comprehensive picture of how the complex dynamics unfolded in and between groups. While Goodall, Boesch, and other founders of primatological field sites had started out on their own, maybe accompanied by a spouse or a field assistant, chimpanzee ethnography and other forms of naturalistic observation had since taken collaborative forms.

By the early twenty-first century, the major chimpanzee field sites accommodated dozens of scientists and assistants who had to make their observations consistent with each other. At Taï, a small group of doctoral students who planned to publish together reverted to a strategy adopted by primatologists at least since the 1960s and anxiously aligned their behavioral classifications by following the same animal for a day, so they could compare their perceptions and reach consensus about what they had seen.[76] Such tests of interobserver reliability also enabled the camp manager to evaluate new field assistants: if their ethogram entries diverged too much from those of a more experienced assistant at their side, they failed. If convergence exceeded 75 percent, they were suspected of cheating. Perfect concurrence seemed impossible, however well researchers had defined their categories in advance. How to integrate variously taken observations had been a problem ever since the emergence of scientific observation as a collective endeavor in the seventeenth century.[77]

Yet present-day ape researchers had not conclusively solved the problem of how to pool the data of many observers. Each field site organized the collaboration of its workers differently. Kevin Langergraber, codirector of the Ngogo Chimpanzee Project in Uganda, worried that assistants were not as conscientious: "It's a good-paying job, but watching the chimps for twelve hours per day is a crazy thing you're asking them to do, from their perspective. The quality of the data I collect is really important to me in a way that it isn't to the guy getting paid $8.00 while it doesn't matter if he misses a lot of stuff." The

Japanese researchers around Tetsuro Matsuzawa also did not trust the records of their African guides enough to use them in publications and consequently frowned on the collective observations of the Taï Chimpanzee Project, which included—although in a rather selective and gradated manner—the work of its field assistants. Boesch recognized the problem as one of cultural difference: "Making the Ivorian assistants understand that they need to do something systematically, because that's the only way we can work in science, is difficult. It's not part of their culture. Even our students have a hard time with this."

The reservations with which scientists at Ngogo, Bossou, and Taï regarded the data produced by local assistants contrasts with the accounts of sociologists and historians of primatology who have emphasized that in comparison with the assistants, "foreign researchers . . . may not even be the most important observers."[78] And yet the latter were the only ones recognized in the moral economy of science, which distributed status, credit, and authorship unequally, these critics implied. The interesting question is this: To what extent did this inequity reflect that certain people's sex, race, class, or nationality was deemed inferior, as historian Georgina Montgomery alleged, and to what extent did it reflect a stratification of inherently epistemic virtues and practices?[79]

Of course, status differences also constituted a hierarchy among the field assistants. The most experienced among the assistants in Taï, Nene Kpazahi Honora, had watched chimpanzees for almost three decades—a craft he had learned under Boesch's exacting tutelage. Honora had continued to exercise the skill daily long after Boesch had returned to Europe. At the time of my visit, he was training a new generation of assistants. Although Honora emphasized that active teaching of this skill was not possible, he had also guided generations of doctoral students, including the project's current director, Roman Wittig, as they learned to see. Wittig was just organizing an event in Honora's honor at the Deutsches Zentrum für Primatenforschung in Göttigen, Germany, where this seasoned chimpanzee watcher would speak about what he had learned during twenty-five years in the field.

Despite occasional doubts about the reliability of individual observers among both assistants and students, such training and coordination, facilitated by inventories of clearly defined behaviors, disciplined everybody's eyes, enabling all to partake in a form of collective empiricism. This collaborative model of knowledge production had come to prevail in Taï, since Boesch no longer ran a mom-and-pop research station in the jungle but had the apparatus of a Max Planck Institute at his disposal. The new approach differed from the way many American primatologists worked out of their anthropology departments. In an interview with me, Langergraber contrasted Boesch's reliance on doctoral students with John Mitani's more individualist approach at Ngogo:

> Christophe develops an idea, puts a Ph.D. student on the project, and he'll be senior author on the papers. That's following the model of big science in the United States where you build a career through grad student labor. In American cultural anthropology, by contrast, grad students come up with their own projects. Mitani works closer to this tradition: he doesn't put his name on his students' publications unless he actually collected data. Biological anthropology is this weird discipline in between the humanities and the science model.

Of course, collective research had never been the exclusive privilege of the natural sciences: in the late nineteenth century, it had been pioneered by historians and philologists before physicists and chemists decided to imitate the group efforts of their colleagues in the humanities.[80] In the early twenty-first century, chimpanzee ethnographers blurred these lines again.

Disentangling Primate Cultures from the Cultures of Primatology

Integrating observations of different observers required keeping their biases at bay. Until the 1970s, primatologists would typically try to record as much as they could or whatever could be most readily observed in the form of "typical field notes," Jeanne Altmann criticized in one of the most widely cited methodological papers of animal behavior research.[81] She worried that such ad libitum sampling allowed unconscious and thus unstated preferences to determine the observer's decisions on what to look at and what to ignore. Despite her commitment to objectivity, Altmann became one of the heroines of Haraway's *Primate Visions*—maybe not that surprising given that both were preoccupied with the problem of observer biases.[82] An outspoken feminist concerned about her mostly male colleagues' one-sided focus on the often more dramatic behavior of male primates, Altmann advocated more disciplined sampling techniques to undo sexist distortions in our image of species studied as models of early humans.

As a corrective, Altmann proposed focal-animal sampling—an approach that had emerged in developmental psychological studies of human children in the 1930s and entered animal behavior research in the 1960s—which required the observer to concentrate on the behavior of one particular member of a primate group for a specified period before turning to a different animal.[83] Now primatologists could make sure that each individual received the same attention. In Taï, every researcher picked a focal animal or "target" that he or she would follow all day. Ideally, they recorded in the first ten seconds of every minute what this individual was doing. This way they would cover not just behaviors that happened to catch their eye but the entire chimpanzee group's spectrum of activities.

Such focal sampling became one of the methodological foundations of cultural primatology. As long as researchers relied on ad libitum sampling, Altmann argued, comparisons between study results left open which differences reflected actual differences between animal populations and which differences reflected predilections of human observers.[84] What appeared to be cultural differences between primates could just as well be cultural differences between primatologists. Focal sampling helped to extricate primate cultures from the cultures of primatology.

Positivism: "Our brain should see through our eyes"

Cultural primatologists also sought to guard against their own flights of fancy by exercising interpretive restraint in their practices of observation. In Loango, for example, a research manual warned: "Never write down what you think, only write what you observe!" Before Boesch's visit to his Gabonese field site, the team studying gorillas in the national park had noted "out-of-sight feeding" and "out-of-sight grooming" on their check sheets if foliage obstructed their view of an ape. These were simply bad observations, Boesch explained to them: "Either you see him eating or you don't—and then you simply don't know what is happening." In the case of the infanticide controversy, the fact that many killings had only been inferred from the sudden disappearance of infants had been grist for the critics' mill.[85] Observation had to be kept free of imagination and prevailing theories. Under the heading "Our brain should see through our eyes," Boesch insisted: "Theories are just tools to help us progress, while only observations will teach us what nature is."[86] Unswerving in the face of more than half a century of postpositivist insistence on the theory-laden nature of observation, Boesch maintained a realist attitude toward observations and an antirealist attitude toward theories, which Ian Hacking had presented as the hallmark of all positivisms.[87]

Although more regulative ideal than reality, the primatologists' separation of observation from theory motivated an effort not to foreclose alternative readings of their data. "We try not to interpret," Samuni told me. "But we do. We put the behavior into a certain context." She gave the following example: if Shogun gently held Kuba's testicles, many observers would infer that Shogun supported Kuba. They would take as a fact that holding the testicles of another male indicated support. But "support" was already an interpretation that went beyond what could be seen. "What if Shogun was actually trying to appease

Kuba?" Samuni asked. Even if no description of a behavior could be entirely free of preconceptions, the statement "Shogun held Kuba's testicles" was more factual than the claim that Shogun supported Kuba. The latter assertion about their social relationship black-boxed what the two animals had actually done. By the standards of science studies, positivism might make a poor description of scientific practice because, as Samuni admitted, the primatologists could not help putting their perceptions in one context rather than another. But, analogous to what historians of science Lorraine Daston and Peter Galison demonstrated for objectivity, the researchers' positivist attitude was not naive or illusionary.[88] Instead it amounted to the cultivation of an epistemic virtue, in this case of interpretive moderation. Positivism did not accurately represent cultural primatology but oriented its scientific culture toward research practices that differed from those engendered by alternative epistemologies.

The chimpanzee ethnographers' positivism also made a virtue of necessity. "No cultural primatologist will secure an interview with a wild ape, but by the same token, no observer of apes will get lied to," remarked McGrew.[89] For him, the reliance of sociocultural anthropologists on what their interlocutors said to "reveal the meanings that underlie and permeate the human condition" was pure "nonsense": "Words are voluntary puffs of air, and so they need not reflect reality in the slightest. Large-brained, intelligent creatures practice deception, and one of the easiest ways of doing so is by telling lies."[90] From this positivist point of view, words could count as valid and reliable data only if they predicted other kinds of action.

Adroitly, Boesch underlined this skepticism toward the methodological value of ethnographic conversation by drawing from the authority of a human scientist who, although distrustful of the "positivistic fetishism of 'data,'" had warned his fellow anthropologists and sociologists not to be taken in by the natives' theories about themselves: "To answer the question of 'what makes us human', we should not forget the great sociologist Pierre Bourdieu's advice to concentrate on what humans do in practice and not what they say they do."[91] What Boesch's one-sided reading of Bourdieu's praxeology omitted was the latter's attempt to reconstruct a "social physics" that observed society independently of the representations of those who lived in it by eventually reintroducing those representations as they structured people's actions from inside.[92] For the purpose of understanding the lived experience of objective structures, Bourdieu had always conducted extensive interviews.[93] It wasn't for lack of interest in how their subjects classified, interpreted, and made sense of the world that cultural primatologists like Boesch and McGrew refrained from more sustained efforts to turn chimpanzee ethnography into a praxeology of primate societies. Unable to interview the apes, they simply lacked the methodological tool kit to determine how the viewpoints of individual animals varied systematically with the positions they occupied in social space.

Ethnography by Ethogram

What chimpanzees did in practice, researchers broke up into neatly divided behavioral rubrics such as vocalization, grooming, food sharing, copulation, or play. They were operationally defined to make sure that different observers arrived at the same categorizations. This will to categorize did not preclude a nominalist sensibility. Jane Goodall's dissertation adviser Robert Hinde had already acknowledged that "each incident of behavior is to some extent unique," but he had also warned his colleagues against "the temptation to believe that every detail must be recorded, for anything not recorded is lost forever."[94] To him the ethnographic excess cherished by many cultural anthropologists appeared fatal.[95] Abstracting operational definitions of behaviors enabled researchers to pool their data and to compare observations across sites—a precondition for chimpanzee ethnology.

Inventories of all types of behavior seen in a particular species were known as ethograms. For chimpanzees a team of mostly Japanese researchers around Toshisada Nishida and William McGrew had compiled the most comprehensive and painstakingly detailed catalog of this kind. Even the most literary description of chimpanzee life could hardly provide more nuance than the almost 1,400 behaviors they meticulously listed and teased apart, in words and with video clips on two accompanying DVDs. Nishida and colleagues even included behaviors so far seen in only one particular individual, which they marked as "idiosyncratic" and considered "important because they are the seeds or origins of local culture."[96] By facilitating systematic behavioral comparison, their book advanced cultural primatology.

Despite these abundant subtle distinctions, Roman Wittig could still complain about the authors' omission of behaviors such as turn taking when chimpanzees shared food, which he had observed in Budongo.[97] He criticized Nishida and McGrew for lumping together pant-hoot variants, such as "roars" produced during traveling and "wails" produced during feeding, which had been described for Gombe, and reminded readers that the book still represented largely the Mahale chimpanzees.

When I interviewed McGrew at a conference in Kyoto, he noted that the increasing awareness of "nuanced variation" offered new insights but had also become an obstacle to the ethnological comparison of chimpanzee cultures. "The grooming hand-clasp started out as one category," he remembered. "But later on we realized that it was a whole constellation of behavioral patterns.

One of the variables has to do with how you actually configure your hands." In 1996, two decades after the initial discovery, he noticed that one community of chimpanzees at Mahale, the so-called K group, folded their wrists around each other, while members of the M group preferred to grasp each other's hands so that their palms touched. McGrew wondered how far they should take such differentiations. For him, the proof of the pudding lay in whether such subtle variations could be related to other variables. Since the operationally defined units of behavior listed in the ethogram were quantifiable, such associations could be tested statistically. McGrew cited his colleague Richard Wrangham, who had just found out that apart from differences between the groups, within the M group individuals showed higher rates of palm-to-palm contact with maternal kin than with nonkin: 37 percent instead of 17 percent. This quantitative difference did not allow inferring any meaning of the action or of a group character that set the M and K groups apart. The variation as such was all that could be observed and counted. No chimpanzee informant had provided any clues about how to interpret it.

In the field, where things happened quickly, it wasn't humanly possible to navigate the increasingly comprehensive species ethogram. Thus, every observer chose from a thinned-out list of behaviors, as Hinde had suggested.[98] At Taï, scientifically untrained assistants collected only basic data on social interactions, nutrition, and so forth, whereas researchers designed their own ethograms to reflect the focus of their investigations. "Research on conflict management will need more detail on aggressive behaviors than a study on food processing, which in contrast will need more elaborate descriptions of, for example, tool use," Wittig noted.[99] Most researchers contented themselves with recording that two individuals groomed each other for a certain amount of time, but a doctoral student who studied grooming behavior also recorded detailed information on how long they spent on different body parts. Giving the ethogram a branchlike structure reconciled the interests of individual scientists and a standardization without which scientific observation would have remained a solitary obsession.

Such ethnography by ethogram generated data on chimpanzee behavior that could be shared between scientists, even if their theoretical commitments diverged. It generated accounts that might not have been entirely devoid of theory but relied on low-level theoretical terms that rarely antagonized other primatologists. Whether or not they believed in chimpanzee culture should not affect their ability to distinguish between palm-to-palm and non–palm-to-palm handclasps. Cultural primatologists maintained their ability to agree about observations in the face of intense theoretical controversy by carefully attending to the difference between what they saw and how they theorized it.

Collecting Data: Notebooks, Handhelds, and Camcorders

Ethnography means "writing people." Cultural anthropologists usually scribble (or punch) their observations into freestyle field diaries (or tablets). Boesch also took field notes, supplemented by drawings and sketches, which he subsequently worked out as ethnographic vignettes that frequently opened chapters of his monographs before he switched to a more systematic discussion of the available literature. His wife, Hedwige, even worked on a whole book of stories about the Taï chimpanzees. By contrast, the PhD students from Boesch's department rarely took out their small notebooks, which they still carried in their leg pockets. The journal articles that would earn them their doctoral degrees provided no forum for anecdotes.

Instead of writing detailed descriptions of chimpanzee life, the new generation of field scientists focused on the collection of ethogram data. Tabulated paper check sheets had been the original medium of this approach. By the mid-2010s, the field assistants of the Taï Chimpanzee Project continued to chase apes with clipboard and pencil in hand. Yet the days when primatologists carried home thirty kilograms of paper records had long passed. The scientists entered their observations into various electronic devices. In the late 1980s, 32K pocket calculator–like machines, occasionally mounted on binoculars, had initiated field computerization.[100] They had since given way to an array of state-of-the-art gadgets, ranging from regular Hewlett Packard smartphones in waterproof cases to shock-resistant Mobile Tout Terrain tablets, especially manufactured to operate under harsh weather conditions and in muddy environments. The CyberTracker GPS Field Data Collection System, a piece of freeware that had originated as a citizen science project for nonliterate trackers in the Kalahari, allowed the scientists to program decision trees.[101] They branched out more or less, depending on how much nuance researchers wanted to give to certain kinds of observation.

Matei Candea argued that handhelds served to confine and make the fieldworkers' behavior more predictable.[102] Deliberately depriving themselves of all poetic license, researchers had to choose from an originally extensive, but ultimately limited catalog of animal behaviors. In opposition to Eileen Crist's analysis of the behavioral sciences, which examined how ethologists discursively constructed animal behavior as determined rather than intentional, Candea emphasized how the scientists' observations were observations of wild animals who did as they pleased.[103] This lack of control over one's object of study distinguished field science from laboratory work. It decreased the so-called internal validity of the results, making the establishment of causal relations between observed behaviors and other factors more difficult, maybe even impossible. But relinquishing experimental interference also protected these results against accusations of artificiality. The power of noninterventionist approaches could be further increased through careful study designs, argued British ethologist Marian Stamp Dawkins: "Even if we cannot control our animals as precisely as an experimentalist would like us to, we can . . . control ourselves."[104] She attributed this self-control to the "computer in the head"—that is, the researchers' own brains—which operated sufficiently alike for different observers to recognize the same patterns of behavior.[105] By contrast, Candea noted that behavioral scientists needed computers not only in their heads but also in their hands to align their neural computations and the resulting observational data (although paper check sheets might previously have done the job just as well). In opposition to radical constructionist accounts in science studies, he maintained the difference between observing and transforming behavior. The observational discipline that handhelds and check sheets imposed on researchers did construct their data in a certain way, but it did not construct the animals' behavior.

This standardization of data collection made the researchers' own behavior more predictable. If it became too predictable, though, it could hinder rather than advance discovery. Similar to cultural anthropologists' receptiveness toward surprises awaiting them in the field, which could potentially undo their culturally charged presuppositions and ethnocentric biases, cultural primatologists privileged their perceptions over books.[106] Boesch called on chimpanzee ethnographers to "be faithful to the animals they study and accept that it is more important to describe exactly what they are fortunate enough to observe and not confine their observations to proving prevailing theories or current orthodoxy."[107]

Yet the media for documenting events out of the ordinary had been marginalized in the prevailing economy of primatological knowledge. If behaviors did not fit into the preprogrammed categories, the handheld software provided a text editor to complement the ethogram data by brief field notes on, say, the social context of an interaction. Moreover, each of the three research camps in Taï Forest kept a so-called event book, a collective but unsystematic field diary of sorts, which allowed researchers and assistants to write down short reports whenever something unusual had happened in their chimpanzee communities—maybe the males had encountered females from another group, or a campfire lit by ecoguards had alarmed the apes. But unlike the rest of the data they collected, these field notes escaped automatic processing and did not lend themselves to subsequent statistical analysis. One of Roman Wittig's students told me that when he is old, he might use the diaries to write a book. Although Wittig worried that an increasingly competitive scientific field appreciated 4,000-word *Nature* publications at the expense of more comprehensive articles, an ambitious midcareer scientist like him had to stay focused on highly cited journals, which did not welcome such anecdotal knowledge.

The students of the Taï Chimpanzee Project collected mostly electronic data not conducive to rich contextualization and narrative. But they recognized the limitations of this approach. Samuni, for example, felt that her ethogram on play schematized the chimpanzees' complex and dynamic interactions all too crudely. But instead of writing, she filmed the animals' behavior. She had adopted this practice from her Budongo mentor Cat Hobaiter. Having come of age with smaller and smaller camcorders equipped with larger and larger memory cards, the young British Lebanese professor and her students belonged to the first generation of primatologists that developed an almost cyborgian relationship with their video equipment. Hobaiter imagined that the chimpanzees must have thought she had lost a limb in a poacher's sling when she showed up one day without camera in hand. In a field distrustful of the contortions of language, filming one's observations was certainly superior to translating them into words, not least because such records enabled other researchers, with different theories, values, and stakes, possibly even blind to the original observer's intentions, to take a second look—and a third, and a fourth. Although video recordings, as any cinematographer knows, do not simply tell it as it is, the medium suited the primatologists' desire for raw sense data better than the field diaries in which Goodall and Boesch had first jotted down chimpanzee culture.

Thin Description

Calling the practice described above chimpanzee ethnography amounted to an "abuse of ethnography," Tim Ingold maintained.[108] In contrast to Marshall Sahlins, who had dismissed the notion of animal cultures as anthropomorphism, Ingold assured his readers that he had no qualms about extending a *genuine* ethnographic approach to nonhumans.[109] What he objected to was breaking up behavior into preestablished ethogram categories. The British anthropologist preferred to talk about "hives of activity," unbroken "flows," and "trajectories of movement."[110] In the same spirit and with reference to Ingold, Nakamura rejected the "Darwinian viewpoint of culture" as an information transfer between individuals and emphasized collectively performed behaviors such as the grooming handclasp.[111] Ingold also objected to the "mere observation of behaviors as cultural traits," which primate, cetacean, and other animal ethnographers extracted from the lived experience of their subjects and which the method of exclusion separated from environmental and genetic factors.[112] True ethnography had to be "sensitive to the intentions and purposes of the people themselves, to their values and orientations, to their ways of perceiving, remembering, and organizing their experience, and to the contexts in which they act—all of which add up to what anthropologists generally mean by 'culture.'"[113]

If an observer actually provided such a thick description, Ingold worried that evolutionary anthropologists would shrug it off as unscientific, just as they ignored indigenous knowledge of animal behavior. The parallel between ethnographic and indigenous knowledge echoed how Ingold conceived of the relationship between the subfields of anthropology: "In the eyes of 'evolutionary biologists', it seems, social and cultural anthropologists are the new savages, locked in a struggle for existence in which they are bound to succumb."[114]

In the face of this stark two-culture divide, Ingold occasionally did find interlocutors among evolutionists. His critique of the antagonism between social and biological anthropology—an antagonism that Ingold both regretted and fueled with his incendiary rhetoric—inspired American primatologist Barbara King to lay out her own vision for an ethnography of African great apes.[115] She translated social anthropologist Marilyn Strathern's account of Melanesian personhood as being always already in relationships into her own language of dynamic systems theory.[116] She wanted to understand the meaning of what the gorillas in the National Zoological Park in Washington did against the background of their relations with each other and with their human caretakers. This led her to dismiss the "checklist approach to chimpanzee culture," which reduced this complex phenomenon to a list of behavioral variants.[117] In the vein of Ingold's opposition to dissecting animal cultures into separable cultural traits, King objected to Whiten and colleagues' article "Cultures in Chimpanzees" because it "abstracted away from the social events and relationships described in the monographs published by these same researchers."[118] She illustrated the possibility of providing richly contextualized accounts of the social construction of meaning with a detailed ethnographic vignette describing how different members of the North Group had responded to the killing of one of their own by a jaguar, which Boesch and Boesch-Achermann had provided in their book *The Chimpanzees of the Taï Forest.*[119] Unlike Ingold, King acknowledged that chimpanzee ethnography included, but could not be reduced to the checklist approach. Thereby, she concurred with McGrew, who responded to Ingold that Jane Goodall's *In the Shadow of Man* had already satisfied all of Ingold's requirements for genuine ethnography some thirty years ago.[120]

Since Goodall's first carefree forays into the world of the Gombe community, the heated atmosphere of the chimpanzee culture wars had certainly made primatologists wary of inferences about the apes' intentions, values, and experiences. Such interpretations invited protests from comparative psychologists who argued that field observations could not possibly support claims about the inner life of other animals.[121] Undeterred, McGrew used book chapters to speculate that chimpanzees might employ their so-far undeciphered soft grunts as kinship terms and took their extended musings as a sign that they might even hold worldviews.[122] Similarly, Boesch considered chimpanzee symbolic cultures as a first step toward the development of belief, myth, art, and discourse. Turning the tables on critics who demanded greater interpretive restraint, he argued that absence of evidence was hardly evidence of absence: "I could provocatively suggest that as long as we remain incapable of communicating with chimpanzees at their level, it seems premature to say that they have no such abilities."[123] In peer-reviewed journal articles, however, cultural primatologists largely refrained from such conjectures about the thickness of chimpanzee culture.

What neither King nor Ingold considered was that it might have been because of, not despite its ethnographic thinness that the checklist approach had made cultural primatology strong. Historian of science Theodore Porter has shown how the aspiration to superficiality—what he dubbed "thin description"—helped the natural sciences gain wider and wider currency in the twentieth century.[124] It was this commitment to the external and observable, while ignoring questions of meaning, that characterized the milestone paper by Whiten and colleagues.[125] When the authors extracted elements from the local primatological knowledge of particular chimpanzee communities and stripped them of their context of discovery to rearrange isolated cultural traits in tables, they made visible the patterns and connections between different study sites that had previously escaped the fieldworkers' ethnographic gaze. This work of abstraction revealed a geographic variation of behavior not easily explained genetically or ecologically and backed the argument that chimpanzees shared our capacity for culture. It created what Latour called an "obligatory passage point," a *Nature* article that, in contrast to Boesch's vignettes, all of those writing about chimpanzee cultures—even if they were hostile to the concept—had to cite.[126] Whatever the biases of Google Scholar, it seems suggestive that according to this search engine, in the not even two decades since its publication, the Whiten article has been cited more often than Boesches' *The Chimpanzees of the Taï Forest* and even Goodall's classic book *In the Shadow of Man*, which provided a close-up view of the Gombe chimpanzees.[127] If the power of epistemic cultures depended on the size of their networks, the cultural primatologists' pooling of collective observations to support a general theory of primate culture, reminiscent of the large-scale structuralist tableaus of anthropologists Claude Lévi-Strauss and Philippe Descola, would grant them a competitive advantage over those miniature painters of biosocial life who reduced cultural anthropology and primatology to ethnography.[128]

Wild or *Pan/Homo* Cultures?

Chimpanzee ethnography emerged as one approach in a whole family that sought to overcome the epistemological and ontological gaps between natural sciences, social research, and humanities, between *Natur-* and *Kulturwissenschaften,* or *sciences naturelles* and *sciences humaines.* King was hardly the only researcher who bridged evolutionary and cultural anthropology. McGrew had previously made an attempt when he wrote a second doctoral thesis in social anthropology.[129] Other scholars colonized the borderland between nature and culture from the humanities side.

At the École normale supérieure in Paris, for instance, philosopher Dominique Lestel sought to combine the empirical richness of ethology and the conceptual resources of philosophy to spell out the consequences of the scientific revolution that had been ushered in by the Japanese discovery of animal cultures.[130] In the new paradigm, ethnographies of animal communities, conceptualized as such in the framework of the social sciences, opened up a future for zoology at the intersection of life sciences and human sciences.[131]

Lestel's account of animal cultures owed much to the kind of chimpanzee ethnography that McGrew and Boesch had developed. But these primatologists remained too committed to the *ancien régime* of classical ethology that maintained the fiction of wild nature and dismissed interactions between animals and humans as a methodological bias.[132] Lestel envisioned an anthropology of human-animal relations instead.[133] Methodologically, it would be based on two symmetrical approaches to interspecies encounters, which he called etho-ethnology and ethno-ethology.[134] Etho-ethnology integrated Jakob von Uexküll's investigations of how each species responded to specific cues in its *Umwelt* and hermeneutic practices interpreting how animals interpreted human behavior, how humans interpreted animal behavior, and how humans interpreted how other humans interpreted both animal and human behavior.[135] Ethno-ethology complemented this biosemiotic approach with an ethnological approach comparing scientific and folk ethologies that acknowledged the validity of what, say, dog owners or hunters knew about animals. If the ethno-ethology of laypeople diverged from that of scientists or if scientists disagreed with each other, argued Lestel and colleagues, their knowledge could not be dismissed as false: just as folk psychology was part and parcel of the psychological functioning of those who believed in it, so anyone's ethology shaped the behavioral functioning of the human-animal community it described.[136] This proposed fusion of realist and constructionist epistemologies, which explained away theoretical differences as reflecting differences between the human-animal communities theorized, could certainly have defused the chimpanzee culture controversy: in their own local context, everybody would have been equally right. Such an "ethology of the singular" amounted to a radical reorientation from a nomothetic to an idiographic approach focusing on particular communities or even individual animals and their caretakers.[137] But it could persuade neither comparative psychologists working hard to establish general laws nor chimpanzee ethnographers privileging wild cultures over human-chimpanzee cultures.

The preference for the wild was partly a matter of taste and personal preoccupations. For example, Boesch could have diverted some of the Max Planck Institute's ample resources to Banco National Park near Abidjan, which had supposedly become a dumping ground for apes that had grown up in expat families until they became too strong and unruly to be kept as pets any longer. But apart from the fact that this dense secondary forest was even more difficult to navigate for humans than Taï, Boesch simply wasn't interested in the emergent culture of a feral population feeding on hotel garbage. Just as earlier generations of cultural anthropologists had favored indigenous cultures less transformed by colonial encounters, Boesch favored chimpanzee cultures less transformed by human contact.

As I have shown ethnographically, the primatologists of the Taï Chimpanzee Project fully realized that their presence did affect the lives of their research subjects and that they could only minimize this effect. Their ethos of detachment made a significant practical but maybe not an ontological difference. Could habituated and possibly "overhabituated" chimpanzees still count as "wild"? Did the simplistic opposition of wild and domestic animals obliterate a much greater diversity of human-animal relations, as Lestel suggested?[138] Such semantic questions might have been more important to philosophers than to practicing scientists.

Of more than semantic significance was the question of whether chimpanzee ethnographers who followed their habituated animals belonged to a "hybrid community" comparable to those created by the ape language projects.[139] Did they participate in the coconstruction of a "*Pan/Homo* culture," as Sue Savage-Rumbaugh and colleagues had argued for the primates and primatologists in her laboratory?[140] Boesch developed his account of wild cultures in opposition to claims that chimpanzees had acquired their cultures from humans. Dutch ethologist Adriaan Kortlandt had thus explained how the Bossou chimpanzees had learned to crack oil palm nuts in an area where humans used similar techniques.[141] Although both Kortlandt and Boesch could only speculate about the history of chimpanzee nut cracking, their disagreement concerned a matter of fact.

Boesch doubted that wild chimpanzees would learn socially from anyone but members of their own community (and even within the community the social status of potential models mattered). They would certainly not learn from a member of another species. Boesch had once put this assumption to the test in an uncontrolled field experiment of sorts. Caught in a poacher's sling trap, a juvenile had been able to break free but had failed to shed the sling around her wrist. Every time she tried to pull it off, it only cut deeper into her flesh. Boesch brought a sling to the forest, put it around his own wrist, and whenever the young chimpanzee observed him, he demonstrated how to remove the sling by pushing rather than pulling the loose end. Much to his chagrin, she would not learn from him. Social learning presupposed that one identified as a member of a group, Boesch explained to me. Even if the animals had been habituated, he did not belong to their community.

An interesting counterfactual consideration is whether this injured juvenile would have learned from Boesch had he cultivated the kind of intimate relationship with the Taï chimpanzees that, for better or worse, Goodall had developed with at least some of the Gombe chimpanzees. Would he have been accepted as a model for social learning if he had engaged in grooming and other social interactions?

Confining social learning to the in-group did not distinguish chimpanzees from humans, Boesch told me. The Hadza, a hunter-gatherer society in Tanzania that Boesch had visited with anthropologist Frank Marlowe, would not learn from outsiders, either. "In Hadza society, you would not believe any old guy. You would attend to the important people in your social group." In Boesch's understanding, the place for genuine social learning was a bounded community, not humanity. He agreed that modern science and scholarship confronted us with the problem of how to organize social learning in much larger collectives. He had also learned much from colleagues he did not know. Historian Steven Shapin examined how scientists adapted to an ever-expanding field of knowledge production by shifting trust from respected people to prestigious institutions.[142] Since their social life had been forced open, Boesch noted, captive chimpanzees such as the subjects of the ape language projects had learned to learn not only conspecifically but also across species boundaries.[143] That was the difference between *Pan/Homo* and wild cultures.

Chimpanzee Ethnography in the War on Cartesianism

In the last chapter, we saw that some cultural primatologists understood chimpanzee ethnography as a method of exclusion: it constituted its scientific object, chimpanzee culture, by eliminating genetic and ecological explanations of observed behaviors.[144] This analysis could not take place in real time but happened only post hoc, after researchers had left the forest and could sit down in front of their computers. Only retrospectively, as primatologists returned to their desks and began to calculate and write culture, did their field observations become chimpanzee ethnography.

Ethnography is not only a method but also a genre. Every cultural anthropologist is expected to have written one. In primatology, however, career advancement depended first and foremost on peer-reviewed articles in the *American Journal of Primatology*, *Behaviour*, or *PLOS ONE*. Japanese chimpanzee researcher Michio Nakamura lamented that the days had long passed when Goodall, Itani, and Nishida had written whole ethnographic monographs, vividly describing monkey and ape societies "in the same manner that anthropologists write about their targets."[145] This genre of writing revealed how every cultural practice was embedded in social interactions that could not be decomposed into a series of neatly categorized behaviors, he claimed. But it had been devalued as "old-fashioned and even nonscientific," worried Nakamura.[146]

Boesch, however, had it both ways. As the author and coauthor of a vast number of journal articles, he also wrote books, which provided him with the freedom necessary to explain his observations and interpretations to his colleagues, especially to those who worked with captive apes and lacked field experience: "The format and length of scientific articles are strictly regulated and there are unwritten rules about what content is permissible," he explained to me in an interview. "But life cannot be reduced to a scientific paper. If you want to provide a detailed account of what the chimpanzees are really doing, you need to write a book." Boesch's books did not take the form of books for a general public that Nakamura missed in contemporary primatological writing, but *The Chimpanzees of the Taï Forest* was certainly an ethnography, and *Wild Cultures* made the transition from chimpanzee ethnography to ethnology that McGrew and colleagues had envisioned.[147]

The epigraphs setting the scene for each of the chapters of *Wild Cultures* reveal a style of thinking and doing that complemented the empiricism of ethnographic observation and ethnological comparison with a practice that had been marginalized in the natural sciences. Boesch not only "wrote culture"; he also read the classics, at least some of them, however spottily. The main text hardly spells out his engagement with Diogenes, Voltaire, and Rousseau beyond the short opening quotations. But Descartes makes repeated appearances, just to be smacked every time for his mechanistic zoology: "Descartes' proposition that animals are mere machines is simply wrong and it is time we start to genuinely look at what animals are capable of," wrote Boesch.[148]

The real target of this anti-Cartesianism was not the seventeenth-century philosopher, though; the real target was experimental psychologists who studied one group of captive chimpanzees and then made claims about cognitive capacities supposedly distinguishing all chimpanzees from all humans. Their belief that "development is only marginally influenced by the environment and social conditions" was a "legacy from René Descartes' equation of animals with machines," Boesch contended. "To reject Descartes' ideas about animals also implies that we cannot simply consider animals living in captivity as equal to those living in the tropical rainforests of Gabon or the savannahs of Senegal."[149] In opposition to this Cartesian lineage, imaginary or real, Boesch followed in the footsteps of eighteenth-century naturalist Charles-Georges Le Roy, who had complained that most philosophers did not realize that to appreciate the intelligence of animals, they would have to live in society with them.[150]

It might be no coincidence that most of the classics Boesch cited were by French Enlightenment thinkers. Not only had he been raised in Paris, but the *philosophes* of the eighteenth century had also fought out the first intense debates over the relationship between humans and apes, triggered largely by Linnaean systematics. They were followed by the nineteenth-century controversies over Darwinism and the still ongoing series of controversies over the primate behaviors that the first field primatologists began to report in the second half of the twentieth century.[151] Social learning and tool use had emerged as new bones of contention, which had not played a central role in the eighteenth century yet—although Rousseau, von Herder, and others did raise the question of whether monkeys and apes shared the human capacity for imitation.[152] In the thick of the chimpanzee culture wars, Boesch's epigraphs recalled the origins of these intellectual conflicts.

Many of the broader cosmological questions about humankind's place in nature returned to prominence with each new cycle of controversy. For instance, the metaphysics of the great chain of being, which put the primates (literally, those first in rank) and especially *Homo sapiens* on top of a hierarchical scale of nature, still lingered in the background of the disagreements between Boesch and Tomasello.[153] The Enlightenment debate over whether apes could speak had first morphed into the ape language projects of the late twentieth century and then into a debate over gestural communication and symbolic cultures in the wild. Since at least the eighteenth century, the metaphysics of apes had not progressed in a linear fashion but appeared to move in circles.

Of course, the invention of new methods and conceptualizations also clarified many issues in each successive articulation and rearticulation of problems. Naturalist and philosopher Comte de Buffon, for example, another author cited in Boesch's epigraphs, had speculated that apes were degenerate human beings, while Rousseau suspected that humans were degenerate apes.[154] By the time Darwin reignited the debate over the ape-human continuum, no serious scholar still doubted that *Homo sapiens* and, say, *Pan troglodytes* belonged to different species. Whether they belonged to different genera was a different question—originally raised by Linnaeus, who had briefly considered calling humans an ape—which bubbled up again in the 2000s when researchers suggested reconciling a supposedly outdated taxonomy and new genetic findings by renaming chimpanzees *Homo niger* or humans *Pan sapiens*.[155]

At the beginning of the twenty-first century, primate nomenclature still bore the marks of early modern science that mixed stories from the ancients and reports of empirical observations. In 1641, Dutch physician Nicolaes Tulp had briefly observed and then dissected a chimpanzee from Angola and presented this *Homo sylvestris* or "man of the woods" as evidence that the Dionysian satyrs described by Pliny the Elder really existed.[156] Eventually, the species would be named after Pan, the Greek god of nature.[157] Ancient cosmologies gave meaning to new discoveries but could also be dismissed as false opinions in light of emerging forms of experiential knowledge. Thus, taxonomic classification merged critical readings of ancient myths with comparative anatomy.

Boesch's references to the classics still endowed his observations with broader significance, but the critical spirit of putting traditional knowledge to the test prevailed. They frequently marked the epistemic rupture between armchair philosophers and field scientists. In Boesch's narrative, the classical thinkers could only speculate about the true abilities of animals at a time when human superiority was dogma. When, in the 1960s, ethologists and evolutionary anthropologists began to actually observe wild chimpanzees, they called into question whether tool use, hunting, and other behaviors were indeed uniquely human.[158]

The modernist gesture of breaking with the past and valuing the new is characteristic of the biosciences, argued anthropologist Paul Rabinow.[159] His account mostly followed Hungarian philosopher Gyorgy Markus's ideal-typical comparison of the temporal structures of natural scientific and humanistic knowledge.[160] Markus diagnosed a confinement to the present on the part of the sciences and an extended engagement with history and supposedly timeless classics on the part of the humanities. Building on a bibliometric analysis by historian of science Derek de Solla Price, he maintained that natural scientific papers cited works of two types: recent literature, mostly from the last five years, and seminal papers from the past fifty years.[161] Philosophical publications, however, referenced a third type: classics that dated back hundreds or even thousands of years. The scientists' abandonment of classical texts entailed that no schools or intellectual traditions assembled around particular authors—except Darwin, maybe.[162] But outside Japanese primatology, evolutionary anthropologists basically knew no alternatives to him. They were "all Darwinians now," as Adam Kuper put it.[163]

And yet Boesch frequently cited the classics. Although no reader would mistake *The Chimpanzees of the Taï Forest* or *Wild Cultures* for the works of a humanities professor, his scattered references to Diogenes, Plato, Herodotus, Descartes, Voltaire, and Rousseau diverged from the bibliometric pattern distinguishing the two cultures in the eyes of Markus and Rabinow. While Boesch's articles appeared to display the historical amnesia characteristic of this publication format, his books mobilized a remote literary past that customarily oriented the interpretations of social and cultural anthropologists. In this structural respect, Boesch's style of writing chimpanzee ethnography converged with the genre of the ethnographic monograph.

This observation carries an important implication for the chimpanzee culture controversy. Markus noted that classical texts organize the humanities socially and intellectually by providing points of reference for antagonistic engagements and confrontations.[164] After all those centuries, Platonists still take on sophists, Hobbesians mock Rousseauians, Nietzscheans disparage Kantians, Wittgensteinians trivialize Heideggerians, and everybody bashes Cartesians. In contrast to the never-ending sectarianism of the humanities, Markus claimed, the natural sciences had developed practices to contain dissent and establish a widely shared background understanding, which made polemics the exception, not the norm.[165]

This account diverges from early social studies of science, which focused their attention on competition and controversies.[166] At the height of the science wars, Cambridge primatologist Robert Hinde perceived this fixation on the divisive as an exaggeration of the differences between schools of thought that failed to present them against a background of their commonalities and the shared goal of unifying knowledge.[167] Although both sides have a point, in my own experience, Markus's observation is consistent with the fact that the vast majority of citations in scientific papers affirm and build on the truth claims of other authors, while more critical engagements prevail in the humanities literature.

Exceptions prove the rule, though. Boesch's recurrent attacks on Descartes as a proxy for experimental psychologists rearticulated cultural primatology in the "polemic-dissensive manner," which Markus presented as the modus operandi of the human and social sciences.[168] *Wild Cultures* staged a clash between the "Cartesian approach" of psychologists ignoring environmental and social conditions in the development of animal cognition and the "Darwinian approach" of biologists who predicted that animals would develop more sophisticated cognitive abilities in more challenging socioecological niches.[169] Boesch ended our last interview by promoting a bellicose vision of science that pitched field observations against laboratory experiments:

> Now you should ask me whether there was anything I would like to say that you didn't ask about. So I would like to say this. The French philosopher Bernard-Henri Lévy says: "Philosophy is not about politics. Philosophy is war. When I discuss with another philosopher my goal is not to compromise. My goal is to convince him that he is wrong and should accept my opinion."[170] I feel science should be like that. Science should not be about doing politics. It should not be about reaching compromises, but about finding the one solution to go forward. Some people in the field of cognition say we need all the different approaches to understand culture and cognition. I disagree. If you want to understand the difference between humans and animals you have to realize that we are biological products and as such we have evolved and adapted to our natural environments. If some people argue that field and captivity are complementary, that's wrong. Removing animals from the wild and putting them into totally artificial situations, sometimes for generations, and then to test them in equally artificial experiments to claim that this was representative of what they could do in the wild is wrong. In that sense I agree with Bernard-Henri Lévy.

When, in 1998, the Max Planck Society appointed both chimpanzee ethnographer Christophe Boesch and comparative psychologist Michael Tomasello as codirectors of the Max Planck Institute for Evolutionary Anthropology in Leipzig, Germany, this combative ethos soon turned the newly founded research center into another battlefield of the chimpanzee culture wars.

4

Controlling for Pongoland

THE CHIMPANZEE culture wars did not rage in the jungle; they raged in universities and nonuniversity research institutions. They were waged at desks and behind lecterns where chimpanzee ethnographers found the time and the comfort to analyze, calculate, and write the cultures they had observed in the field. They were also waged at lab benches and behind Plexiglas walls where comparative psychologists tested whether chimpanzees were cognitively capable of culture in the first place. The Max Planck Institute for Evolutionary Anthropology in Leipzig, Germany, had become one of the hotspots of this controversy.

Tensions between the Department of Developmental and Comparative Psychology and the Department of Primatology ran high. Led by Michael Tomasello, the psychologists conducted experiments with human children and captive chimpanzees, bonobos, gorillas, and orangutans in Pongoland. This ape enclosure of the Leipzig Zoo and the developmental psychology laboratory on the premises of the Max Planck Institute itself represented tightly controlled research settings. At first glance, it seemed as if they could not have contrasted more sharply with the African rainforests where Christophe Boesch and his field primatologists observed great ape and occasionally human communities in their natural habitats. This chapter examines how the different research settings and approaches of psychologists and biologists informed their disagreements. While the experimenters argued that uncontrolled naturalistic observations failed to provide access to the workings of other minds and could therefore not shed much light on the cognitive capacity for culture in nonhuman animals, the fieldworkers doubted the validity of zoo experiments that did not control for their own artificiality.

This controversy over epistemic practices gained its anthropological significance and savage tone from perennial philosophical quarrels over the place of humans in the natural world. In Germany, it acquired a distinctive early twenty-first century character as two philosophical traditions, the Frankfurt School and contemporary heirs of philosophical anthropology, adopted

Tomasello as an honorary philosopher. Both valued his emphasis on how different humans were from other animals. But they also appreciated the analytic form his conceptual work took. What gave Tomasello's experimental psychology its philosophical appeal? Could the emphasis on the ape-human continuum of Boesch's field biology inspire alternative philosophical projects? A philosophical primatology, maybe?

Anthropology Returns to Germany

After everything that had happened, physical anthropology would find a new home in Germany. Not in Berlin, to be sure, where the Kaiser Wilhelm Institute for Anthropology had regularly received human body parts from the concentration camp in Auschwitz for use in studies supporting eugenics and other race-related biopolicies under National Socialist rule.[1] In the wake of World War II, the Kaiser Wilhelm Society was renamed the Max Planck Society and became the foremost research organization of the Federal Republic. Following the reunification of East and West Germany in 1990, it set up several new institutes in the not quite blossoming landscapes of the former German Democratic Republic—a scientific recovery program, of sorts. Operating outside the university system and its established, at times sclerotic disciplinary landscape, the Max Planck Society used its significant financial and symbolic capital to give new impulses to German science and scholarship. In a climate of growing national self-confidence and among calls for what philosopher Jürgen Habermas warily eyed as "a new German normality," the time appeared ripe for reintroducing evolutionary anthropology as a field that was rapidly advancing in the United States, other European countries, and Japan, but that had been thoroughly discredited in postwar Germany.[2] The Max Planck Society picked the East German city of Leipzig as the location for the new institute. But what to call a research facility dedicated to the question of what makes us human?

Concerned about evoking the very spiritual continuity with the Nazi past that the Federal Republic had assiduously fended off for half a century, the Max Planck Society hesitated to establish another institute for anthropology. Maybe it was no coincidence that it would be the first of the society's institutes run exclusively by foreigners. As founding directors, the society hired French Swiss field primatologist Christophe Boesch, American comparative psychologist Michael Tomasello, Swedish paleogeneticist Svante Pääbo, and British linguist Bernard Comrie. When it asked them to propose a designation that did not contain "anthropology," unencumbered by German sensitivities, the codirectors collectively decided that "precisely because anthropology is no longer what it used to be, the word should be in the name," Boesch remembered in an interview. "Fifty years after his death, Hitler should not be allowed to dictate what we could and could not do," Pääbo told the committee. And he added that the institute should not be a place where one "philosophized about human history."[3] "We wanted to distinguish ourselves from the field of sociocultural anthropology," Boesch said, "so we chose 'evolutionary anthropology.'" Thus the Max Planck Institute for Evolutionary Anthropology was born in 1998.[4]

Soon the researchers could move from temporary lab spaces and offices to a luxurious steel and glass building at Deutscher Platz, with a fifty-four-foot climbing wall, a sauna on its roof, clean labs for paleogenetic analysis in the basement, and one of the world's fanciest primate facilities in the Leipzig Zoo, known as Pongoland. The institute became home to four hundred people and one of the internationally leading research centers for evolutionary anthropology. The dreaded public outcry never came.

Controlled Comparison

As an anthropologist of science, I grew interested in the Max Planck Institute when I discovered that it had become the site of a heated controversy as relations between two of its founders soured. As they became codirectors, Boesch and Tomasello sought to resolve their earlier dispute over Tomasello's contention that cultural learning was uniquely human.[5] In an article titled "Chimpanzee and Human Cultures" published in *Current Anthropology*, they suggested that both humans and chimpanzees had culture, but that culture wasn't monolithic. While Tomasello tolerated "the biological point of view" that "cultural transmission" had contributed to differences in behavioral repertoires between chimpanzee communities, Boesch acknowledged "the psychological point of view" that these differences were not the product solely of social learning but also of individual adaptations to local environments.[6]

This episode of mutual recognition was short-lived. One year later, Tomasello distanced himself from the concept of chimpanzee culture again. In his book *The Cultural Origins of Human Cognition*, he questioned whether chimpanzee tool use was really culturally transmitted and returned to his earlier claim that culture was a "uniquely human adaptation."[7] Boesch reciprocated with a scathing methodological critique of his colleague's experiments on captive apes, which Tomasello and his coworker Josep Call in turn rejected as a "disturbing misrepresentation" of their work.[8] Ten years after the founding of the Max Planck Institute for Evolutionary Anthropology, Pääbo's original sense that the codirectors were "truly connected to one another" had proved an illusion.[9]

Beyond their strained interpersonal relations, which were of only local concern, the disagreement between Boesch and Tomasello suggested that differences between disciplinary perspectives and methodologies drove the chimpanzee culture controversy. At the Max Planck Institute, field biologists committed to naturalistic observations clashed with psychologists conducting laboratory experiments.

When I first noticed this in-house controversy, it seemed to provide an opportunity for a "controlled comparison." Ethnological comparisons could count as controlled if they examined differences and similarities between geographically and ecologically proximate cultures, American anthropologist Fred Eggan had argued.[10] This approach contrasted sharply with juxtapositions of the West with a culture that was as different as possible to demonstrate that real alternatives to the European and American ways of life existed.[11] Boesch had conducted a controlled comparison in cultural primatology when his group contrasted nut-cracking techniques of three neighboring chimpanzee communities in Taï Forest.[12] Analogously, I wanted to observe two neighboring scientific communities. By focusing on the Max Planck Institute I could get a close-up perspective on evolutionary anthropology as a field torn by a vehement disagreement between laboratory and field scientists about the nature and even the very existence of ape culture. Looking at the scientific endeavors of two directors of one and the same elite research institution—who were roughly equal in status and age, were both Westerners and foreign nationals in Germany, and were both studying chimpanzees—came as close to the control exercised in a laboratory as was possible at an ethnographic field site. The many commonalities would bring their most relevant difference into relief: Boesch observed wild chimpanzees naturalistically whereas Tomasello studied their captive conspecifics experimentally. I hoped to find a key to the controversy over chimpanzee cultures in the codirectors' discrepant scientific objects and methodologies, a simple and elegant research design that should have pleased cultural primatologists and comparative psychologists alike—or so I thought.

Running against Plexiglas Walls

In January 2013, I asked both Boesch and Tomasello whether I could come to the Max Planck Institute to familiarize myself with their research practices, so I would better understand their differences. Boesch suggested that I visit Leipzig to discuss my plans in person. Eventually, he and Roman Wittig, who had taken over the direction of research projects in Taï National Park, not only opened the doors to their department but also invited me to observe their fieldwork in Côte d'Ivoire and Gabon. Considering how many people had lined up to use their logistical infrastructure to conduct research on the exceedingly well habituated Taï chimpanzees, the offer was generous and a sign of respect and trust.

Michael Tomasello, by contrast, immediately rejected my request: "I'm sorry, but I have consistently said no to philosophers wanting to come observe how our lab works." The explanation surprised me because his books signaled openness toward philosophy and the humanities. Tomasello promoted a peculiar integration of the humanities and the natural sciences. In the foreword to the German edition of *The Cultural Origins of Human Cognition*, Tomasello compared his own approach with that of Wilhelm Wundt, a professor of philosophy at the University of Leipzig often credited with having founded the world's first experimental psychology laboratory in 1879.[13] But Wundt had assumed that some psychological phenomena, especially those of cultural origin, escaped laboratory methods.[14] They had to be approached by way of historical and cultural comparison. At the beginning of the twenty-first century, the neat division of labor between such folk psychology (*Völkerpsychologie*) and experimental psychology, or between humanist and natural scientific research, was no longer tenable, Tomasello maintained.[15] He described his own approach as applying methods of the natural sciences to classical research topics of the humanities, especially the cultural activities of humankind. Experimentally, his work explained "what enabled human beings to conduct a process of interpretation in the first place, i.e., the kind of understanding that Dilthey and others had regarded as essential for doing interpretive social science (e.g., for being able to identify the thoughts and feelings of human beings from another age)."[16] By comparing human infants and great apes, Tomasello hoped to reveal a basic cognitive capacity that distinguished our entire species and was professionally cultivated by humanities scholars who made sense of people in other times and cultures.[17]

In the late nineteenth century, Wundt and his students, many of whom also found employment in philosophy departments, had sparked the so-called psychologism controversy over whether philosophical questions could be answered by empirical, especially experimental means. By the 1910s, pure philosophers such as Gottlob Frege and Edmund Husserl had decided this conflict in their favor, ousting psychologists and forcing them to establish their own departments. This disciplinary breakup set twentieth-century philosophy on a path largely apart from empirical research, argued historical sociologist and philosopher of science Martin Kusch.[18] In recent decades, however, some natural science–oriented philosophers and philosophically oriented natural scientists had sought to reconnect, and Tomasello's revamped psychologism actively contributed to this reconciliation. So why wouldn't he allow "philosophers" to visit his laboratory?

Tomasello suggested that instead of coming to Leipzig, I should study the laboratory of his academic adversary Frans de Waal. Since my original research design could no longer be realized, I followed this advice and contacted the Dutch primatologist at the Yerkes National Primate Research Center in Atlanta. But de Waal wrote back: "An extended stay with us will unfortunately not work. Seeing experiments being carried out will not be possible, we don't let outsiders in close proximity." Contacting Andrew Whiten at the University of St. Andrews in Scotland, I again hit my head against a Plexiglas wall. Apart from Jared Taglialatela from the Ape Cognition and Conservation Initiative in Iowa, who simply ignored my request, everybody agreed to interviews, but nobody wanted an anthropological observer in their laboratory.

When I mentioned my difficulties to a historian of the behavioral sciences, he responded: "You are not the first person I know that has had difficulties getting access to psychological sites. There is a lot going on. Fears about animal liberation, anxieties around experimenter effects. In primatology, there is definitely the long shadow of Donna Haraway's *Primate Visions*, which most people know and actively hate!" An animal welfare activist dedicated to great apes suspected that the laboratory researchers had much to hide and were worried about whistle-blowers. A North American doctoral student familiar with the Max Planck Institute imagined that Tomasello was acutely aware of how harmful an outside observer could be because Vanessa Woods, the spouse of his former postdoc Brian Hare, had damaged Susan Perry's reputation by publishing her book *It's Every Monkey for Themselves: A True Story of Sex, Love, and Lies in the Jungle* about the cultural primatologist's field site in Costa Rica.[19] Of course, it was also conceivable that the scientists were worried that a stranger would distract their test subjects from their experimental tasks. I decided to remain agnostic toward these speculations. Although the guarding of social boundaries is ethnographically significant, there was simply no way of telling where the laboratory researchers' reservations came from—only that Tomasello was not alone. At least I had learned something about how these experimentalists were perceived by other players in a very tense field, and it seemed as if this image was what they were trying to control.

Experimental Philosophy Today

The next surprise followed suit. When I mentioned Tomasello's explanation that he generally didn't allow philosophers into his lab, one field primatologist laughed: Tomasello's lab was "full of philosophers," I was told.

During my visit to Boesch's department in the summer of 2013, I quickly came to realize that evolutionary anthropologists used the word *philosopher* quite liberally. Boesch never introduced me to his colleagues as a fellow professor of anthropology because they would have assumed that I studied human beings or other primate species from a natural history perspective. Whenever I made the mistake of presenting myself as a *cultural* anthropologist, they expected that I could answer their questions about hunter-gatherers. The idea that an anthropologist would study other anthropologists, being more interested in modern knowledge cultures than in foraging or chimpanzee cultures, was too outlandish to fit into the thirty-second pitch of what I was doing at the institute. Boesch circumnavigated the anticipated confusion by introducing me as "the philosopher who is interested in how we study culture and cognition."

I couldn't help feeling like an impostor when I first met Tomasello's postdoc Richard Moore, who actually held a doctoral degree in philosophy and who had already heard about "the philosopher" visiting Boesch's department. I don't know whether other philosophers had worked with Tomasello before him. There were definitely PhD students with some training in philosophy, to which the field primatologist's hyperbole that the laboratory was teeming with lovers of wisdom might have referred. But even if Moore did not represent a whole platoon of empirical philosophers in the psychologist's lab, his presence indicated a sustained intellectual exchange.

Before my fieldwork in Leipzig, I had studied collaborations between analytic philosophers of mind and neuroscientists, wondering about the nature of philosophy as an academic discipline and a human activity that had long preceded the disciplinary division of scholarly labor. I had examined how these philosophers related to the empirical research that natural scientists conducted in the laboratories that the philosophers visited. Additionally, I sought to understand how the philosophers' enterprise related to the empirical research that anthropologists and sociologists of science conducted in such labs. Although both neurophilosophers and science studies scholars had developed a keen interest in brain research since the 1970s, they looked at it very differently and I tried to explain that difference, partly out of sheer ethnographic curiosity, but also because I felt that as an anthropologist of science, I could learn from these distant relatives.[20]

Since Moore had begun to make a career of doing philosophical *and* experimental work in a primate and child psychology laboratory, he represented the kind of empirical philosopher that I had previously engaged with in neuroscientific contexts. Having fairly recently entered the world of primate research as an outsider himself, he generously shared his astute perceptions of people and practices and quickly became one of my best ethnographic informants—a window through which I could gain at least a cursory glance at the work in Tomasello's laboratory. More specifically, Moore provided a perspective on one of the aspects I was most interested in: How did the lab straddle a natural scientific approach to primate minds and humanistic questions about culture and history? And why did its work attract the attention of philosophers?

Richard Moore's story of how he, a native of Sheffield, England, had found his way to Leipzig, Germany, began at age four when he and his identical twin brother developed a secret language with each other that only they could understand. As the two grew older, Richard came to understand that their argot had reflected the things that mattered to them but not to other people. Both brothers became philosophers, and when Richard first read Wittgenstein he found echoes of this insight in Wittgenstein's claim that human agreement rested on shared forms of life. Ever since, he had pursued the question of how people managed to communicate and make sense of each other.

As a student, Moore cherished so-called continental philosophy and immersed himself in Derrida, Deleuze, Heidegger, and Wittgenstein. In 2005, he began doctoral work at the University of Warwick, struggling to relate Wittgenstein and Heidegger to the psychological literature on how cognitive development enabled preverbal children to communicate—and so he entered Tomasello's turf. Moore noted that historically, continental philosophers like Maurice Merleau-Ponty had blazed a trail for such cross-disciplinary projects. Initially, the novelty of the rediscovered territory between the conceptual and the empirical called for less meticulous and more speculative temperaments willing to venture out of their area of expertise. "Because continental philosophers engage more superficially with topics, they have an attitude that allows them to dip into things liberally," Moore explained. By contrast, "analytic philosophers really want to do everything properly and spend a lot of time engaging with the details. They're less willing to speculate." In an eruption of self-doubt three years into his dissertation research, Moore came to realize that Heidegger and Wittgenstein would not get him through a PhD thesis in the philosophy of language. In 2010, he resurfaced from this bookish adventure as a freshly baked analytic philosopher with doctoral degree in hand.

Determined to not just dabble in the published literature, Moore asked Tomasello, whom he had previously met at a conference, whether he could come to the Max Planck Institute for postdoctoral research. Tomasello accepted the philosopher—under the condition that Moore would not just write about their work (as a visiting anthropologist of science might have) but would participate in it and run his own experiments. He could not just exploit and depend on the scientists' data or take on the role of armchair critic, carefully analyzing and ascertaining the meaningfulness and coherence of the scientists' concepts, as Max Bennett and Peter Hacker had done for the neurosciences.[21] He would have to make positive contributions to Tomasello's project by finding things out himself.

Although Moore experienced the practical challenges of experimenting with apes and toddlers as humbling, he flourished in the new environment. The cooperative nature of laboratory life allowed him to escape from his solitary work at the desk: "In philosophy, when you get stuck, no one can help you. In psychology, you sit down with someone who also has a stake in the answer. Group work is so refreshing." Thus Moore ran experimental studies suggesting, among other things, that two-year-olds understood communicative intentions even without language, gestures, or gaze, whereas the orangutans of Pongoland struggled to understand pointing gestures.[22]

And yet Moore still understood himself as a philosopher whose job it was to clarify the concepts that Tomasello and his colleagues used, especially those borrowed from philosophy, before they put them to work in experiments: "If you don't have a philosophical account of what you're looking for, you can't look for it. You can't test whether or not a certain behavior is a communicative act unless you have an account of what a communicative act is. To specify those constraints, you need to do philosophy." Thus Moore helped the comparative psychologists to think more clearly about philosopher Paul Grice's concept of communication or their own psychological concepts such as imitation.[23] Although empirically oriented, this form of philosophizing would never dissolve in science. For science constantly required conceptual therapy as it progressed. Moore compared his job to that of a masseur: "It's like, if you go and lift weights a lot, your muscles get stronger but you'll continue getting knots in your back. Science is extending the knowledge, but philosophers need to massage the knots out, so the muscles develop properly and stay focused on what the real issues are."

Compare this low-key conception of philosophy as conceptual massage therapy for scientific strongmen to Derrida's desire to deconstruct the key categories of Western thought or Heidegger's ambition to reveal the fundamental nature of being. The goal of getting readers to see the world in an entirely new light by offering alternative conceptual frameworks, bringing about a cognitive gestalt switch similar to the paradigm shifts in the sciences described by Thomas Kuhn, appeared unrealistic and outright crazy to Moore.[24] He felt reservations even about his colleague Kim Sterelny's more modest proposal that philosophy's contribution to science was showing how everything hung together by integrating knowledge from different disciplines.[25] That might have helped to resolve the chimpanzee culture controversy, at least as it played out between the biologist Christophe Boesch and the psychologist Michael Tomasello. But Moore did not claim to solve the big questions of primatology. "My work is trying to see how the small differences between apes and humans best fit into a continuity or discontinuity story," he explained. He had adopted the ethos of analytic philosophy, as he understood it.

In Moore's eyes, the controversy between the Max Planck Institute's field primatologists and comparative psychologists was driven partly by empirical findings and partly by the different perspectives that their respective methodologies provided. Disagreements ensued when observations in the wild suggested capacities that could not be demonstrated in the laboratory. "It's very difficult to make strong claims on the basis of field data," Moore argued, "because you can't control the parameters as you can in the lab. Dismissive rebuttals of each other's work don't change the fact that this is part of the scientific process. Yet it causes tensions and maybe even resentment between our departments. It's a shame that this antagonizes people, really."

But the controversy was also motivated by theoretical commitments and speculations, Moore pointed out. His own response was an orientation toward details, both empirical and conceptual. It exceeded even that of vocational scientists. Sociologist Max Weber had already described how scientists focused all their attention on specifics that bordered on the insignificant.[26] Over time, however, these trifles added up, enabling an incremental scientific progress. "I think more than either Mike or Christophe, I'd be interested in a nuanced story, not one that makes big claims based on theoretical assumptions," Moore explained. For Boesch's taste, Moore's conceptual microtome sliced the world up too thinly. One man's conceptual rigor was another man's pettiness, while the latter's theoretical boldness might be the former's flights of fancy. Ullica Segerstråle had already noticed a similar clash of temperaments between the sociobiologists (mostly naturalists with a penchant for the big picture) and their critics (often experimentalists skeptical of any claim not supported by a controlled experiment).[27] These differences in epistemic virtues were not universally accepted differences in the quality of scholarship, but differences between scholarly styles of thinking and doing.

It struck me that empirical philosophers tended to collaborate with laboratory researchers, not field scientists. Maybe this, too, was a matter of temperament. Desk work and benchwork probably drew in more like-minded people than desk work and fieldwork. Although Moore himself would have loved to spend time in the forest, he noted: "If you think of the stereotype of the philosopher as someone who sits in a chair and reads, that's less adventurous than fieldwork. Maybe philosophy attracts bookish types."

But there might also have been historical reasons for the philosophers' lack of interest in the field. When the widespread disciplinary agreement that psychologism had been refuted began to crumble, it was easier to reconstruct the old intellectual connections between philosophy and experimental psychology (including its late twentieth-century reincarnation as cognitive neuroscience) than to forge new ties with, say, field biology.[28] Philosophers and psychologists regularly met at joint conferences to which no naturalist had been invited.

Despite Aristotle's zoological writings, philosophers had always been more interested in the human mind than in animal behavior. Although Boesch also made claims about how chimpanzee minds worked in the wild, Moore had adopted the experimental psychologists' sense that fieldwork pertained primarily to behavior, not to cognition. Naturalistic observations hardly supported the fine-grained conceptual discriminations that analytic philosophers in particular sought to make, Moore explained, because multiple cognitive possibilities were consistent with the same behavior. Consequently, the revival of an experimental philosophy à la Wundt in Tomasello's laboratory and elsewhere had not been accompanied by the emergence of a field philosophy.[29] Comparative psychology was hardly the intellectual milieu to realize Rousseau's dream of a Montesquieu or a Diderot who would study Pongoes and Engecoes outside Pongoland.

Yet the philosophical reception of Tomasello's work was not limited to analytic philosophers who ventured beyond the confines of conceptual analysis. In Germany, his project, which had gained its first contours at the Yerkes National Primate Research Center in America's Deep South, was about to give new impulses to one of the internationally most influential schools of continental philosophy.

The Making of a True Philosopher

In 2009, the mayor of Stuttgart took the unusual step of handing Michael Tomasello the Hegel Prize. With this prestigious award, Georg Wilhelm Friedrich Hegel's hometown and the International Hegel Society had honored important philosophers, social researchers, and humanities scholars such as Jürgen Habermas, Hans-Georg Gadamer, Niklas Luhmann, Paul Ricoeur, and Richard Sennett. But no natural scientist had ever received the prize. In his eulogy, Habermas made Tomasello an honorary member of his guild. In case anyone wondered why, Germany's most eminent postwar philosopher pointed out that "the intellectual profile of Michael Tomasello is explanation enough: he *is* a philosopher, although not because of his discipline, but because of his questions and the way he thinks."[30] Moore found this gesture "fucking insulting" because it disregarded the scientist's lack of desire to be a philosopher in the first place.

One reason for the questionable compliment was the form in which Tomasello presented at least part of his work. For Habermas, what counted more than the three hundred peer-reviewed journal articles that Tomasello had written "with the usual collective authors" was that the psychologist also wrote single-author monographs: "The book format reveals the constructive effort of a theoretical synopsis of the investigated details. For humanities scholars it is reassuring to see that also in the natural sciences theoretically constructive achievements still have to go through the synthetic energy and the power of expression of a single mind."[31] Despite all the talk about the demise of the book, especially in its printed form, the medium's capacity to connect the dots and facilitate critical thinking by decelerating both reading and writing had kept it at the center of the moral economy of the humanities.[32] Through monographs such as *The Cultural Origins of Human Cognition* Tomasello partook in this economy.[33] In Habermas's opinion, they earned him the title of "true philosopher."[34]

Of course, books alone don't make philosophers. Despite his monographs, it would not have occurred to Habermas to raise to his peerage Gerhard Roth, a neuroscientist (who even held a doctoral degree in philosophy) with whom he had clashed over free will, or Edward O. Wilson, a sociobiologist whose genetic determinism was anathema to Habermas's conception of the human. What distinguished Tomasello from these equally prominent book authors was not that his research spoke to philosophically relevant questions and theirs did not, but that he gave the right answers. In the preface to the German edition of *Cultural Origins,* Tomasello had explicitly rejected the reduction of human cognition and social life to neurons and genes.[35] Habermas and Axel Honneth, the most prominent representatives of the second and third generation of the Frankfurt School, had realized that Tomasello could empirically support their conceptual analysis of the social constitution of the human mind.[36]

The main argument that Habermas backed up with references to Tomasello—that humans relate to other humans not just as "social objects" but as "intentionally acting beings"—had been one of the long-standing tenets of the Frankfurt School.[37] As nine-month-old prelinguistic children began to take the perspective of caretakers and learned to direct their gaze to an object of then-shared attention, they adopted the communicative role of the second person toward others, the philosopher had learned from *Cultural Origins.*[38] The book "gives a natural science undergirding to some of Habermas's concerns," Tomasello noted in an interview with me. In his *Theory of Communicative Action,* published two decades earlier, Habermas had drawn from turn-of-the-century philosopher and social psychologist George Herbert Mead to make much the same point.[39]

While Habermas used both Mead and Tomasello to emphasize the importance of intersubjective relations in the development of human cognition, Honneth distinguished himself from his mentor by interpreting both authors a little differently. To take Habermas's paradigm of intersubjectivity beyond communicative action, he emphasized the priority of recognition to cognition: ontogenetically, an empathic engagement with an attachment figure preceded the neutral grasping of reality. While Mead and Habermas conceived of the triangular relationship between two human subjects and an object in a largely emotionless space, Honneth called on Tomasello for scientific support of his persuasion that "emotional identification with others is absolutely necessary to the taking over of another person's perspective, which in turn leads to the development of the capacity for symbolic thought."[40] That only a child's love for his parents enabled his subject-formation and object relations Honneth might actually have learned from psychoanalyst Donald Winnicott, though.[41]

Ultimately, Honneth needed these psychologists to provide "independent evidence" instead of merely invoking the authority of his philosophical ancestor Theodor W. Adorno to demonstrate "that the human mind arises out of an early imitation of a loved figure of attachment."[42] In an aphorism written in 1945, this founder of the Frankfurt School had described anthropogenesis from what he understood, rather idiosyncratically, as "the primal form of love": "The human is indissolubly linked with imitation: a human being only becomes human at all by imitating other human beings."[43] Although Tomasello's experimental research said nothing about love, it was meant to serve as a naturalistic justification of Adorno's, Habermas's, and Honneth's shared philosophical conviction that the human mind was not the origin but the product of social relations. It seemed as if the Frankfurters read Tomasello mostly to provide further confirmation of what they had known all along. Why then did they promote the psychologist so aggressively?

In Tomasello's story of how sociocultural forms emerged, Habermas sympathized with two distinct threads. First, it offered a historical perspective on culture, which could account for the modern experience of faster and faster change. "Today, an 80-year old who remembers the social conditions and technical aids and appliances of his grandparents and tries to anticipate the everyday life of his grand children must feel dizzy in the face of the exponentially accelerating developments," noted the philosopher. The behavior of *Homo sapiens* had already transmuted dramatically in the millennia preceding the twentieth century, especially after the agricultural revolution about ten thousand years ago. Primatologist Carel van Schaik and historian Kai Michel concluded that as a result of cultural evolution, it would be misleading to speak of human nature in the singular: the transition from nomadic to sedentary lifestyles endowed our kind with a second and even a third nature.[44] Historically modern humans appeared to be almost a different species. Building on Robert Boyd and Peter Richerson's dual inheritance theory, which postulated two interacting evolutionary processes, genetic and cultural evolution, Tomasello's *Cultural Origins* provided an answer to the question of how it was possible for an animal to change its habitat, nutrition, group size, and social behavior so fundamentally, although very few genetic changes had taken place in such a brief evolutionary time span.[45] His answer to this question was the "ratchet effect" of cumulative culture: humans had acquired the ability to pass on their knowledge from generation to generation, while constantly making better what their forebears had achieved.[46]

In the first of two special issues of *Deutsche Zeitschrift für Philosophie* that had been dedicated to Tomasello's work (another unprecedented honor for a natural scientist), philosopher Wolfgang Welsch objected to this optimistic conception of culture as essentially progressive: its one-sided emphasis on the conservation and accumulation of improvements left no room for the kind of destruction and innovation that Kuhn had theorized as paradigm shifts.[47] Philosophical anthropologist Joachim Fischer noted that the accretion of knowledge over generations had certainly not happened in philosophy and sociology: after all, the insights of philosophical anthropology itself had disappeared into a maelstrom of linguistic turns.[48] In other words, Tomasello's notion of culture seemed to reflect the comparative psychologist's own experience of normal science—a positivist accretion of facts undisrupted by scientific revolutions.[49]

Habermas followed Tomasello's gradualist intuition that it was precisely this ability to pass on what had been learned and invented collectively that made us human: "Chimpanzees, too, use simple tools, but only in hominids do we observe their continuous improvement, for example, the technical progress from the stone tools of the Oldowan culture to the paleolithic hand axes. What sets man apart from the apes is the kind of communication, which enables both the intersubjective aggregation and the transgenerational inheritance and reworking of cognitive resources."[50] In other words, it was precisely the human capacity for continuity that had introduced a radical discontinuity in natural history.

Boesch, McGrew, and other chimpanzee ethnographers, however, doubted Tomasello's claim that the ratchet effect distinguished human from ape culture.[51] The chimpanzees of Loango, for example, combined up to five tools to extract honey from underground beehives.[52] In theory, this ability might have developed within one generation. But the fieldworkers deemed it more likely that the Loango population had accumulated this complex tool kit over many lifetimes. Unfortunately, no chimpanzee historian had recorded what Rousseau's Engecoes had or had not done in the eighteenth century, and archaeologists had come to know rainforests as notoriously poor places for the conservation of cultural artifacts, especially those made of wood.[53] Whether chimpanzees stood on the shoulders of giants as they extracted honey and fished for termites remained a matter of speculation.

Habermas's claim that intersubjectivity had driven the development of human but supposedly not chimpanzee culture pointed to the second and more recent thread in Tomasello's natural historical narrative adopted by the philosopher. In *Why We Cooperate* and more comprehensively in *A Natural History of Human Thinking,* Tomasello shifted the emphasis from the diachronic aspect of cultural transmission toward the synchronic aspects of social coordination and collaboration.[54] This transition had occurred in the mid-2000s as new evidence called into question his original account of culture.[55] Previously Tomasello had presented the understanding of others as intentional agents as the uniquely human foundation for imitative learning, which in turn enabled cumulative culture.[56] When experiments of his own laboratory suggested that chimpanzees actually shared this sociocognitive capacity with humans, Tomasello had to find a new explanation for why the "ratchet effect" had enabled *Homo* but not *Pan* to develop increasingly complex cultures.[57] Eventually, he proposed that modern humans came into being as their apelike ancestors moved from individual to increasingly cooperative problem solving, from predominantly competitive relations with others, which Tomasello still saw in chimpanzees, to collaborative foraging with a partner and eventually the larger and more permanent shared life of a cultural group.[58] Thereby, group cohesion was rendered possible on a bigger and more complex scale. From this angle, culture appeared to have emerged as humans acquired the ability to put their heads together with others, solving problems, conventionalizing languages, producing norms, and building institutions.

"The systematic comparison of children and chimpanzees sheds light on that period of evolution during which the subjectively confined consciousness of hominids broke out of its isolation," Habermas paraphrased. "In coping cooperatively with a surprising environment, it switched to joint intentions."[59] It was no coincidence, he suggested, that the Hegel Prize laureate's account resembled Hegel's *Phenomenology of the Spirit,* in which the philosopher got even with "the mentalistic conception of a self-referentially closed subjectivity that delimits itself against its environment." Tomasello vindicated Habermas's conviction that the inherently social makeup of the human mind distinguished *Homo sapiens* from other animals: "Chimpanzees don't seem to be able to break out of the confines of their egocentric perspective, which is driven by each individual's own interests. . . . They can't enter into interpersonal relations with others."[60] Without interpersonal relations, Habermas seconded Tomasello, they could not develop culture:

> The kinds of traditions, rituals, and tool use that are also encountered among chimpanzees do not reveal an intersubjectively shared cultural background knowledge. Without intersubjective understanding, there can be

> no objective knowledge. Without the reorganizing "connection" of the subjective mind [*subjektiver Geist*] and its natural substrate, the brain, to an objective mind [*objektiver Geist*]—that is, to symbolically stored collective knowledge—there can be no propositional attitudes to a world that is distanced in a way that allows for objectivation. Nor can there be any of the technical achievements in coping with an objectivated world of that sort. Only *socialized brains*, linked up with a cultural milieu, become bearers of those highly accelerated, cumulative learning processes that have become uncoupled from the genetic mechanism of natural evolution.[61]

Habermas emphasized that he did not pose the question of what distinguished humans from animals to polemically demarcate the higher from the lower but to obtain an evolutionary explanation of sociocultural forms of life that wasn't reductionist.[62] This project was equally philosophical and political. The tradition of critical theory in which Habermas worked had always been concerned with not just how things were but how they could and should be. For the Frankfurt School, the formative historical experience was the Holocaust, which its members understood as an excess of modernity's tendency to reify human beings. In the political atmosphere of postwar Germany, which associated biological accounts of human life with Nazi racism and eugenics, their opposition to scientism and their antibiologistic rejection of the emerging alliance between Anglo-American philosophy of mind and the neurosciences, which reduced mind to brain, won over large parts of Germany's leftist intelligentsia.[63] The corresponding marginalization of biological anthropology, a field stained by the Kaiser Wilhelm Institute's collaboration with Auschwitz, coincided with the dissociation of most members of the Frankfurt School from philosophical anthropology.[64] Adorno's "negative anthropology" refused to answer the question "What is man?" because it was impossible to anticipate how human beings would change, and any positive determination would inevitably entail their reification as caricatures of contemporary humans.[65] After 1945, Germans knew all too well where looking at humans as objects could ultimately lead.

Habermas's opposition of chimpanzee egocentrism and human sociality revealed that he had shed his mentor's qualms about a positive anthropology and even welcomed the return of evolutionary anthropology to Germany—as long as it provided support to the Frankfurt School's insistence on a primacy of the social. In an increasingly naturalist climate, the breakdown of fruitful communication with the natural sciences since the mid-twentieth century, which Habermas had noticed, put the traditions that had informed his thinking at a competitive disadvantage.[66] In the 2000s, a growing number of humanities and posthumanities scholars eagerly aligned themselves with the life sciences while the tradition of critical theory, which Habermas, Honneth, and their allies sought to preserve in a reunified Germany, had positioned itself as a last bastion against the neoliberal ideology underlying genetic and neural explanations of human behavior.[67] "With the innovative research of the laureate this constellation might change," Habermas hoped in his Hegel Prize speech on Tomasello.[68] "His work pursues philosophical questions by empirical, but non-reductionist means." Finally, the Frankfurt School had found a powerful ally in the natural sciences who—like the Russian anarcho-communist geographer Pyotr Kropotkin—emphasized cooperation rather than competition.[69] In the spirit of mutual aid, Tomasello came to enjoy a philosophical reception in Germany, which he attributed to Habermas's advocacy and which had failed to materialize in the United States.

Although Tomasello had come to Leipzig from a very different part of the world, the experience of the calamitous effects of racism had also shaped how he had come of age politically. Having grown up in the American South in the 1950s and 1960s, which he conceived of as an apartheid system, Tomasello knew firsthand how the human disposition to direct aggression against other groups could corrode the social integration of multiethnic societies. In the face of an increasingly globalized world inhabited by a fast-growing human population of unprecedented size, he rearticulated the anthropological concern with human sociality in a form both reminiscent of and very different from its eighteenth-century origins. In his Hegel Prize speech, Tomasello wondered whether our evolved capacities for cooperation in small groups scaled up successfully to large-scale modern civilization: "We are still here. But of course we are only a few nuclear bombs or a few more decades of rampant environmental degradation away from not being here."[70]

Considering the almost apocalyptic zeitgeist of the Anthropocene, which foresaw the disappearance of humanity at the very moment of its greatest expansion, Tomasello spun a surprisingly optimistic narrative. He drew political hope from natural history. While evolutionary psychologists had worried that our modern skulls housed a Stone Age mind that had not evolved to cope with the exigencies of a globalized industrial world that was home to more than seven billion people, Tomasello's antireductionist account of cultural inheritance and social coordination suggested that *Homo sapiens* had acquired mechanisms of behavioral adaptation many orders of magnitude faster than organic evolution. This made him confident that the very capacities that had led to the problems humanity was now facing also enabled the political practices and institutions that would solve them: "New prosocial norms for being careful with our environment and for recognizing the dignity and value of all peoples from all ethnic groups seem to be spreading in influence, not receding, and we are continually finding new ways for creating more cooperative and open arrangements for communication and coalition-building in large-scale societies, as Professor Habermas has argued."[71] Thanks to human culture, Tomasello seemed to suggest, the future of humanity might be brighter than the vision of the world's nuclear destruction by intelligent baboons, which writer Aldous Huxley had imagined in his dystopian novel *Ape and Essence*.[72] This was the heyday of Barack Obama's presidency—years before populist movements all over Europe and the United States would ferociously attack the very prosocial norms that had seemed on the advance in 2009.

The contrast between chimpanzees that fended only for themselves and humans who were born helpful and cooperative and—unless corrupted by neoliberalism and racist ideologies—developed cultures to maintain their prosocial nature in the face of new challenges endeared Tomasello to the Frankfurt School. Some of his colleagues in primatology, on the other hand, looked less favorably at Tomasello's account of human nature. The idea that group-living animals who—not in the Leipzig Zoo but in the wild—organized military-style campaigns against their neighbors would not be able to cooperate, and that, at least in Habermas's pointed interpretation, these highly social primates lived in mental isolation from each other, led chimpanzee researcher Volker Sommer to decry Tomasello as a wing chair primatologist.[73] In a newspaper article titled "No Feeling of Togetherness in Pongoland," the University College London professor recommended that, if not in the forest, Tomasello could look for "a shared we" at least in the corridors of his own institute. After all, his fellow Max Planck Institute director Christophe Boesch maintained that wild chimpanzees coordinated the allocation of roles as they formed groups to hunt monkeys.[74] If they indeed cooperated, then Tomasello's new explanation of why only humans were capable of cumulative culture would fall apart like the previous one. Sommer concluded his polemic: "That the eulogy of the newly minted Hegel Prize laureate was given by Jürgen Habermas might not have helped establish the truth. For at best he knows the urban jungle." Fieldworkers critical of Tomasello resented the revived liaison between experimental psychology and philosophy.

A Philosophical Anthropology for the Twenty-First Century

The Frankfurt School's anointment of Tomasello as philosopher marked only the beginning of an extraordinary reception in his adopted country. In 2014, German philosophical anthropologists, another twentieth-century school of thought that sought to modernize, announced him the first winner of the newly created Helmuth Plessner Prize of the city of Wiesbaden (followed by

Peter Sloterdijk). Professor of philosophy Christian Thies explained their interest against the background of this research program's marginalization since the 1960s.[75] Unlike the Frankfurt School, the tradition established by Max Scheler, Arnold Gehlen, and Helmuth Plessner had not produced a creative mind like Habermas to lead the new generation. At the same time, the empirical knowledge about human beings had grown so fast that it required a collective effort to process it philosophically. From the start, the Frankfurt School had built institutions to integrate advances in the social sciences: in the 1930s, the philosophers made a home for themselves in the Institute for Social Research in Frankfurt; in the 1970s, Habermas codirected the Max Planck Institute for the Study of the Scientific-Technical World; between 2007 and 2017, the Cluster of Excellence "Normative Orders" kept the integration of critical theory and social research alive in a more competitively organized system of higher education. By contrast, philosophical anthropology had associated itself with neither social scientific nor natural scientific institutions. Thies daydreamed that the Max Planck Institute for Evolutionary Anthropology, especially its Department of Developmental and Comparative Psychology, could become for philosophical anthropology what Habermas's Max Planck Institute had been for the Frankfurt School.

In his history of German philosophy, Herbert Schnädelbach interpreted the birth of philosophical anthropology in the 1920s as a response to the identity crisis that philosophy experienced as it faced the ascent of the empirical sciences.[76] Before the 1830s, German philosophers had understood their scholarship as *Wissenschaft*, but by *Wissenschaft* they meant knowledge of the universal, of the necessary, and of what eternally is. In the nineteenth century, however, science came to be identified with empirical research. The natural sciences broke away from natural philosophy while history began to emancipate itself from the philosophy of history. The sciences no longer represented a system of truths but defined themselves as a never-ending process of methodical research that constantly revised earlier findings. This upsurge of empiricism forced philosophy to justify its existence and to redefine its role vis-à-vis the sciences in this new knowledge economy.

The question of what human beings were had been so central to philosophical reflection since antiquity and especially to Enlightenment thought that anthropology might appear to have always been philosophical. In 1800, Immanuel Kant referred the questions "What can I know?" and "What ought I to do?" and "What may I hope?" to the question "What is man?"[77] What set the twentieth-century tradition of philosophical anthropology apart from these earlier anthropologies in philosophy was the formation of the empirical human sciences such as cultural and evolutionary anthropology, argued Schnädelbach.[78] Scheler, Plessner, and Gehlen gave different answers to the question of what remained genuinely philosophical about philosophical anthropology. But Schnädelbach reconstructed the common denominator of their projects as translating the question of "What is man?" into "Who are we?" At a time when what it meant to be human no longer seemed self-evident, they shifted the emphasis from the species characteristics of *Homo sapiens* to understanding our identity as humans. At this point, the identity crisis of philosophy and the identity crisis of philosophers as members of the human species converged: philosophical anthropology worked through doubts about traditional conceptions of humankind *and* doubts about the traditional means of philosophy to tell us who we were.[79] The historical events, technological innovations, and scientific findings that have occurred in the meantime have not rendered *anthropos* any less problematic to himself. Nor have philosophers resolved their troubled relationship with the empirical sciences.[80]

In his laudation for the Plessner Prize, Joachim Fischer adopted the same strategy as Habermas, which Schnädelbach had dubbed "identification with the aggressor," and declared Tomasello's empirical science a philosophy.[81] He wondered whether the experimental psychologist would emerge as a "protagonist of twenty-first century philosophical anthropology." Although they came from very different traditions, Tomasello and the German philosophers shared important intellectual orientations. First and foremost, both sides wanted to show in a nonreductionist fashion the special place of human beings in natural history. Tomasello's opposition to primatologists such as Volker Sommer (or Christophe Boesch, for that matter) advocating continuity between human and ape behavior aligned him with the theoretical orientation of philosophical anthropology. Fischer noted that the scientific controversy between these two camps reenacted the disagreements between Arnold Gehlen and ethologists around Konrad Lorenz—with Tomasello in the role of the philosophical anthropologist.

This return to the origins of philosophical anthropology indicated that the psychologist's historical role was not to revolutionize but to preserve an endangered philosophical tradition. Fischer concluded from his attempt at relating Tomasello to German philosophers, which the American had never read, that Tomasello's research and theory "did *not* amount to a paradigm shift in the cultural and social sciences because the paradigm already exists; it is not a new paradigm, but an (indirect) paradigm renewal."[82] For example, Scheler, Plessner, and Gehlen had already discussed the psychological experiments that Wolfgang Köhler had conducted with chimpanzees in the 1920s. Karl-Siegbert Rehberg revisited their reception in light of Tomasello's work to update this line of philosophical research.[83] The comparative psychologist's job was to help with the reconstruction of an empirical philosophy that would position "the cultural and social sciences in the age of (evolutionary) biology."[84]

However, in contrast to the affirmative reception of Tomasello's work by the critical theorists of the Frankfurt School, the philosophical anthropologists also spelled out their differences. In Fischer's eyes, the psychologist overemphasized the social constitution of human interiority and underestimated how much the unfathomable depths of our inner worlds contributed to the exceptional status of *Homo sapiens* in the animal kingdom.[85] He also objected that Tomasello's focus on social cognition ignored the importance of social emotions. Finally, the primal scene of Tomasello's developmental psychology, one subject pointing another to an object (that might be attainable only through cooperation), did not account for a supposedly more fundamental triangulation involving a third subject that enforced norms independent of "you" and "me." This institutional dimension of human cultures was unknown to ape communities.

While Fischer's critique of Tomasello was fed mostly by intuitions that philosophical anthropologists had harbored since the early twentieth century, Thies also cited other primatologists—Christophe Boesch, Volker Sommer, Julia Fischer, and Frans de Waal—to call into question one of the deepest convictions of his own philosophical tradition: "Tomasello probably overdraws human-animal differences."[86]

The eagerness with which both the Frankfurt School and philosophical anthropologists sought to enroll Tomasello in their projects might say more about the challenges faced by German philosophy than it says about comparative psychology. But this cross-disciplinary reception was part of a broader effort to make sense of what evolutionary anthropologists learned about the cognitive capacities of humans and other primates. It wove the bare-boned accounts of experimental results into the rich fabric of early twenty-first-century Euro-American cosmology.

Captive Experiments

Pongoland, that great ape habitat where the chimpanzees were said to lack a community spirit, was accessible to any downtown naturalist willing to pay for a ticket to Leipzig's zoological garden. Massively modernized since the 1990s, the zoo, with the help of the Max Planck Society, had built a $14 million ape enclosure that also housed the Wolfgang Köhler Primate Research Center. That the Max Planck Institute for Evolutionary Anthropology sought to explain the process of hominization rather than the superiority of the Aryan race was illustrated by a cave painting at the visitors' entrance, suggesting an evolution from *Austrolopithecus afarensis* to *Homo sapiens*. The latter was the only extant hominid not on display in the zoo.

In 1906, the Bronx Zoo had promoted the idea that we were just another ape species by adding a human exhibit to its Monkey House. In line with nineteenth-century theories that envisioned present-day Africans as remnants of earlier stages of hominid evolution, the New York Zoological Park had made human-animal continuity more palatable to its audience by putting on public view Ota Benga, an Mbuti (known as pygmies at the time) whom an American anthropologist had purchased from African slave traders. The show came to an early end as enraged black clergymen successfully protested both the racism and the Darwinism that had inspired the display.[87] The Leipzig Zoo made no white supremacist or any other attempt at undermining human exceptionalism.

The one role in which humans did figure in the zoological exhibits of Pongoland was as scientists. When the Leipzig Zoological Garden opened its gates, the researchers had already conducted the bulk of their experiments. But occasionally, visitors could still observe through Plexiglas walls how humans tested apes. Alongside a small exhibition in the foyer, this staged transparency was part of the Max Planck Institute's public relations work, assuring visitors that—unlike the highly controversial invasive research on macaques at the Max Planck Institute for Biological Cybernetics—their animal experiments were benign and even served as a welcome enrichment of life in captivity.

The shell game that I observed a Spanish postdoctoral researcher playing with the chimpanzees to test their memory was harmless enough. When I visited Kumamoto Sanctuary in Japan, however, I met Hungarian chimpanzee management consultant Michael Seres, who painted a very different picture of Pongoland. Seres was a chimpanzee consultant who had worked for Frans de Waal at Zoo Atlanta and for Tomasello at the Leipzig Zoo before Tetsuro Matsuzawa hired him to help with the resocialization of chimpanzees that the Japanese pharmaceutical industry had previously used in biomedical research. He told me that he had quit his job at the Max Planck Institute in 2003 after Tomasello had refused to intervene when Seres and his wife showed him video footage of zookeepers beating chimpanzee infants into submission. Whatever was happening in 2013, a peek through the Plexiglas window in the ape house could provide only a very partial perspective on the relationships between primates and primatologists in Pongoland.

Boesch's reservations toward such captive research were palpable. They were ethical, epistemological, and in a way, ontological. But the situation was too complex to allow a simple dismissal of zoo experiments. Behind bars and Plexiglas, the Leipzig chimpanzees had it good compared to what they had

FIGURE 10. Psychological experiment in Michael Tomasello's laboratory, Leipzig Zoo, Germany. Photo by author.

endured before. Jane Goodall had successfully lobbied the Dutch government to free them from a biomedical research institute in Rijswijk. They had found a better home in the new ape enclosure in Germany.

But when they first arrived, these chimpanzees did not dare to enter the outdoor compound. "I think it was the first time they saw the sun and they might never have stepped on grass before," Boesch told me. "Instead of exploring the outside, they stayed on the iron mesh of the air conditioning system right next to the door. Now it's more than fifteen years and the zoo director has just told me that some individuals still don't go on the grass. Can you imagine what these chimpanzees must have gone through? And Tomasello claims that, if these animals can't do something, it means that chimpanzees as such cannot do it!"[88]

This anecdote illustrated what Boesch took to be epistemologically and ontologically wrong about Tomasello's captive experiments: How could the behavior of such psychologically deformed animals represent *the* chimpanzee while Leipzig kindergarten children represented humankind? In the eyes of a field primatologist who had made a career of questioning the false universalization of Goodall's woodland-living chimpanzees in the name of cultural and ecological diversity, the psychologist's approach appeared fundamentally misguided.

Boesch contra Tomasello

In 2007, the lingering disagreements between the two codirectors of the Max Planck Institute for Evolutionary Anthropology erupted in a head-on controversy over the relative importance of field observations and controlled captive experiments in cognitive cross-species comparisons—at least, that was how Christophe Boesch sought to frame the issue. But underneath their methodological disagreements, he saw a deeper theoretical rift. It was the old rift between universalism and particularism, which primatologists had inherited from anthropologists. Against the background of what Boesch considered the single most important finding of the last two decades of chimpanzee fieldwork, namely the behavioral diversity within the species, it took the form of a heated debate over how biographical experience and living conditions shaped the development of cognitive capacities.[89]

Methodologically, Boesch called into question whether Tomasello's research design actually allowed him to make valid claims about the difference between how humans and chimpanzees thought. Since the early days of comparative psychology in the 1910s, when Louis Boutan had tested human children against his gibbon Pepée, the validity of cross-species comparisons rested on the similarity of the respective experimental and psychological conditions.[90] But Tomasello tested human children and adult chimpanzees under different conditions. While the children were "free-ranging," as Boesch put it, and came to Tomasello's laboratory at the Max Planck Institute in the company of a parent, the chimpanzees lived in captivity and for most experiments were isolated from their group.[91] The children were tested by conspecifics, the chimpanzees by another primate species. Since the experimenters did not want to risk entering an experimental booth with a potentially violent chimpanzee, they interacted with test animals through a thick Plexiglas wall, while no such wall separated experimenters and children. While the toddlers were told what to do in their native language, verbal instruction could not be given to the apes. In many studies testing whether chimpanzees could imitate—a bone of contention in the chimpanzee culture controversy—they were presented with a human model. The question answered by these experiments was really whether the apes imitated members of another species, while the corresponding human experiments tested whether children imitated members of their own species. Such differences between experimental conditions allowed critics to call into question the proposed causal relationship between the experimentally manipulated variable and the observed effect (e.g., maybe it wasn't being a chimpanzee, but being presented with a model belonging to a different species that predicted the inability to imitate). Every uncontrolled variable enabled alternative explanations and compromised what philosophers of science as well as Tomasello and Call themselves called the internal validity of their experiments.[92]

Boesch also challenged the *external* validity of Tomasello's findings by calling into question whether they applied beyond the walls of his laboratory. The behaviors tested—pulling ropes, opening doors with keys, playing shell games—were not among the things that chimpanzees ever did in the wild. For German children, however, these activities were nothing out of the ordinary. Moreover, the participating children had been raised by middle-class parents sufficiently invested in science to enroll their offspring in a psychological study, while many of the Leipzig chimpanzees had been raised in cages by employees of a Dutch pharmaceutical company, presumably not under the most nurturing conditions. In Boesch's eyes, their cognitive capacities and social behavior were about as representative of wild chimpanzees as the cognitive capacities and social behavior of the retarded children rescued from Romanian orphanages after the fall of the Iron Curtain were representative of humankind.[93]

At first glance, the chimpanzee culture controversy appears to follow a playbook similar to that of the infanticide controversy, which sociologist of science Amanda Rees laid out.[94] Between the 1970s and 2000s, primatologists had fought over whether the killing of infants was a natural behavior or a pathological response to particular social and environmental conditions. Now they debated whether culture and cooperation belonged to the natural behavioral repertoire of chimpanzees and just remained hidden under the abnormal conditions of the laboratory, or whether the inability of comparative psychologists to verify these capacities in controlled experiments suggested that fieldworkers had to interpret their observations more conservatively. In other words, did the absence of experimental evidence provide evidence of absence? The major difference between the two debates is that the infanticide controversy occurred largely between field scientists, whereas the chimpanzee culture controversy also pitted field against laboratory scientists.

Studying cultural transmission mechanisms under laboratory and zoo conditions amounted to studying "culture outside of culture," Boesch claimed as a fieldworker.[95] The experimental findings couldn't be compared with naturalistic observations because they did not take into consideration the massive differences in enculturation between captive and wild apes. Boesch accused Tomasello of not controlling for the effects of the radically different environments to which these populations had adapted and in which they were tested.[96] And since most laboratory chimpanzees came from West Africa while most fieldwork had been conducted in East Africa, it wasn't even clear whether behavioral differences between these populations represented cultural or ecological differences between captive and wild communities, or whether they had to be attributed to genetic differences between the subspecies *Pan troglodytes verus* and *P. t. schweinfurthii*. In a bellicose tone, Boesch alleged that "such captive studies would be akin to studying the culture of the Aka Pygmies in Central Africa with Nigerian prisoners in German prisons!"[97]

Philosophically, Boesch situated his challenge to Tomasello in the campaign against Cartesianism, which had united the most disparate schools of thought at least since the early twentieth century.[98] But what did he mean when he accused Tomasello and other psychologists as well as some unnamed circles in cultural anthropology of Cartesianism? Boesch freely translated Descartes's claim that animals were mere machines while humans had been endowed with mind into the proposition that "humans are born resembling a white sheet upon which everything can be written, whereas animals are genetically rigidly fixed."[99] Of course, it wasn't the rationalist René Descartes but the empiricist John Locke who had suggested that, at birth, the human mind resembled a blank slate on which life would subsequently inscribe its experiences.[100] And neither of these seventeenth-century philosophers knew anything about genetics. At the beginning of the twenty-first century, however, evolutionary psychologist Steven Pinker connected the dots and popularized genetic explanations of behavior, previously championed by sociobiologists, against the "modern denial of human nature": far from being a tabula rasa, the human mind had inherited adaptations to our ancestors' environment that genes, not experience, had instilled in us.[101] Although Boesch generally embraced evolutionary interpretations of human and animal behavior, he felt that the genetic determinism of evolutionary psychologists underestimated the behavioral variability of primates—not just of humans but of all apes.[102] His argument worked in the opposite direction of Pinker's. Boesch's concern was not that the human mind had been wrongly considered a "white sheet," but that contemporary "Cartesians" like Tomasello failed to recognize that the chimpanzee mind was quite a white sheet, too.

Captive experiments presupposed that "individuals' cognitive abilities were not influenced by experiences during development and that individuals developed full-fledged species-specific cognitive abilities under a whole array of different environmental conditions," claimed Boesch.[103] In his eyes, this was equally false for humans *and* chimpanzees. He pointed to the wealth of evidence provided by cultural psychologists suggesting cognitive differences between white middle-class Westerners and people who had been brought up under different conditions. "Our hardcore psychologist needs to remember that not all humans are westerners and that the socially and contextually integrative, less individualistic African and Asian way of educating children is observed in very large segments of humanity," Boesch reminded Tomasello and colleagues. Boesch also cited studies demonstrating how detrimental life in captivity was to the development of primate cognition.[104] It followed that if different life experiences transformed ape minds as profoundly as they did human minds, then comparisons between Leipzig children and zoo-living chimpanzees did not warrant claims about species differences between humans and chimpanzees in general. Boesch even contended that "whatever deficits we find in captive populations are more likely to be artifacts of captivity than characteristics of the species."[105]

The methodological linchpin of Boesch's critique was an alleged lack of experimental control. Historian of science Jutta Schickore showed that late nineteenth-century physiologists began to call for control as they realized how difficult it was to manage the variability and complexity of living organisms in the laboratory.[106] Acknowledging that the ideal experimental situation could not be attained in practice, they emphasized the importance of keeping conditions constant as much as possible. They also recognized that there were no universal standards in biology to which an experimental condition could be compared, and they introduced experimental controls as a pragmatic alternative: a second experimental condition that differed in only one crucial respect. Boesch's attack on Tomasello exploited the fact that captivity introduced an array of additional variables, so that comparative psychologists would either have to bring these in line with each other, or demonstrate experimentally that these variables did not affect their results.

A controlled cross-species comparison would require keeping the environment constant, for example by testing forest-living humans against forest-living apes, Boesch insisted.[107] And that is what Boesch and colleagues subsequently did when they compared how Aka and Mbendjele people and Taï chimpanzees cracked nuts.[108] The rationale was not to relegate contemporary foragers to the human past—an accusation that anthropologist Tim Ingold had leveled against McGrew's comparison of the material cultures of Gombe chimpanzees and Tasmanian hunter-gatherers.[109] In Boesch's eyes, these human groups allowed for an ecologically valid comparison because their present environment resembled that of Boesch's Ivorian chimpanzees more closely than the city of Leipzig did. Philosophically, Boesch sought to extend an empiricist philosophy of mind from *Homo* to *Pan* and demanded that the quest for human nature go through the eye of both human and chimpanzee diversity.

Michael Tomasello and his right-hand man Josep Call pushed back in a cool tone that betrayed nothing of Boesch's thymotic anger. They welcomed the methodological reflection that his critical assessment stimulated but insisted that their human-ape comparisons were perfectly valid, both internally and externally. Although it was true that they tested humans and chimpanzees under different conditions, identical conditions were neither possible nor desirable. Species differences required adjusting experimental designs in a way that made them "functionally equivalent."[110] For example, human children needed a parent in the room in order to be as calm and secure as the adult apes were naturally (positing an accommodation of differences in age rather than species because in the absence of their mothers chimpanzee infants would get anxious, too). Or, the chimpanzees could be motivated with food rewards but did not care much for the toys incentivizing human children. At the same time, the latter were unimpressed by grapes because they could count on their parents for a steady supply of snacks. (Had Tomasello worked with Leipzig children right after World War II or with children from Somalia, which had just been ravaged by famine, his human subjects might have preferred the food rewards as well.)

Whether related to genetics or ecology, Tomasello and his coworkers saw no way of eliminating these differences in experimental conditions between humans and chimpanzees. And they emphasized that they didn't have to. After all, they did not compare the performance of the two species directly. Instead they tested children and apes against a control condition *within* the respective species. Ideally, a control condition kept all variables constant apart from the independent and dependent variables tested, but Tomasello did not aspire to such consistency between experimental setups for humans and chimpanzees. Only informally did his laboratory compare the results of these controlled experiments across species. For example, chimpanzees understood false beliefs in one condition but not in the other, while children understood them in both.[111] When I asked Tomasello why he avoided direct comparisons, he explained that absolute differences between the behavior of chimpanzees and children depended on too many factors he could not control. Hence, they weren't telling. But as a cognitive scientist, he wasn't as interested in how well his test subjects performed as in the factors that enabled their performance. If he could show that children did significantly better after having observed a demonstration while the performance of chimpanzees did not change, he had learned something about the cognitive mechanism underlying the respective performances of the two species—unless these experimental outcomes had been confounded by differences in human and chimpanzee research designs. Whether or not their supposed functional equivalence validated such an informal comparison remained a matter of practical wisdom and prudent judgment—and here Boesch and Tomasello harbored very different intuitions.

Boesch alleged that the parents' presence introduced a Clever Hans effect into their children's performance.[112] Like the famous German horse that seemed to successfully perform arithmetic calculations in front of gasping audiences but was eventually shown to respond to unconscious bodily cues provided by its owner, the children received unintended help from their parents. The primatologist continued to call into question the "fairness" of such asymmetrical cross-species comparisons and dismissed Tomasello and Call's defense that testing under more similar conditions was simply not possible. In Japan, Tetsuro Matsuzawa had carefully developed relations of trust with the chimpanzees in his laboratory so that they allowed him into their experimental booth. There he could test a chimpanzee infant in the presence of its mother, just as Tomasello tested human children. While Tomasello's experiments usually showed that adult apes had failed to solve a cognitive task that human children could master, one of Matsuzawa's young chimpanzees had just surpassed the performance of a human control group in a numerical memory task.[113] When tested under less disadvantageous conditions, Boesch declared, "chimpanzees either are equal to or outcompete humans."[114]

In defense of their approach's external validity, Tomasello and Call argued that the cognitive abilities they tested in children were basic enough to show no significant cultural variation.[115] As far as chimpanzees went, they claimed that no data supported Boesch's suggestion that the cognitive skills of wild apes were superior to those of captive animals. In fact, captivity presented chimpanzees with a particularly challenging environment and, if combined with enculturation, could even stimulate them to develop skills that had never been seen in their forest-living conspecifics. The Japanese chimpanzee memorizing Arabic numerals could have been a case in point, but Tomasello and Call drew from another ape language project: Sue Savage-Rumbaugh's human-reared bonobos had developed the ability to share attention by pointing, and they had learned to imitate each other. "If enculturated chimpanzees do resemble human children more than their conspecifics in imitative learning, it would say a great deal about the power of cultural environments to encourage human-like skills of social cognition," noted Tomasello, Savage-Rumbaugh, and Ann Cale Kruger.[116] Tomasello considered imitation a cognitive prerequisite of culture, which only *Homo* could bestow on *Pan*.[117]

Although they had made no systematic methodological effort to understand how different environmental conditions shaped the cognitive capacities of captive apes, Tomasello and Call insisted that denouncing them as genetic determinists amounted to a disturbing misrepresentation of their theoretical position.[118] Tomasello's argument that chimpanzees and bonobos could socially learn new forms of social learning if raised by humans—acquiring a *Pan/Homo* culture of culture, so to say—certainly ruled out the idea of genes and genes alone dictating chimpanzee behavior. His trust in the human potential for adjusting our social norms and practices to the exigencies of life in technologically modern large-scale societies on an increasingly overcrowded planet, which Tomasello had expressed in his Hegel Prize lecture, revealed an even less determinist anthropology.[119] Not over his dead body would Habermas have supported a genetic determinist. Or would he?

Boesch's allegation aimed at the structural core of Tomasello's approach: the attempt to define what set humans apart from other animals. When I interviewed Tomasello, he pointed out that his pretheoretical commitment to demonstrating that humans and chimpanzees were different distinguished him from Boesch and de Waal: "They don't like talking about difference. If you beat them with a stick, you get *language* out of them." Boesch had indeed written that "the one thing that is unique to human culture is language, which acts to amplify cultural transmission and development."[120] But although our linguistic capacities represented a crucial milestone in the process of hominization, Tomasello did not want to stop there. He turned from such extraordinarily complex achievements of cumulative culture to inapparent but more fundamental differences in cognition, which had allowed these majestic markers of human exceptionalism to evolve. For instance, at first glance, the difference between imitation (defined as the ability to precisely copy a model's behavioral strategy to achieve a desired outcome) and emulation (aiming at the same outcome without paying detailed attention to the behavioral strategy), which Tomasello had presented as a decisive anthropological difference, appeared to be nothing more than a hairline crack between two closely related types of social learning, of interest solely to a few experts on primate social cognition.[121] But if imitation indeed enabled humans, but not chimpanzees, to first copy and then refine the behavioral strategies of others, it would open a deep ontological fault between these genetically so closely related species that would become visible only over time, as an increasingly complex culture emerged from this particular kind of social learning.

In Tomasello's eyes, the difference between a species that used stone hammers and anvils to crack open nuts and one that built Boeing 747s was not merely quantitative but qualitative: a jet airliner consisted of so many more parts that no single individual could develop such a technology within his or her lifetime. Only a species capable of cumulative culture could develop such complexity. Tomasello conceived of the underlying cognitive capacity as so fundamental that no normal human being failed to develop it (only autistics did). Yet it would always remain beyond wild chimpanzees. Although Tomasello granted that human-reared apes could learn some aspects of cultural learning from their evolutionary cousins, they still did not acquire the ability to create a culture de novo.[122] In other words, they could be shaped by human culture but they could not actively contribute to a shared *Pan/Homo* culture.

Tomasello's assumption that early humans had taken a small evolutionary leap, which soon opened a giant chasm, and Boesch's insistence on a more gradual phylogenetic emergence of culture rehashed the logic of the mid-nineteenth-century debate over the origins of language. Back then, one camp, most prominently represented by Oxford-based Sanskrit scholar Max Müller, had declared language an impassable barrier, which the first humans had erected quite abruptly between themselves and all other animals. The other camp gathered around Charles Darwin and argued that language had evolved in a step-by-step process, which connected *Homo sapiens* to our natural historical cousins and forebears. The Darwinians' insistence on gradual changes was part of their rejection of the geological doctrine of catastrophism. In line with the Christian belief in God's ex nihilo creation of life, the Deluge, and so forth, catastrophists held that large-scale changes could occur rapidly.[123] Tomasello sought to identify the small-scale cognitive changes that had enabled human culture and thus the rapid and massive transformations it has brought upon the world. Like a biblical flood, it threatened to drive countless species extinct (with the biodiversity programs of zoological gardens taking the role of Noah's ark). In the end, Tomasello's natural history of human exceptionalism amounted to cognitive catastrophism, at least from the perspective of staunch Darwinian gradualists like Boesch.

In response to the philosophers who had contributed to the special issue of *Deutsche Zeitschrift für Philosophie* on *The Cultural Origins of Human Cognition*, Tomasello and his postdoc Hannes Rakoczy (also a former student of philosophy) demarcated their research program from philosophical anthropology: "Argumentatively, this is not about an attempt to distill the core of human nature, to answer the philosophical question 'What is Man?' (by pointing to the capacity for cultural learning or the like). This line of argument is not philosophical at all and does not deal with 'What is . . . ?' questions; instead it is rooted to a down-to-earth cognitive science debate."[124] They also dismissed the idea that one specific difference set humans apart from other animals ("of course, nobody believes in such a thing"). Instead they sought to explain the numerous cognitive differences between human and nonhuman primates as economically as possible. Tomasello wanted to show how a few small but enormously consequential biological adaptations had enabled cultural transmission, which set *Homo sapiens* on an altogether different track.[125]

Boesch suggested that the theoretical presupposition of a genetic difference between *Pan* and *Homo* informed Tomasello's laboratory method of comparing children and chimpanzees irrespective of their histories and living conditions and put it at odds with an ecologically oriented chimpanzee ethnography.[126] Tomasello and Call, on the other hand, denied that there even was a debate over field observations and controlled experiments—or in any case, they refused to have one.[127] When I addressed the tensions between field and laboratory researchers, Tomasello's response was conciliatory and almost self-effacing: "I've said to my fieldworker friends for many years: natural observations come first; they are primary. We are the cleanup operation." The two approaches simply fulfilled different but complementary functions: fieldwork told primatologists *what* animals did in their natural habitat while laboratory experiments revealed *how* exactly the animals did it cognitively.

Yet Tomasello remained adamant that behavioral observation alone did not suffice to answer the latter question because "the exact same behavior may be underlain by very different cognitive mechanisms."[128] Seconding this division of epistemic labor, Richard Moore saw it as the main reason philosophers interested in how the human mind worked engaged with an experimentalist like Tomasello rather than a field biologist like Boesch. This attempt to grant fieldworkers the first but not the last word on animal cognition failed to pacify this methodological front line in the chimpanzee culture wars. "If we want to understand the specificity of cognitive abilities in humans and chimpanzees we have to take into account what they do in real life," Boesch charged. He refused to acknowledge that cognition was beyond the fieldworker's pay grade.[129]

The Laboratory as Field

The primatological literature provides many examples in which field and laboratory research complemented each other.[130] But when controversies erupted, benchworkers and fieldworkers handled them differently. Hence, the controversy between Boesch and Tomasello could also be read as a controversy over how to deal with controversy.

In the 1980s, when controversy studies became the bread and butter of the social studies of science, sociologist Harry Collins argued that replicability served as the "Supreme Court" of the scientific value system: it operationalized the norm of universality, which demarcated scientific knowledge from other forms of knowledge by proving its independence from whoever had originally established it.[131] In the face of this epistemological dogma, laboratory scientists rarely replicated their colleagues' results in an exact manner—if not for lack of skill and tacit knowledge then because the institutions of science, especially journals, did not reward the confirmation of what one's competitors had already found. Researchers didn't get credit for checking old facts but for discovering new ones. If an experiment was repeated, the experimenter usually sought to introduce a significant and thus publishable difference. Yet any deviation from exact replication posed a problem: if a controversy broke out and the warring parties exploited a conflict between two experimental findings for their respective purposes, only a third experiment could validate one or the other—and validating this third experiment required a fourth, and so on. Collins dubbed this need for tests of tests of tests "experimenter's regress."[132] No experiment can ever put a logical end to this infinite movement, he claimed. Only people can. Consequently, it takes sociologists, anthropologists, and historians of science to explain how a scientific controversy came to be resolved. This social constructionist account left open one important question: What role did the actual behavior of nonhuman primates play in the resolution of primatological controversies?[133]

Analogous to experimenter's regress, field researchers attributed differences between their observations to differences between research sites, as Amanda Rees argued in a study of the infanticide controversy between field primatologists.[134] But such "fieldworker's regress" diverged from experimenter's regress in that fieldwork, by definition, takes place in uncontrolled environments. Whereas laboratories, at least in theory, enable researchers to keep all factors but independent and dependent variables constant, the whole point of naturalistic observation is to avoid such manipulation. Historically, primatologists initiated field studies in the 1930s to provide the external or ecological validity that laboratory research lacked.[135] The price to be paid for the aura of naturalness was the logical and practical impossibility of exact replication. While laboratory researchers maintained replicability at least as a regulative ideal, fieldworkers acknowledged that all field sites were different and that even at the same field site, you could not step into the same jungle stream twice.

This uncontrollability made field sciences both more and less vulnerable than laboratory sciences. The necessity to contextualize all findings enabled opponents to challenge any generalization. But it also meant that if others did not corroborate what a fieldworker believed to have seen, she could always blame it on the different times, places, and objects of her critics' observations. Fieldworkers could compare but never really replicate their colleagues' observations. Consequently, they tolerated competing claims to a much greater extent than laboratory scientists.[136] If their animals did not behave like their colleagues' animals, they could sell their observation as a new discovery rather than a failed replication. Chimpanzee ethnographers and experimental psychologists did not abide by the same epistemic norms and expectations.

As a student of the environmental and cultural diversity of chimpanzee communities, Boesch showed little tolerance toward universalist claims if they were at odds with his own field observations. Doubting that psychologists studying chimpanzee cognition actually exercised the high levels of control that Tomasello and other experimenters asserted, he read reports from different laboratories as if they came from different field sites—just as a laboratory ethnographer would.

Under the heading "Culture and cognition in chimpanzees," Boesch compared the histories and living conditions of three captive groups at the University of New Iberia, Ohio State University Chimpanzee Center, and Leipzig Zoo to explain behavioral differences these communities had shown in experimental tests.[137] He claimed that Daniel Povinelli's New Iberia chimpanzees had experienced the poorest socioecological conditions and that they had shown the poorest performance in cognitive tests.[138] So poor that natural selection would have eliminated them in the wild, Boesch suggested: "The cost of misjudging physical properties in real life is too high to allow for the type of errors that the New Iberia chimpanzees made."[139] By contrast, Tomasello's Leipzig chimpanzees enjoyed the best housing. But they had endured probably the worst hardship during their upbringing in a biomedical research facility. Some of the Ohio apes had been enculturated in human homes and did better in cognitive tests than members of the other groups. But none of the settings compared favorably with the wild, where survival posed a constant cognitive challenge. Nor did the captive chimpanzees' tool-using and tool-making skills or social cognition match the abilities of the Taï community. The contrast Boesch evoked between the "small, stable, and restricted man-made environment" inhabited by the New Iberian chimpanzees and the luxurious "free grassy outdoor area" that their Leipzig cousins had access to suggested that the ecologies of the laboratory populations differed as much as Taï Forest differed from the savanna roamed by the chimpanzees of Mont Assirik, Senegal.

Boesch used the resulting inconsistency between experimental findings to cast doubt on the ability of captive research to represent chimpanzee cognition in general: "In a disturbing comparison, all studies done by one group (Tomasello's) concluded that apes are strongly limited in their ability to imitate, while all studies in another group (Whiten's) concluded the exact opposite. This highlights the challenges of interpreting the work by research teams that have adopted different definitions and work with captive individuals with very different backgrounds and experience."[140] While Boesch left no doubt that he sided with Whiten, he mostly dismissed the focus on imitation as arbitrary because even "in humans culture can develop without much reliance on imitation."[141] Why then set the bar so unreasonably high for nonhumans?

Apart from disagreements over how broadly or strictly to define culture, Boesch had pointed to a second possible explanation for why Tomasello's and Whiten's findings diverged. Each captive group comprised individuals that had experienced very different upbringings: some had been born in the wild while others had been reared by their own mothers, by zoo staff in a nursery, or by a human family in an ape language project.[142] These disparate biographies contrasted sharply with the uniform life histories of the laboratory rats and mice that researchers could purchase like other industrially produced equipment from specialized companies. While experimental psychologists looked at the data collected from chimpanzees that had come from all walks of life as if they were Wistar rats, field primatologists had customarily documented their research subjects' unique life courses ever since long-term observation of individual animals had become state-of-the-art practice in the 1960s.[143]

I wondered whether Boesch's comparison of different captive groups amounted to a chimpanzee laboratory ethnology of sorts. But when I asked him whether he saw cultural differences between these communities, he pushed back: "I can't tell you that, no. There have been some attempts to look at cultural differences between captive groups. Jane Goodall pushed that forward. But, to my knowledge, no publication came out of it, probably because most people, like me, would ask: How can you tell that a behavioral difference is cultural and not a response to the artificial environment?"

Although Boesch refused to speculate about chimpanzee laboratory cultures, he insisted that comparative psychologists take into consideration their animals' distinct histories and interpret experimental findings against the background of varying contexts of discovery.[144] Lacking the necessary controls, each laboratory produced a form of local knowledge that was informative only insofar as it gave rise to comparisons with the local knowledge generated by other labs—or field sites, for that matter. Boesch concluded that "we need to adopt a population-sensitive approach in chimpanzees if we want to answer questions about how flexible they are, how culture affects their cognition, and how this compares to humans."[145]

In an interview with me, Tomasello dismissed Boesch's claim that every captive group was particular because of its different upbringing and environment: "I don't know on the basis of which data he is saying that. There's nothing on that." No primatologist had studied possible behavioral differences between laboratory chimpanzee groups, Tomasello insisted.[146]

Considering how much attention cultural primatologists had paid to behavioral differences between wild chimpanzee groups, I began to wonder why nobody had developed a positive research program for doing fieldwork in laboratories. Weren't they interested in documenting and explaining the ways of life that different zoos and laboratories had fostered? Or would field primatologists seeking access to laboratories encounter the same resistance that I had experienced because experimental psychologists worried that their findings could be reduced to merely local knowledge? If chimpanzee ethnographers had to rule out that behavioral differences were not the outcome of learning individually how to deal with different laboratory environments, why couldn't they apply the method of exclusion just like their colleagues in the wild? Why had no chimpanzee laboratory ethnography emerged?

While Boesch looked at different laboratories through the eyes of a fieldworker, he never suggested that experimental psychologists switch to naturalistic observations. Unlike Sue Savage-Rumbaugh and colleagues, who had marginalized themselves by turning from controlled experiments to an ethnography of the *Pan/Homo* culture of her lab, Boesch did not give up on the distinct logics of benchwork and fieldwork.[147] At a time when French philosopher Vinciane Despret proposed to make the Clever Hans effect part of a new approach that deliberately "attuned" animals and researchers in experimental settings, Boesch sought to minimize such human interference in the field and demanded that comparative psychologists check how they affected the behavior of both apes and infants.[148] He held Tomasello accountable for what he perceived as a breach of the experimental method. Basically, Boesch charged Tomasello and Call with not controlling for Pongoland. Accusing the comparative psychologists of systematically ignoring the evidence supporting his claim that wild apes had more sophisticated cognitive skills than captive apes, which hundreds of field researchers had supposedly collected over fifty years, Boesch asked indignantly: "Why should we grant them tolerance for their violation of the experimental paradigm, when they decline any tolerance for field data?"[149]

After the Field: How Statistics Revolutionized Chimpanzee Ethnography

In the first two decades of the twenty-first century, the mathematization of field primatology overtook that of experimental research in comparative psychology—a development that added to the already dramatic transformation that chimpanzee ethnography had undergone since the early days of Jane Goodall's participant observation in Gombe. In the 1950s, field primatologists still struggled with the continuing reputation of doing qualitative rather than quantitative research. The solution to this problem seemed to be the increasing application of experimental practices in the field.[150] By the 1970s, however, they had replaced the descriptive ethnographic approaches of anthropology with the quantitative, but not experimental methods offered by biology. Biological anthropologist Karen Strier drew attention to the historical irony that the comeback of primate ethnography in the 1990s had been mediated by a highly quantitative population-oriented perspective that revealed intraspecific population-wide behavioral variation among chimpanzees.[151]

In the forest, the desire to mathematize ethnography materialized in the form of check sheets and handheld computers, which enabled researchers to categorize and thus count observed behaviors. In 1974, mathematician-turned-baboon-watcher Jeanne Altmann reviewed sampling methods relevant to primatological field research that would generate quantitative data sets for subsequent statistical analysis.[152] According to her colleague Shirley Strum, the increasing interest in quantification had grown out of a concern over how to interpret naturalistic observations: "The problem was how to decide who was correct if the description of the baboons in one place did not match the description of the baboons in another place."[153] Unless fieldworkers had collected their data from samples of individuals that represented the whole range of actors in their respective groups, differences between their accounts could reflect either real group differences or differences between the fieldworkers' attention biases: maybe they noticed the brouhaha of male baboons more readily that the quieter behavior of the females, or they didn't realize that they spent less time observing the well-habituated animals than the shy ones. Unless behavioral differences were purely qualitative, the question of whom and what to count was of crucial importance to the project of chimpanzee ethnology, which sought to reveal real behavioral differences between wild chimpanzee communities, not between their ethnographers.

Although the choice of suitable sampling methods and their application in the field prepared the ground for subsequent analysis, the real work of numbers happened only after researchers had left the field. How central it had become to primatological fieldwork in general and chimpanzee ethnography in particular had already struck me during my first week at the Max Planck Institute for Evolutionary Anthropology. Upon their return to Leipzig, I overheard people in the Department of Primatology constantly asking: "Where is Roger?" "Do you think Roger will have time for me?" "I'm really worried about my meeting with Roger today." The person in such high demand was the institute's statistician—mere technical staff, as far as the human resources department was concerned, but few projects could do without his input.

Roger Mundry's curriculum vitae stood out at an elite research institution like the Max Planck Institute. At age nineteen, this West Berlin native had flunked out of public school, worked as a postman and gardener, dabbled in organic farming, and organized peace marches. In his mid-twenties, he attended night school to qualify for student grants and discovered that he actually enjoyed learning. He moved on to study biology at Freie Universität and took up bird-watching. For his dissertation, he conducted extensive fieldwork across Europe on individual birds that had learned their songs from other bird species. Mundry taught himself the statistics necessary to construct models from mining large numerical data sets. But he did not publish much, nor did he make an effort to build the social network necessary for obtaining one of the few postdoctoral positions in ornithology. So he applied for a job as a statistician instead. Soon after, he found himself at one of the world's foremost research institutions for evolutionary anthropology.

For Mundry, the application of statistics to biology was key. "If it wouldn't be about understanding life, if I sat here with data from insurance customers, I would have quit after a month," he told me. But his interest in the subject matter not only motivated him to do his job but also shaped the way he worked. When researchers came to him, they first discussed what they had seen in the wild. Since he had never been to Taï or any other chimpanzee field site, the statistician had to learn about their experience in the field. Mundry explained: "You can't do statistics without understanding the problem."

Understanding the problem meant generating what historian of science Theodore Porter called a thin description against the background of a thick description.[154] Statistics—Porter's prime example of a research practice that stripped observables of their circumstances—required careful thinking about the situation in which the data had been collected, Mundry maintained.[155] That was why he refrained from mindlessly formulaic data analyses and drew from his own experience as a biologist when discussing with fieldworkers what exactly they had done and what they had observed at their research site. Even statistics maintained a craft character at odds with the claims of the scientific method as applicable by anyone anywhere.

These conversations about how to make sense of their numbers helped fieldworkers, especially students and postdocs, to think more carefully about their hypotheses. For example, Lydia Luncz's comparison of nut-cracking techniques in three neighboring communities in Taï Forest sought to identify cultural differences between groups whose genetics and ecology were almost identical.[156] Many students first expected to find stable differences in the overall behavior between groups. In Luncz's study, the chimpanzees disappointed that hope. At the beginning of the nut-cracking season, all three communities did the same: they cracked *Coula* nuts with stone hammers. Only later in the season, as the shells grew softer, did behavioral differences emerge: while the South Group continued to use mostly stone hammers, the North and East Groups began to rely more heavily on wooden hammers. Initially, the nuts were so hard that the problem of opening them afforded only one possible solution, and no cultural differences could emerge. As the nuts became easier to crack, chimpanzees could either stick to their always reliable but scarce stone tools or use abundant but often rotten pieces of wood. Statisticians referred to such a change in behavior in response to a change in ecology as an interaction. In a laboratory, researchers would have kept the environment constant to avoid such a complication. In the field, however, complex interactions between large numbers of factors were the rule rather than the exception. Mundry taught students who had returned from the field to think less in black and white and to recognize and quantify interactions.

Rendering cultural primatology complex came at a price. "If you end up with a four-way interaction, you don't have a sexy story anymore because it gets complicated," Mundry admitted when I asked him whether more straightforward answers could be published in higher-ranking journals. Since controlled experiments provided simpler answers, that might well be one reason why benchwork stood a better chance than fieldwork to make it into *Science* or *Nature*. Considering that field research took longer and was more exacting, the primatologists experienced this preference for laboratory studies as unjust. Because of Mundry's statistical scrupulousness, many students saw him as the bearer of bad news who downsized their claims to what was statistically defensible.

Mundry shrugged off this reputation: "Are simpler stories easier to sell? We argue over this every day. Why are we doing science? To have a career and write sexy papers that will make it into *SPIEGEL ONLINE* [Germany's most popular news portal]? The German state pays us to provide knowledge. If our findings turn out to be complicated, they only reflect the complexity of life. Why would you even expect something simple?" Mundry had come a long way from high school dropout to vocational scientist, but throughout this development he had remained true to his disregard for extrinsic rewards like money, degrees, or prestigious publication venues.

Despite this austerity, Mundry was much sought after among field primatologists because he knew how to do what nobody else in the department could do. He had taught himself to code and he was able to write a program in ten minutes that would save researchers two months of dull mouse clicking. He was a proficient user of "R," an open-source software suite for statistical computing that offered a broad range of advanced methods. And he understood these advanced methods better than most people in evolutionary anthropology. Consequently, he taught workshops as far away as Japan and provided advice by email to colleagues all over the world. His high demand reflected changes in the epistemic landscape of field primatology.

Until the 1990s, both field and laboratory scientists had relied mostly on relatively simple techniques such as nonparametric tests and analyses of variance (ANOVA), which they had learned during their own training. These methods helped researchers to distinguish signal from noise in their data. "Chimpanzees have different personalities, physiologies, genes, histories, and experiences," Mundry explained. "If you test the same animal twice, it won't do exactly the same thing." And maybe it would not show the expected behavior at all. Considering the wayward nature of great apes, this posed a problem even in the laboratory, where researchers exercised a relatively high level of control over their test subjects. Thus, it occasionally happened that some cells of the data collection sheets could not be filled—for example, if certain individuals could not be tested under all experimental conditions. Previously, this had meant that either those individuals or the conditions under which they had not been tested had to be removed from analysis. In the field, where manipulating animal behavior was anathema and complex interdependencies could not be controlled, such incompleteness haunted data collection at almost every step. Fifteen years ago, Mundry remembered in an interview in 2013, it was not practically feasible to analyze social interactions in a whole group of chimpanzees, which involved repeated observations of how particular individuals interacted with recurring as well as with changing partners. In the extensive matrix of all possible permutations, too many cells would remain open. Consequently, there were questions that chimpanzee ethnographers could ask but could not expect to answer.

In recent years, advances in statistics and statistics software packages such as "R" made modern but demanding statistical tools that processed complex data structures available to ordinary scientists. Generalized linear mixed models had opened new possibilities for experimenters, but especially for fieldworkers. "Today, people come up with more complicated research designs. Fifteen years ago they would have said that it would be nice if they could test this, but they wouldn't know how to analyze the resulting data," Mundry explained to me. "Now they say: 'Roger is going to figure it out, somehow.'"

Even in Europe, where the division of scientific labor in research teams was more common than it was in American anthropology departments, few other institutes could afford their own statistician to keep up with the rapid advances in the field. Yet their scientists eagerly adopted the new research designs. Many tried to analyze the data of the 2010s with the methods of the 1990s, complained Mundry. The patchy distribution of statistical expertise also challenged the scientific quality control demanded by journals: even senior people in the field often had difficulties reviewing article manuscripts based on the latest quantitative analyses. The fast-paced scientific change confronted the discipline with a structural problem. Evolutionary anthropology clearly wasn't what it used to be.

The new statistical tools also emboldened fieldworkers in relation to experimenters. "In an experiment, you exclude all variability and that makes it unnatural," Boesch explained to me in an interview. "In the wild, we don't control by removing, but by including variability. We try to collect quantitative data on all the factors that might affect the behavior we study. Twenty years ago this was still a problem, but today we can control for them statistically. If people say, we don't need to remove animals from the wild anymore to be able to control for everything in captive experiments, that's partly a result of this progress in statistics."

Can Fieldworkers Tell Apart Cause and Effect?

The sense of radically new possibilities found its most enthusiastic expression when a postdoctoral researcher in Boesch's department discovered a recent paper from the Institute of Behavioural and Neural Sciences at the University of St. Andrews. The other powerhouse of European evolutionary anthropology also sought to expand the field's mathematical tool kit. Kevin Laland and colleagues boldly claimed that a new statistical method enabled field researchers to establish causal relations.[157]

Until then, the consensus had been that only experiments could demonstrate that an observed effect had been brought about by a particular cause, namely the preceding experimental treatment, rather than some confounding factor. Controlled experiments sought to achieve this goal by testing the experimental against a control condition, which would ideally be identical with the experimental condition apart from not receiving the experimental treatment. As fieldworkers did not control the environment in which they made their observations, they were unable to exclude alternative explanations when they noticed a correlation between two processes. For example, the Gombe chimpanzees' use of sticks to fish for termites and the Taï chimpanzees' lack of any tool use for this purpose could have been brought about by different traditions of social learning and thus represent a cultural difference, but it could also be attributed to genetic or ecological differences because the two chimpanzee populations differed along all three axes. The method of exclusion could render such alternative explanations less plausible, but it could never rule out all alternatives. As one of Luncz's coauthors, Mundry conceded: "Just because we can't find any ecological causes of a presumably cultural behavior doesn't mean that there aren't any."[158] Uncontrolled naturalistic observations could be correlated with each other (e.g., continued use of stone hammers tallied with belonging to the South Group), but it was not possible to determine which factor was cause and which one effect, or whether a third factor caused both to vary while the original two factors did not affect each other. Only experiments could discern causal relations and their directionality—or at least that was what most scientists had believed.

What got some of the younger field primatologists so excited about Laland's paper was that he proposed that analyzing so-called causal graphs with the help of d-separation, a relatively new statistical method developed by artificial intelligence researchers in the mid-1980s, enabled field scientists to move from discerning mere correlations between observational data to establishing causal relations. If true, this argument would have major implications. For example, it suggested that cause-effect relations in animal behavior could be established without experimenting on captive animals. It promised to elevate the field sciences in relation to the more prestigious laboratory sciences that promised control and repetition of generalizable results.[159] Maybe the time would come when *Nature* and *Science* would no longer be brimming with laboratory studies while many excellent field studies were relegated to second-rate journals. The department decided to collectively reassess the value of experiments in comparison with naturalistic observations in its Journal Club, a weekly meeting discussing recent publications. In an email announcement, the organizers wrote: "A recent paper by Kevin Laland, not the least intelligent person in this world, proposes that there are new ways, besides experiments, to reliably infer the presence and direction of a causal link between variables. Do we believe him?"

Laland's claim about statistically discerning causal relations turned out to be a mere aside in a short paper that advocated nothing short of a revolution in evolutionary biology. It presented a narrative in which the controversies over animal culture, cooperation, and niche construction constituted battlegrounds where traditionalists and radicals clashed over the "dominant scientific paradigm."[160] This paradigm had been established by biologist Ernst Mayr, who had proposed to resolve a number of controversies in the life sciences by distinguishing between two kinds of causality that researchers from different subfields of biology often mistook to be mutually exclusive, although they really offered complementary explanations of any given phenomenon.[161] Proximate causes comprised factors in a particular organism's immediate environment, which affected its living processes in a mechanistic fashion. By contrast, ultimate causes had historically shaped the organism's DNA and thereby its behavioral predispositions without necessarily acting on the organism in a given situation.

Mayr illustrated the distinction by asking: "What is the cause of bird migration? Or more specifically: Why did the warbler on my summer place in New Hampshire start his southward migration on the night of the 25th of August?"[162] The ultimate causes were ecological and genetic. As an insect eater, the warbler had to migrate because it would not find enough food in New Hampshire during the winter. The evolutionary history of its species had selected for individuals that responded to particular environmental stimuli by migrating to warmer climates. By contrast, the proximate causes were physiological. The shortening of days prepared the warbler for its departure, and when a cold air mass passed over New Hampshire on 25 August, the drop in temperature led the bird to fly off. There was no point in arguing over whether warblers *really* migrated because of their species' natural history or because of a spell of cold weather, Mayr suggested. Students of natural history and of bird physiology simply explained the same behavior on two different levels of causality. Laland and colleagues conceded that this "meta-theoretical conceptual framework" had indeed helped to understand bird migration. But it proved an obstacle to resolving some of the most heated debates in contemporary biology.[163]

What the controversies over animal culture, cooperation, and niche construction had in common was that proximate causes doubled as ultimate causes and contributed to the dynamics of selection. While the cold-induced migration of Mayr's warbler did not change the climate of New Hampshire, which had driven extinct all warblers that had not left for warmer climates, the creation of a more livable ecological niche or of a cooperative culture that benefited its members reduced selection pressure on the very individuals engaged in these activities. In Laland's eyes, bird migration was a well-chosen but exceptional case of animal behavior. As one of the most influential ambassadors of gene-culture coevolution to primatology, Laland suggested that studies of how socially learned behaviors allowed animals to construct new niches for themselves provided better paradigm cases for a biology that did not feature organisms as passive objects of natural selection but as active makers of the social and ecological worlds they inhabited. In the case of *Homo sapiens*, for example, agricultural practices selected for gene alleles that helped us to digest starch, carbohydrates, and proteins. Here, the distinction between proximate and ultimate causes collapsed, Laland and colleagues argued, and had to be replaced by a notion of reciprocal causation.[164]

A division between experimental sciences, which offered classical causal explanations, and observational sciences, which had to contend with such a motley of potential causes that they could offer only descriptions, now appeared obsolete to Laland and colleagues: "Modern causal modeling methods overcome Mayr's concern that biological complexity would make impossible an accurate description of causality as traditionally defined."[165] Following biologist and statistician Bill Shipley, Laland maintained that new statistical methods had invalidated the truisms that correlation did not imply causation and that only experiments could demonstrate causal relations.[166] "Statistical methods now exist that allow researchers to translate a causal hypothesis into a corresponding model and thus to distinguish between competing causal hypotheses by using observational data," Laland and colleagues wrote.[167] "We can use this to test whether our data are consistent with a specific causal hypothesis, or to determine which of several hypotheses best fit the data," William Hoppitt and Kevin Laland explained in a more extended exposition of their causal modeling approach.[168] Of course, the evidence would not be demonstrative but remained probabilistic: "This approach allows a researcher to rule out some causal hypotheses as improbable, which in turn increases their knowledge about which hypotheses are likely."[169]

Although Laland's contention seemed to entertain any naturalist's wildest epistemological dreams, Boesch and Mundry met the proposal with skepticism. In the Journal Club, the conversation initially clarified some of the general tenets of the field primatologists' theory of knowledge. "I'm surprised by this discussion," Boesch opened the debate. "In some cases, things are quite clear. Let's assume we considered all factors affecting the travel distance of chimpanzees and found that it only correlated with temperature and fruit availability. Do we really think that chimpanzees, if they walk longer distances, affect how many fruits are available in the forest? No. There is an obvious causational chain here."

Tobias Deschner, one of the senior researchers in the department, responded: "Somebody who is into experiments will say: 'Okay, you have measured temperature, fruit availability, and travel distance, but how can you be sure that other factors aren't involved? How can you be sure that there is a causal relationship between the two variables you recorded rather than between those two variables and an unknown third variable?'" Controlled experiments, however, were not the solution to this problem, he contended, because they controlled only for confounding variables the experimenter considered relevant, not for all possible confounding variables. "Laboratory researchers also introduce a new bias by creating an artificial environment and group structure," seconded a postdoc. "Thus you run into exactly the same problem as in fieldwork," Deschner concluded. "Just the means of control are different. In the case of laboratory experiments, you try to control for confounders by changing the structure of the cage; in the case of fieldwork you use statistics. But the question of whether you have controlled for all the relevant variables remains the same."

"There are still huge differences between naturalistic observations and manipulating the conditions in an experiment," countered Mundry. "Of course, there are many factors you can't control like the internal states of individual animals, their past experiences, etc. But you can randomly allocate these individuals to different treatments." Mundry's objection introduced an important element in the historical epistemology of experimentation into the debate: the transition from controlled to randomized controlled trials. In 1926, British statistician and population geneticist Ronald Fisher proposed: "Randomisation relieves the experimenter from the anxiety of considering and estimating the magnitude of the innumerable causes by which his data may be disturbed."[170] While controlled experiments allowed ruling out that the effect was really the cause, they left open the possibility that one of the innumerable uncontrolled and often unrecognized causes had simultaneously impacted both of these variables without them impacting each other. But if the experimental treatment had been allocated randomly, researchers could be certain that the randomization was the only factor that had determined its allocation. No uncontrolled variable could have impacted the treatment, which had to be the cause of the observed response. And yet Fisher recommended controlling randomized experiments as much as possible, establishing the paradigm of randomized controlled experiments.[171]

Of course, randomization entailed the possibility that the experiment had confirmed a cause-effect relationship between treatment and outcome by pure chance. But researchers could calculate the probability for the experimental result to be a matter of bad luck. Fisher proposed considering a result statistically significant only if such an error was less likely than 5 percent.[172] In the course of the twentieth century, the vast majority of scientific journals had adopted this so-called p-value below 0.05 as the arbitrary cutoff line to decide whether the claims of an article manuscript counted as valid. In Mundry's eyes, this had led to an overemphasis on statistics in the quality control process: "We dichotomize what can't be dichotomized. The tiny difference between 0.049 and 0.051 can decide what gets published."

Although randomized controlled experiments were considered the most powerful approach to discern causal relations, much scientific research adopted other methods. For ethical, financial, or practical reasons, many situations and questions did not allow for either randomized or controlled experiments. And epistemologically, no behavioral scientist considered experiments a *non plus ultra*. Their artificial nature drastically reduced their external validity: whether findings applied outside the experimental situation could be determined only by observing how animals behaved when not experimented on—ideally in their natural habitat.

Such naturalistic observations remained equally limited—even if supplemented with the statistical tools that Shipley along with Laland and colleagues proposed to assess causal hypotheses on the basis of observational data.[173] Although Mundry agreed that causal modeling could determine which of several competing explanations was the likeliest, he objected that "the fact that one model was more probable than the other does not tell us anything about causation." He argued that it still did not eliminate the possibility that the two variables considered cause and effect were correlated only because both were impacted by an unrecognized third variable.[174]

"That unknown factor is at the heart of good science," Boesch seconded. "Our job is to develop a nose for all possible factors that might affect an observed phenomenon. That's how you come up with a good hypothesis, not with statistical tricks. Statistics is just a tool. Biology guides what we should do." His bottom line was that even the most advanced statistics could not make up for good intuitions developed during years of observation in the field. That was why Boesch emphasized that ape cultures had to be experienced personally, that they could be understood only by a fieldworker familiar with the context in which events unfolded.[175] His thick descriptions of chimpanzee life in Taï found an outlet in the books he wrote. These qualitative accounts not only provided indirect cues to what things might mean to the apes but also informed the hypotheses that Boesch and his group tested statistically against their collective observations—even if they published the results in the abstract and anemic genre of scientific journal articles. Ethnographic thickness in the sense of a rich contextualization of behaviors offered hints to those "third variables" and "unknown factors" that, according to Mundry, no statistics could identify or eliminate. Only an experienced fieldworker could make an educated guess about what might have caused an observed event.

The Endurance of Human Nature

Human nature had been the cardinal concept of Enlightenment anthropology. In the nineteenth century, the rise of statistical thinking displaced it with normalcy, argued philosopher of science Ian Hacking.[176] The quest for what made us human gave way to quantitative analysis of how human traits were dispersed across a normal curve. Instead of asking what *differentia specifica* distinguished humans in general, statisticians now studied what distinguished normal people from those who were above or below average. A debate ensued over whether the normal amounted to the right and the good or whether it was merely mediocre and in need of improvement, as statistician and eugenicist Francis Galton suggested.

The replacement of human nature by normalcy satisfied the narrative conventions of the twentieth-century history of science, which buzzed with stories of such dramatic ruptures: scientific revolutions, paradigm shifts, epistemic breaks. But rarely does an emergent style of thinking and doing repeal and replace an earlier one. Usually, the new addition does not eliminate but transforms and restructures an existing mesh of concepts, norms, and practices.[177] And that's how it went when normalcy joined human nature. The Max Planck Institute for Evolutionary Anthropology can serve as a case in point: this large research facility analyzed its field observations and experimental findings with the help of the latest advances in statistics to extract from distributed data points new answers to anthropology's original question: What makes us human?

While Christophe Boesch focused on differences *within* the species, Michael Tomasello sought to understand differences *between* humans and apes that could explain our exceptional place in the natural world. Of course, he also recognized that not all chimpanzees or bonobos or humans were the same. Every experiment revealed variation among the test subjects' individual performances, which had to be analyzed statistically. But their application of statistics came to an abrupt end at the species boundary. Comparative psychologists Tomasello and Call excluded cross-species comparisons from quantitative analysis: "We mostly make our species comparisons not by statistically comparing the performance of the two species directly, but rather by statistically comparing the experimental and control conditions within each species separately and then comparing the pattern informally (e.g., one species is higher in experimental than control whereas the other is not)."[178]

By contrast, McGrew's cultural comparison between Tanzanian chimpanzees and Tasmanian humans had enabled direct statistical comparison between a human and an animal culture.[179] For this purpose, he had quantified the two populations' respective technological complexity along one and the same axis. Tomasello's approach, however, did not allow for curves in which the highest chimpanzee scores matched or exceeded low human scores—a perspective that had earned McGrew the accusation of reviving the open racism of an earlier era of anthropology.[180] Technically, Tomasello avoided comparing the numerical results of human and chimpanzee experiments directly because they had been obtained under different experimental conditions. And the conditions had to be different because humans and chimpanzees were different. This research design of qualitative but not quantitative comparison materialized Tomasello's philosophical conviction that humans and apes did not differ just in degree but in kind.

In Tomasello's eyes, this qualitative distinction recurred within *Pan*: Savage-Rumbaugh's enculturated bonobos seemed to differ from their conspecifics in that they had developed the capacity for imitation, which Tomasello considered a cognitive cornerstone of cumulative culture.[181] This shifted the unbridgeable gulf between human and animal only a little further because Tomasello stressed the difference between responding to a culture and creating a culture anew.[182] And yet he did conceive of the possibility that cross-fostering would endow *Pan* with capacities otherwise reserved for *Homo*.

As we have seen, this assessment of anthropogenic changes in primate cognition and behavior was one of the points on which Boesch differed sharply. While Tomasello saw human-made environments, at least potentially, as so stimulating that apes could acquire almost supernormal powers, Boesch believed that captivity largely diminished chimpanzee capacities. That was one of the reasons he rejected comparisons between Tomasello's zoo chimpanzees and free-roaming human children.[183]

A decade after this in-house controversy, Boesch and colleagues presented their alternative: a direct qualitative-quantitative comparison between wild chimpanzees and humans in their natural habitats.[184] To avoid the pitfalls of juxtaposing African forest dwellers and European urbanites, the fieldworkers controlled for the environment by observing both species in tropical rainforests. Since humans in West Africa had abandoned life in the few remaining patches of forest, Boesch compared his Ivorian chimpanzees with Aka in the Central African Republic and Mbendjele in the Republic of Congo. Among these peoples' cultural practices was one they shared with the Taï chimpanzees: they used tools to crack *Coula* and *Panda* nuts. Generalized linear mixed models enabled Boesch to compare the efficiency of chimpanzee, Aka, and Mbendjele nut cracking while taking into account which individuals used which types of hammers and anvils to open the two species of nuts on different days of the season in nut-cracking sessions of varying duration. Of course, such a complex statistical analysis produced complex results. I will just highlight that applying the same efficiency measures—nuts opened per minute and number of hits per nut—to both species revealed that humans and chimpanzees did not differ significantly. In certain respects, the chimpanzees worked even more efficiently: hammers of different weights allowed them to crack nuts with fewer hits. The human foragers, however, could exploit more types of nuts because they used more specialized tools (e.g., axes and bush knives as anvils), which consisted of more parts than the chimpanzees' natural tools (stones and rocks).

Note the language of "more" and "less." Boesch postulated no categorical differences between "man the toolmaker" and animals lacking such technical intelligence, or between humans who had culture and chimpanzees that did not. The differences described in his article were in degree, not in kind. Applying the same quantitative measures and the same statistical tools across species boundaries served as the epistemological correlate of a monist ontology that recognized no qualitative chasm between *Homo sapiens* and other animals (apart from language).

This gradualism set Boesch's cultural primatology apart from philosophical anthropology. From the start, the latter had sought to pin down the specific difference between humans and animals. In Tomasello's work, Habermas, Fischer, and other heirs of this tradition found experimentally supported answers to the question of human nature. Boesch's emphasis on continuity must have been much less appealing to philosophers whose job was first and foremost concept work. Conceptual distinctions are qualitative distinctions: they provide clarity by telling things apart not in a graded, but in a categorical manner. What if primate species indeed turned out to differ more in degree than in kind? Wouldn't that require abandoning the quest for human nature and inventing a new philosophical anthropology—a philosophical anthropology that could do without discerning essential qualities, maybe scattering cognitive capacities once deemed exclusively human across the whole hominoid family?

On the other hand, the breathtaking destruction of biodiversity at the hands of one species, *Homo sapiens,* is currently earning us an exceptional place in natural history. It has brought back the Enlightenment question of anthropological difference with a vengeance: How did humans rather than chimpanzees or baboons come to transform the face of the Earth in such dramatic fashion? Wouldn't answering this question require the restoration of an anthropology that accounted for the unequaled place of modern humans in the animal kingdom? An anthropology that enabled direct and quantitative comparisons between *Homo sapiens* and other primates but that will eventually return to a qualitative account, narrating and conceptualizing whatever historical break might have set us on this peculiar course? An anthropology that explained how differences in degree had been transformed into differences in kind and that forged new concepts to capture these qualitative differences?[185] Such an equally empirical and philosophical anthropology would rearticulate the eighteenth-century question of human nature as the pressing twenty-first-century question of how we have transmuted into what primatologist Shirley Strum called a superdominant species.[186]

5

Japanese Syntheses

PHILOSOPHER DOMINIQUE LESTEL noted that the question of animal cultures *within* human cultures remained the blind spot of primatological studies of culture.[1] While ethologists like Boesch focused on animal cultures outside human cultures, comparative psychologists like Tomasello studied how captive animals behaved in a human-controlled space to assess the capacity for cultural cognition of the species in general. Yet neither faction of the chimpanzee culture wars studied what Lestel called the formation of "hybrid communities," which bring together primates and primatologists.[2]

In Leipzig, I could catch only glimpses of what Sue Savage-Rumbaugh and colleagues might have thought of as the *Pan/Homo* culture of Tomasello's laboratory.[3] While European and American psychologists studying chimpanzee cognition kept their doors shut, Tetsuro Matsuzawa immediately invited me in 2015 to visit the Kyoto University Primate Research Institute (KUPRI). In Japan, behavioral scientists did not have to worry about animal rights activism, and the doyen of early twenty-first-century Japanese primatology set himself apart from the Kyoto School in that he aggressively internationalized the field. A steady stream of French interns, Portuguese and Chinese doctoral students, Indonesian professors, American science writers, and Brazilian and German anthropologists flowed through Matsuzawa's laboratory, located in a small town some 150 kilometers from Kyoto. Many chimpanzees, especially the world-famous Ai, after whom the primatologist had named his entire research program, had grown so used to the presence of visitors that the apes interrupted their work on experimental tasks only to pose for a quick photo with a guest.

While comparative psychologists at the Max Planck Institute had to go to the other end of town for their experiments at the zoo, the KUPRI chimpanzees lived in a large outdoor enclosure right outside the researchers' offices. If turmoil broke out, the primatologists often interrupted their work to rush to a gallery from where they could see what their test subjects were up to. When Matsuzawa stepped outside he frequently impressed visitors by greeting the chimpanzees with a pant-hoot to which the animals would respond in kind. With vast amounts of state funding he had created an unusual space of human-chimpanzee cohabitation very different from Pongoland and Taï Forest. Whether or not the two species had actually formed a common culture, they definitely shared a life.

Matsuzawa conceived of his work as a fusion of observation and experiment: "My research approach has focused on synthesizing these two different approaches."[4] So-called participation observation in the laboratory was one practice, which integrated both methodologies. The other one was field experimentation in the wild. In Bossou, Guinea, Matsuzawa had developed it into the second column of his research program (we will turn to it in the next chapter). It was the holistic philosophy inspiring this methodological integration of laboratory and field research, rather than Kinji Imanishi's anti-Darwinian evolutionary theory, that Matsuzawa proudly presented as what was Japanese about Japanese primatology. If Andrew Whiten was right that the chimpanzee culture controversy was fueled by the use of different methods, then Matsuzawa's Japanese synthesis offered an opportunity to resolve some of the most vexing disagreements. After Leipzig, the Orient promised to be a place of pacification.

From Kyoto to the Nihon Rhine

Kinji Imanishi had used his contacts with friends in high places to institutionalize primatology in Japan. Although the Kyoto School, which he had founded, originated from the main campus of Kyoto University, the Japan Monkey Centre and the Kyoto University Primate Research Institute were built in Inuyama. This town today is home to seventy-four thousand residents and is twenty-five minutes by train from Nagoya. It was this train line that elevated Inuyama to one of the world's most important places for primate studies. In the mid-1950s, when there was hardly any public funding for research on monkeys and apes, Imanishi persuaded Motoo Tsuchikawa, a fellow graduate of Kyoto University and vice president of Meitetsu, that this private railroad company could profit from opening a monkey park and a primate zoo. These tourist attractions would incentivize visitors to buy tickets for Meitetsu's trains to Inuyama.[5]

Even before this local development initiative, Inuyama had already drawn day-trippers from Nagoya and other cities. A samurai castle from the fifteenth century, supposedly the oldest in Japan, offered panoramic views of the Kiso River. In the early twentieth century, the "Japanese Rhine" and the nearby "Japanese Alps" had become popular among painters and mountaineers who followed the European Romantics as they embarked on national self-discovery through artistic and recreational engagements with the landscape.[6] Meitetsu realized and systematically developed the touristic potential of the place. It built an open-air museum for historic architecture from the Meiji Era and the Little World Museum of Man, an anthropological amusement park where visitors could wear traditional ethnic costumes from over twenty-two countries. Before it began to lose most of its visitors to Tokyo Disneyland in the mid-1980s, the Japan Monkey Park attracted up to one million people per year. After riding a roller coaster and a Ferris wheel overlooking the Kiso River, visitors could cross the parking lot to the Japan Monkey Centre, which maintained a larger collection of nonhuman primate species than any other zoo in the world.

The exhibit opened in 1956. At the same time, Imanishi established an onsite research center where he could place his students. "When Itani and Kawai got their first jobs," Matsuzawa told me, "they were employees of Meitetsu." Funded by the railway company but under tutelage of the Ministry of Education, Science, and Culture, the Japan Monkey Centre began in 1957 to publish *Primates*, the oldest English-language primatology journal. Soon Imanishi mobilized another friend, physicist and Nobel Prize laureate Shinichiro Tomonaga, who lobbied Prime Minister Hayato Ikeda—who, like Tomonaga and Imanishi, was an alumnus of Kyoto University—to support the creation of a purely academic center for primate studies. Considering the already available resources, Inuyama appeared the ideal location. In 1967, Kyoto University launched its Primate Research Institute at the Nihon Rhine.

KUPRI was more than an outpost of the Kyoto School, though. Imanishi had brought on board Toshihiko Tokizane, a neurophysiologist from the University of Tokyo. From the start, the new research facility housed both biosociological fieldwork and invasive laboratory experiments on monkeys. Imanishi disdained "people in white smocks" who "have never once been out of the laboratory."[7] He preferred researchers who studied "nature in its entirety, with flowers blooming, butterflies fluttering, and birds singing." But he accepted that the institute's first two directors would come from Tokyo because, as Matsuzawa put it, "they had the money and the power." Only in 1978 did Masao Kawai take over.

As director of KUPRI, Kawai reportedly suffered from a sense of inferiority toward Junichiro Itani, whom Imanishi had made his successor in Kyoto. But the institute's geographical distance from the main campus and the exchange with Tokyo-trained experimental scientists soon endowed Inuyama with some intellectual autonomy, as KUPRI-trained primatologist and science studies scholar Osamu Sakura told me.[8]

When Matsuzawa got his first job at the Primate Research Institute in 1976, he arrived with a degree in psychology from Kyoto University. At KUPRI, the psychology department was the poorest and least accomplished—"a colony of neuroscience," as Matsuzawa remembered. He neither came from the Tokyo lineage nor belonged to the Kyoto School of primatology. Emphasizing his own independence, he described Kawai as a "miniature-scale Imanishi." If Japanese primatology had fallen behind American, English, and German primatology, he explained to me, it was because every generation was smaller than the previous one. Although Matsuzawa occasionally did portray himself as a "spiritual descendant of Imanishi and Itani"—possibly to underline his claim to leadership—he understood himself less as a follower than a founder.[9]

Japanese Primatology after Imanishi

Today, Matsuzawa is known internationally for his synthesis of laboratory and field research. But he was trained as an experimentalist. How to get by in the wild he learned from climbing eight-thousander peaks as a passionate mountaineer. In the 1970s and 1980s, he took part in four Himalayan expeditions with the goal of reaching a summit that nobody else had climbed before. It was not as a primatologist, but as the fourteenth president of the Academic Alpine Club of Kyoto that Matsuzawa considered himself an heir to Imanishi, who had cofounded the organization and who had written popular books about mountain climbing. Advocating an "isomorphism of mountaineering and science," Matsuzawa sought to remain true to Imanishi's spirit of first ascent by deliberately not going where his predecessor had already been: "You have to go to places where no one has ever reached before. You have to think of things that no one has ever thought before. You have to find the route that no one has ever tried before."[10]

Matsuzawa set himself apart from the Kyoto School by shifting the focus from the evolution of primate societies to that of primate minds. In this endeavor, the most prominent points of reference were American and European authors, ranging from Gustav Theodor Fechner and Noam Chomsky to David Premack and Jane Goodall. Nevertheless Matsuzawa thought of his project as Japanese primatology because of what he described as its holistic or synthetic spirit. In an interview, he characterized the "Japanese way of thinking" thus: "We don't like to segregate, we want to assemble things."

Matsuzawa's hybridization of Japanese and Euro-American primatology did not replace the Kyoto School, which continues to be a strong force in the Japanese field—even though some of its representatives complained about their international marginalization.[11] Yet the concentration of disciplinary power within and beyond Japan in the hands of Matsuzawa—in 2016 he was the president of the International Primatological Society, general director of the Japan Monkey Centre, coordinator of the international graduate program

Primatology and Wildlife Science, and editor in chief of the journal *Primates*—suggests that his project changed the landscape and opened a genuinely new chapter in the history of Japanese primatology after Imanishi.

Despite Matsuzawa's desire to distinguish himself, he also used every opportunity to situate himself and others in lineages. One day after lunch, he took me to a small museum next to the canteen, which displayed memorabilia from the Imanishi era. Being Japanese, he explained, means thinking in family continuities—and he began to recite the Japanese emperors who all belonged to the same house. The samurai had been ready to commit hara-kiri because they knew that the son would carry on the line. In this spirit, Matsuzawa assured me that he did not care as much about himself as about the next generation, which would continue his work—work that had not just broken new ground but also kept up a tradition.

While Matsuzawa frequently emphasized that he was no product of the Kyoto School of primatology, he considered the famous founder of the Kyoto School of *philosophy* his "spiritual great-grandfather." Kiyoko Murofushi, head of the Section of Psychology at KUPRI, who supervised the twenty-six-year-old assistant professor Matsuzawa's research, had trained under psychologist Taro Sonohara, who in turn had been a student of Kitaro Nishida's. Nishida had been a friend of Imanishi's and the founder of a philosophical tradition that used Western philosophy to rearticulate East Asian and especially Mahayana Buddhist thought. Matsuzawa had started out as a philosophy major but soon realized that he preferred the outdoors to the conceptual thicket of philosophical writings. Bored after only a few pages, he had put aside Nishida's famous book *An Inquiry into the Good*, never to look at it again.[12] But mediated by the lineage of Sonohara and Murofushi, Matsuzawa believed, he had inherited Nishida's holistic way of thinking—not on an ontological but on an epistemological level. He never theorized a transcendence of the subjective and the objective, nor did he postulate any fundamental indecomposability of social interactions. Instead he practiced his own understanding of holism by synthesizing benchwork and fieldwork. Matsuzawa dubbed this endeavor the Ai Project.

A Talking Ape: "Is there anything new?"

The Ai Project began in 1976. It was named after a freshly captured one-year-old chimpanzee infant. The Primate Research Institute had purchased her from a West African animal dealer four years before Japan's ratification of the Convention on International Trade in Endangered Species of Wild Fauna and Flora (CITES) prohibited further trading of wild apes. Ai became the foremost test subject of the first and last Japanese ape language project.

As its principal investigator, Kiyoko Murofushi asked three young assistant professors, Toshio Asano, Shozo Kojima, and Tetsuro Matsuzawa, to explore the linguistic capacities of chimpanzees. A quarter century after the Hayeses had taught their home-raised chimpanzee Viki to utter "papa," "mama," and "cup," and a decade after the Gardners had begun to teach Washoe American Sign Language, the general idea was derived from the American ape language projects.[13] In the mid-1970s, they were still in full swing—although Herbert Terrace's disillusionment with Project Nim would soon throw the field into controversy.[14] At KUPRI, the neuroscientists supported the psychologists' project because they wanted to study the brain processes underlying language acquisition in apes. Since Asano and Kojima were on sabbaticals in the United States when the research began, it was left mostly to the twenty-six-year-old Matsuzawa, who had been hired that very year, to train and take care of Ai and two more chimpanzee infants, Akira and Mari.

In the course of the 1980s, Ai learned twenty-six Roman letters and twenty-five signs of an artificial language called the Kyoto University Lexigram System (instead of Japanese kanji because the latter would have required too many pixels to be displayed on the computer screens then available). This native of Guinea also used Arabic numerals to count up to six and to state how many objects of a particular kind and color she saw.[15] However, when Matsuzawa spent his own sabbatical at the laboratory of American psychologist David Premack, the latter showed himself unimpressed by Ai's cosmopolitan language skills: "The important point is that *you* taught the chimpanzee to do these things and the chimpanzee learned them," Matsuzawa remembered his recently deceased mentor saying in an interview in 2015. "Is there anything strange, anything new? I don't think so."

Premack must have felt that the young Japanese professor's first *Nature* publication could not teach him anything he did not know about chimpanzees because Matsuzawa had not given them the chance to be "their own agents," as Premack and his coworkers would put it elsewhere.[16] His critique aimed at what he took to be a behaviorist research design, which conditioned animals instead of affording them an opportunity to speak for themselves. In the language lessons of their own chimpanzee Sarah, David and Ann Premack attended to whether her learning was confined to training examples, or whether she could apply it to new cases.[17] Unless a behavior had fully undergone the transition that freed it of environmental control, they argued, it had not become a completely spontaneous act that the animal could exercise at will, just as humans did.

Whether apes could be taught to communicate in a human language, however, did not really concern Matsuzawa. As a reader of German Estonian biologist Jakob von Uexküll and primed by research he had done as a student on stereoscopic vision in rats, he found his attention caught by a different question: "How do chimpanzees perceive this world? Do they perceive it like we do?"[18] The original Ai Project used ape language research to address this question by way of psychophysics.

When Theodor Fechner coined the term *psychophysics* in 1860, he described the object of this new field of study as the relationship between mind and matter, body and soul. While the natural sciences studied matter by way of outside observation and psychology studied the mind through introspection, the *relationship* of mind and matter remained a matter of philosophical speculation.[19] Fechner sought to capture this abstract object, not given to immediate experience, by developing mathematical formulas to describe how a physical stimulus and the resulting sensation, as experienced by a test subject, correlated with each other.

Human participants in psychophysics experiments usually became active when asked to express their sensations in either words or numbers. Using psychophysics to understand how other animals perceived the world, however, turned out to be a problem because they had no means to share their experience of a given stimulus with human researchers. At this point ape language research came in handy. If Matsuzawa could teach his chimpanzee subjects to use a language-like medium to communicate with him, they could provide access to their inner experience as they verbalized the kind of introspections that Fechnerian psychophysics required.

By the 1970s, American ape language research offered two alternative approaches. Viki's sorry abilities to talk to the Hayeses had taught primatologists that the chimpanzee vocal tract did not lend itself to speech. The Japanese psychologists also decided against the Gardners' approach of teaching their chimpanzee infants American Sign Language because the gesturing between humans and apes offered ample opportunity for social cuing. To make ape language research more objective, Duane Rumbaugh and his wife, Sue Savage-Rumbaugh, had just introduced some of the first computer technology into the field. They trained their test subject Lana to match artificial symbols—so-called lexigrams—on a computer keyboard to various stimuli like objects or colors. As they presented these samples, Lana could not see the human experimenters and therefore could not receive any cues from them. A computer recorded her responses, unpolluted by the kind of human interpretation required to decipher a gesture.[20] Thus the Rumbaughs implemented mechanical objectivity to reduce human interference in chimpanzee behavior. This was two decades before Savage-Rumbaugh's methodological change of heart. To study the cognitive capacities of great apes she switched from an experimental to an ethnographic approach.[21]

At the Primate Research Institute, Imanishi's student Masao Kawai advocated an empathetic understanding of other primates, his so-called feel-one method, or *kyokan*, which he claimed to be "the most striking aspect of Japanese primate studies."[22] The neuroscientists and experimental psychologists at KUPRI, however, were more closely aligned with the medical school in Tokyo than with the Kyoto School. Matsuzawa did not identify with either of these two camps dividing the institute. But in his experimental work he cherished objectivity more than empathy. Moreover, his more experienced coworker Asano had previously used computers for his behaviorist experiments with monkeys. So their group decided against the sign language approach and adapted Rumbaugh's paradigm for the Japanese ape language project.

The Ai Project's first goal was to understand how chimpanzees perceived and categorized colors. Matsuzawa and his colleagues started out with the behaviorist paradigm of discrimination learning. As the test subject sat on her own in a box roughly two meters by two meters, they showed her through a small window a stimulus such as different hues on Munsell color chips, conforming with the Japanese Industrial Standard. In response, the chimpanzee had to match this sample to the correct lexigram on the keyboard in front of her. If she got it right, a chime sounded and an automatic food dispenser provided a raisin or small piece of apple. If she failed to name the color correctly, the machine played an unpleasant sound and the subject came away empty handed. Thereby, the chimpanzees learned to discriminate and name eleven colors.[23]

The behaviorist paradigm based such matching-to-sample tasks on an understanding of the organism as tabula rasa, which researchers could set with any associations they desired. By reinforcing the chimpanzees' use of a particular terminology to designate a given palette of colors while punishing deviations from this human-invented system of symbols, the Japanese experimenters sought to form the animals' behavior in the image of their own species—or culture? Presumably, that was the reason Premack felt that he had not learned anything new when Matsuzawa taught Ai how to apply Arabic numerals to quantities of objects just as he had taught her to name the colors of these objects.

However, the chimpanzees were not putty in the experimenter's hands. When learning to name the eight different objects presented to them, Ai took 57 days while Akira needed 90 days, and Mari struggled for 120 days. Matsuzawa related these very pronounced differences in learning rate to the chimpanzees' personalities: Whereas Ai proved to be a patient student, the young male, Akira, often tried to solve cognitive problems by sheer force, hitting the computer and other experimental equipment as hard as he could. Mari, on the other hand, grimaced and screamed every time she failed in a trial. Looking back at his life's work in his last lecture series at Kyoto University in 2015, Matsuzawa explained with some regret that a teacher should adapt his teaching style to his students. In Mari's case, for example, errorless learning without the unpleasant sounds of the buzzer would have been more appropriate. "But I was a rigorous scientist keeping to the same method," he said.

It was not just because the three chimpanzee infants did not fully comply with their behavioral engineering that it would be inappropriate to consider the human experimenters the only relevant actors. Unlike Asano, Matsuzawa was no behaviorist. "I use the techniques, but I do not rely on the philosophy of Skinner," he maintained. "I don't believe in the idea that every behavior can be explained by operant conditioning. In that sense, I'm closer to ethology and fieldwork." The most important part of the experiment began only once the Skinnerian conditioning was complete. It examined what Matsuzawa took to be a naturally occurring behavior rather than an artifact of the laboratory. Once the chimpanzees had learned to discriminate and name eleven standardized colors, he presented them with color chips they had never seen before. "Now my question was: how about this color?" Matsuzawa explained this second step. "You may say red, pink, orange, or brown. It's completely up to you. In contrast to discrimination tasks, there is no correct or incorrect answer." The key question was not whether chimpanzees could use symbols to designate objects—the ape language projects had provided ample and uncontroversial evidence for that—but how the apes would generalize from what they had learned. Matsuzawa believed that such stimulus generalization, as Skinner had called it in his book on language acquisition, was not a direct product of conditioning but would be determined by innate features of the chimpanzee mind.[24]

From the chimpanzees' categorizations of new colors, Matsuzawa and his colleagues created a map, showing which hues the chimpanzees grouped together. They adopted this approach from anthropologist Brent Berlin and linguist Paul Kay's book *Basic Color Terms*, which at the time was Matsuzawa's bible.[25] In opposition to the linguistic relativism of Edward Sapir's and Benjamin Whorf's earlier work, which had argued that differences between languages caused differences in their users' ways of seeing the world, Berlin and Kay's comparison of color lexicons across languages arrived at a universalist conclusion: although differently developed, the color terminologies of all human groups evolved toward one and the same categorization.[26] Berlin and Kay speculated that species-specific biological structures were not only behind the universality of syntax, as postulated by Noam Chomsky, but could also be found in the realm of semantics.[27]

The Chomskyan emphasis on deep cognitive structures generative of more contingent behaviors left a deep mark on Matsuzawa's thinking. This postulate of innateness contradicted the linguistic relativity proposed by Sapir and Whorf as well as Skinner's assumption that animals acquired their behaviors by way of association learning. Combining the method of operant conditioning with a belief in the universality of the chimpanzee mind, the Ai Project had fallen into the tense space opened by the late-1950s controversy between Skinner and Chomsky over whether mental faculties such as language developed as inborn human capacities or were gradually instilled by rewards and punishments.[28] Matsuzawa explained: "Many people think Ai is a genius. But for me Ai is a representative of chimpanzees, of *all* chimpanzees. I'm very much skewed toward species-specific universals." However, although Matsuzawa stressed the innate over the conditioned, his adherence to Skinner's concept of stimulus generalization prevented him from completely abandoning behaviorist thought and from climbing onto the bandwagon of the Chomskyan revolution.

Matsuzawa's cartography of the chimpanzees' color space showed that it largely corresponded to the map that Berlin and Kay had proposed for humans. This finding confirmed earlier psychophysiological studies suggesting that the trichromatic color perception of chimpanzees resembled our own.[29] Although in accord with Chomsky's emphasis on inborn universals, Matsuzawa's account diverged from the structure of Chomsky's argument in that it provided evidence *against* species specificity: "These results suggest that there is a common basis of color classification not only across human cultures but also across primate family lines, Hominidae and Pongidae."[30]

Matsuzawa described the approach underlying this early work as the prototype of the Ai Project: a single subject conducting matching-to-sample or discrimination tasks in automated experiments—"and they have to answer my questions." Forcing the subject to choose among two or more sample stimuli and to match them to a so-called comparison stimulus, these discrimination tasks asked which stimuli corresponded to each other and conditioned the animal to pick the right one. Only the subsequent stimulus generalization experiments provided opportunity for spontaneity on the chimpanzee's part. In Matsuzawa's eyes, this did not indicate individual agency, but the spontaneity of innate propensities—what ethologists used to call *instincts*—shared by the entire species.

At first glance, the Ai Project could serve as a paradigm case for Lestel's project of studying hybrid communities, in which "apes are genuinely integrated in a human culture."[31] When I visited the Primate Research Institute, Gabriela Bezerra de Melo Daly, a Brazilian doctoral student supervised by

Lestel and French anthropologist Philippe Descola, was studying human-chimpanzee interactions in Matsuzawa's laboratory from exactly that perspective.[32] But such integration did not necessarily conform with the primatologists' notion of culture as the product of social learning. While Washoe, Nim, and other apes who had learned to communicate in American Sign Language had acquired those gestures by copying their human teachers, Ai had been deliberately "isolated from a social context and worked with a computer" to learn by trial and error how to use lexigrams.[33] In the language of comparative psychology, it was individual rather than social learning that had endowed her with linguistic skills—even if she adopted a system of symbols that humans had devised. Lestel, however, did not follow the primatologists' "minimalist" conception of culture as social learning but defined it as a society of subjects who gave different meanings to the world and acted in accordance with reasons rather than being determined by genetic and environmental causes: "The natural history of culture is first and foremost a natural history of subjectivity, signification, and freedom."[34] But how free were the captive apes in Matsuzawa's laboratory?

Freedom and Control

At the Primate Research Institute, the chimpanzees lived in a densely planted outdoor enclosure and two giant outdoor cages connected to indoor rooms where they spent the night. Because of tensions between two males, Akira and Gon, they had been split into two groups. Their wild conspecifics lived in fission-fusion societies that broke up into parties of varying size and composition (for example, for foraging), but eventually they would come together again as one community (for example, for sleeping). Since the veterinarians worried about injuries that could result from violent conflicts between the rivalrous males, the primatologists had socially engineered a similar way of life by allowing the females to go back and forth between the two groups and the three habitats. The males had to stay put, however. Thus, the groups had become something in between temporarily divided parties and permanently divided communities. In the wild, communities did not exchange females on a regular basis but only when they reached sexual maturity. Matsuzawa would have preferred to reunite the groups of Akira and Gon—even if this came at the cost of a few bitten-off fingers: "Fighting is part of chimpanzee nature," he explained to me. But since the veterinarians were in charge of animal welfare, he sought to make the best of this artificial demography. It still appeared more natural than keeping a community of captive chimpanzees like a baboon troop where everybody was always together, as most zoos did.

FIGURE 11. Outdoor enclosure at the Kyoto University Primate Research Institute, Inuyama, Japan. Photo by author.

The keepers could control traffic between groups and spaces through a system of passageways, waiting rooms, hatches, and locks. Within these constraints (and they were significant constraints), the chimpanzees decided where to go. Getting them to the places where humans wanted them to be took much time and effort. Every morning caretakers and researchers spent up to one and a half hours enticing particular individuals to come to the laboratories for which they had been scheduled. A good deal of my conversation with Matsuzawa—otherwise an extraordinarily busy man—happened in the basement of KUPRI, where we sat on children's chairs to appear smaller and less threatening to the reluctant test subjects.

Some chimpanzees did not come to the laboratories at all, presumably because they felt too uncomfortable when isolated from their community. Others, especially males, showed up only occasionally because they prioritized competing over estrous females and rising through the ranks. Even when they participated in experiments, they often appeared distracted. Whereas field primatologists employed sampling techniques to avoid paying undue attention to the flamboyant males, the experimental psychologists at KUPRI tested more females because they proved to be more disciplined test subjects.

The animals' main motivation for participating in experiments was food. They received three regular meals a day and could feed another four times, if they joined the test sessions in the morning and afternoon. Again Matsuzawa sought to model their laboratory life on life in the forest, where chimpanzees engaged in six to seven feeding bouts of thirty to forty minutes per day. Visiting different KUPRI labs—apart from Matsuzawa's there were also those of Masaki Tomonaga, Ikuma Hadachi, and Yuko Hattori—was like visiting different fruit trees. Even the quantity and size of food rewards—an apple cut into three hundred pieces with edges eight millimeters long—mimicked the fruits of *Antiaris africana*, one of the Bossou chimpanzees' favorite foods.

But extra calories were not the only incentive. In the sultry heat of the Inuyama summer, the cool of the institute's basement and the air conditioners turned on by some faculty provided additional reasons for coming to their laboratories, but not necessarily for working on any cognitive tasks. Frequently, chimpanzees would lie down in front of the computer, cross their legs, stare into space, and enjoy a break from the tropical outside temperatures while the researchers were working hard to get them to generate data.

Other chimpanzees were said to come for human company, especially if they could expect to meet people they liked or who offered special treats such as frozen Jell-O shots. Of course, sympathies could also go the other way. In an interview, Oxford zoologist Dora Biro, who had studied with Matsuzawa, remembered: "At the Primate Research Institute, you have to be friends with the chimps. If you did something nasty to them, or you didn't give them the right food, they won't come to your experiment."[35] Researchers never had to quit their job because the chimpanzees did not like them, Matsuzawa's coworker Misato Hayashi told me, but their data would be poorer and they usually had to leave the institute upon completion of their degree.

Although the chimpanzees could decide whether and when to go down to the labs, they could not decide when to exit. Once they were in, a gate closed behind them and only a human could open it again. This disciplinary measure improved their compliance with experimental protocols.

Yet Matsuzawa repeatedly expressed his regret about keeping Ai and her conspecifics under lock and key: "I feel terribly sorry about the chimpanzees here because I know the wild chimpanzees. I don't like any captive situation, including this one, which is one of the best. The size of their habitat at the institute is one thousandth of their range in the wild where they would be free to roam all the way to Inuyama Castle. They cannot go to [the department store] Ito Yokado for shopping, they cannot take a walk when they want. I can only maximize their freedom to eat, to move, and to interact in their social group."

In the face of his fast approaching retirement, Matsuzawa's last major effort to give greater liberties to the captive chimpanzees under his tutelage was to build the WISH cage (the acronym stood for Web for the Integrated Studies of the Human Mind). This giant steel construction was twenty meters long, twenty meters wide, and fifteen meters high—as high as Japanese building standards allowed. Matsuzawa emphasized the importance of providing vertical space to animals that, in the wild, would spend more than half of their days in treetops. He had won a large grant from the Ministry of Education, Science, and Culture for a project comparing the social cognition of chimpanzees, bonobos, and humans, which had allowed him to build such WISH cages for $5 million each at the Primate Research Institute in Inuyama and at a primate sanctuary in Kumamoto.

When the Inuyama cage opened to the chimpanzees later in 2015, it not only offered a third habitat to emulate a fission-fusion lifestyle in captivity but also contained small computer-operated walk-in booths. Each contained workplaces for up to three animals to do cognitive tasks at touchscreen monitors and thus to feed whenever they felt like it. They would no longer have to wait until human employees came to the institute at 9:00 a.m. but could—just like their wild conspecifics—find their first food as soon as they got up at daybreak. Video cameras and a face recognition software adapted to chimpanzee physiognomy would identify and grant access to individual animals. Keeping the experiments controlled, a computer system would present them with the trials they had been enrolled for, and an automated food dispenser would

FIGURE 12. Cognitive test battery: chimpanzee working at a touchscreen. Photo by author.

deliver the prepared food rewards. If test subjects wanted to leave, though, they were free to go at any time and to resume working on their test battery at a later point.[36]

Matsuzawa compared this new approach to their field experiments in Bossou where the chimpanzees also lived in their own community and visited the outdoor laboratory as they pleased. In the Japanese high-tech lab, computers were meant to give the captive animals a little bit of the freedom that their Guinean cousins enjoyed. The chimpanzees would no longer depend on human caretakers to allow them in and out of the experimental spaces where they fed. They could decide on their own. Referring to a bright annular icon appearing on the touchscreen, which the chimpanzees pressed when they wanted to start a new trial, Matsuzawa noted: "The white circle is the symbol of freedom here."

Unsolicited Behaviors: Field Observations in the Laboratory

The Ai Project exceeded a series of controlled experiments. It went along with forms of fieldwork, constantly situating experimental results in broader contexts. Around 1980, before tighter and tighter regulations precluded such carefree excursions, Matsuzawa took Ai for walks outside the institute. He brought colored objects as well as the corresponding lexigram symbols and occasionally asked Ai to name the color of a woodblock or spoon to see whether she generalized from her laboratory experience to situations of daily life. One day as they sat in an orchard and Matsuzawa was writing down the results of such an informal test in his notebook, Ai picked a dandelion and handed him the lexigram for yellow. "It was her who spontaneously used her vocabulary to describe the colors of her world to me," he remembered.

For several reasons, this and many episodes like it never made it into Matsuzawa's scientific articles. First, it is in the nature of spontaneous behavior that it occurs suddenly and without apparent external cause, which makes its methodical investigation difficult. But if reports were limited to opportunistically collected singular events, scientific journals tended to dismiss them as anecdotal. Even in the field sciences, the standard of repeatability that so profoundly informed the epistemology of the laboratory sciences had come to determine what counted as genuinely scientific. Matsuzawa did publish these stories in his popular Japanese books but felt that he was not fluent enough to write monographs in English. Moreover, he wanted to set himself apart from American ape language researchers who had put much weight on such episodes because demonstrating the ability of spontaneous language use was key to showing that chimpanzees were capable of linguistic communication, understood as a creative rather than a conditioned combination and recombination of signs.

Such field observations were not confined to the outdoors. Matsuzawa also conducted them within the Primate Research Institute. Although his publications presented mostly experimental results proper, he systematically collected data on factors other than the displayed stimuli, which affected the chimpanzees' behavior in the trial. "I'm doing my fieldwork here in the laboratory," he explained. "This is the chimpanzees' semi-natural or artificial habitat."

The fully automated experiments allowed him to carefully observe and take notes in a field diary of sorts on what the animals did in the laboratory setting while a computer system recorded their test performance. From the beginning of the Ai Project, film recordings provided these field observations in the laboratory with an objective basis. When I visited KUPRI in 2015, three video cameras captured every single trial from different angles, generating an enormous archive of contextual information that could be used to interpret the quantitative data from the tests.

One example of how such field observations offered additional insights came from Matsuzawa's most famous experiment, demonstrating the cognitive superiority of a young chimpanzee over human subjects in a test of visual short-term memory. After learning to discriminate and order Arabic numerals from 1 to 9, a touchscreen presented chimpanzees with up to five numbers, which they had to tap in ascending order. In this so-called limited-hold memory task, the numbers appeared for only a blink of an eye—often a mere 210 milliseconds—before being masked by white squares. While human subjects regularly failed to memorize the numbers in their spatial distribution in such a short time, Ai's five-year-old son, Ayumu, excelled at remembering them and outperformed all nine university students serving as controls.[37]

A paper by two American psychologists, Alan Silberberg and David Kearns, claimed that with adequate practice university students had been able to outperform the chimpanzee, but Matsuzawa dismissed it as "misleading": "Some people really hate that chimpanzees can do better than us."[38] After this challenge, Matsuzawa conducted additional experiments, further reducing the presentation time to 150 milliseconds. Ayumu maintained his performance, he claimed, while the performance of human test subjects deteriorated. When I asked Matsuzawa why he had never published these results, he responded: "We don't like to fight people who have such strong prejudices." Whatever the truth was about human and chimpanzee visual short-term memory, Matsuzawa exhibited a very different ethos than Boesch and Tomasello.

There was another finding Matsuzawa had not published. Located in the basement of KUPRI, his laboratory rooms were not soundproof. As in the wild where chimpanzees often forage out of sight of each other, the community, which had grown to fourteen individuals altogether, communicated through long-distance calls, so-called pant-hoots, while participating in different experiments. On one occasion, Ayumu had begun tapping the masked numbers on the touchscreen in front of him when—unforeseen by the experimental protocol—some brouhaha in the group distracted him. As a young male, he got passionate about chimpanzee politics. For a moment—subsequent analysis of the video footage revealed that it was for exactly ten seconds—he turned around and listened intently. Then he looked at the screen again and touched the remaining masked numbers in the right order. Matsuzawa concluded: "It is very clear that there is a photographic memory in chimpanzees, which lasts for at least 10 seconds."

The fact that this episode had been filmed and that other researchers who had not witnessed it themselves could look at it again and again, even though it happened only once, raised its epistemological value beyond that of mere anecdote. Matsuzawa gave Ayumu's shift of attention from procuring food by solving cognitive tasks to obtaining valuable information about the social dynamics of the group an ethological interpretation: as is typical of chimpanzees of his sex and age, he responded more strongly to external stimuli enabling him to work his way up the dominance rank hierarchy.

By contrast, the last instance of a field observation in the laboratory I wish to present confirmed Matsuzawa's conviction that chimpanzees were not mere automatons translating stimuli into responses, as both behaviorists and early ethologists had imagined animals to be. It occurred during my own fieldwork at the Primate Research Institute. When the chimpanzees came to the laboratory for thirty- to forty-five-minute sessions, different tasks awaited them. To avoid Clever Hans effects, the researchers could not see the chimpanzee touchscreen, but the monitor of the laboratory computer mirrored the presentation of the stimuli. Every time an animal completed a task, a researcher went to this computer to enter new parameters for the next task. Until a few weeks before my arrival, the chimpanzee screens blacked out during this time. But when a doctoral student set up a new monitor system, she forgot to include the switch that decoupled the chimpanzees' screens from the humans' screen. Now, every time Ai and her conspecifics completed a task, they saw the Windows desktop with a dialog box appear. Well versed in her experiments, Matsuzawa's student always changed the parameters within seconds. But when she could not come to the laboratory one day, Matsuzawa adjusted the settings himself. Unfamiliar with this routine, he was still double-checking his entries when Ai impatiently pressed the button with the Japanese kanji for "Start"—and the experiment began. "The first time, I thought it was by accident, but the next day she did it again," Matsuzawa recounted a few weeks later. At that point, he always told Ai *not* to touch the screen before he was done. He had grown convinced that "she has meta-knowledge of the system. She knows that, if she presses this button, the experiment starts. And she knows the right timing. She is waiting until I have entered all the necessary information."

Henrika Kuklick and Robert Kohler have pointed out that the uncontrollability of the field sets it apart from the laboratory.[39] However, many historians and anthropologists of science, including me, as well as chimpanzee ethnographers like Boesch, have argued that in reality even laboratories are not as controlled as the ideal-typical conception of experiments carving out the effects of independent on dependent variables would require.[40] "You can often learn more about how chimpanzees think and see the world when experiments do not work out as planned," a field primatologist from Britain who was visiting Matsuzawa's lab told me. Although Matsuzawa's style of experimentation aimed at making the animals provide direct answers to confined questions, his view of the laboratory as field provided space for the observation of spontaneous behaviors that went beyond what the experimenter had asked.

While most primatologists emphasized the conservative dimension of culture—the ability to copy and traditionalize—Lestel foregrounded its creative side, especially in hybrid cultures of humans and other animals.[41] But creativity was not what cultural primatologists found most impressive about nonhuman cultures. McGrew noted: "Innovation is not rare in wild primates but inventions that catch on are. For every behavioural pattern that emerges and is passed on to others, scores probably never spread beyond the inventor, and thus fade away."[42] This was also true for Ai's new behavior: however ingenious and unpredictable her breach of experimental protocol might have been, she had learned it on her own and nobody in the laboratory thought of this behavior as cultural.

Participation Observation

Matsuzawa's Ai Project intercalated experimentation, field observation, and a peculiar version of participant observation—a term but not a practice he had borrowed from cultural anthropology.

In the course of the twentieth century, participant observation had become almost synonymous with ethnographic fieldwork.[43] Although Bronislaw Malinowski and Margaret Mead have often been credited, they did not invent the immersion in unfamiliar forms of human life as a research method, nor did they come up with the expression *participant observation*.[44] It was American educator and social researcher Eduard Lindeman who first described the new role of "participant observers" who "are participating in the activities of the group being observed."[45] Their goal was not just to describe what the group was doing but also to understand what the group thought it was doing. Malinowski and Mead laid out the principles of participant observation in ethnographic fieldwork and successfully promoted a practice that anthropologists had developed at least since the 1870s to understand other cultures from the point of view of their participants.[46]

Many field primatologists harbored a general sense of participating in the social life of the primate group: they grew attached to the observed animals, buried individuals that had died (although not only for sentimental reasons but also to be able to exhume the bones at a later time, if needed), and developed from such immersion an empathetic understanding of a very different form of life.[47] By 2015, however, primatologists studying wild apes usually made every effort to stay out of the social lives of their subjects—not only because any intervention potentially altered the observed behavior but also because becoming involved in often violent chimpanzee politics could easily threaten a researcher's life. Instead they based their close-up observations on habituation. Ideally, the animals became so accustomed to the presence of human observers that they behaved as if the at times strenuously disengaged researchers did not exist: "The best situation is when the chimpanzees don't pay attention to me," said Matsuzawa about his fieldwork in Guinea. In contrast to laboratory experiments and participant observation, the goal of unmanipulated field observations was not to interact but to capture animal behavior that had not been elicited by humans.

However, the original setting of Matsuzawa's scientific work could not have been further from the naturalistic research sites at which Jane Goodall, Junichiro Itani, William McGrew, and Christophe Boesch had come to know chimpanzees. The Ai Project had departed from the North American ape language projects, which attempted to socialize nonhuman primates into human language use, sometimes by raising these hominoid infants like human children in human families.[48] In 1983 and 1984, Matsuzawa briefly cross-fostered the newborn Pan with his own daughter—until he decided to hand Pan over to KUPRI staff to go on a mountaineering expedition in the Himalayas. Even though Ai and the other chimpanzees at the institute had grown up in the laboratory rather than in his home, Matsuzawa engaged much more with them than he or any other primatologist did with research subjects in the field. It might be difficult to work with an orphaned one-year-old chimpanzee without being solicited as an attachment figure and a playmate. From the start, Matsuzawa's experimental work had been accompanied by some sort of participant observation.

Direct interactions in a shared space presupposed an intimate familiarity between researchers and apes, which Matsuzawa did not find in European and American laboratories. They maintained a strict division of labor between scientists and caretakers, or, as Matsuzawa put it: "The researchers don't take care and the caretakers don't do research." Although KUPRI keepers did not conduct research either, doctoral students, postdocs, and even junior faculty spent a significant amount of time on feeding, cleaning cages, and other menial jobs that did not produce scientific data. "I love people who make an intensive effort to become the friends of chimpanzees," Matsuzawa told me. One of these people, his former student Satoshi Hirata, spelled out the rationale: "A division of labor in terms of research and husbandry may appear to be more efficient, but it is our contention that the researchers increase their knowledge and understanding of the chimpanzees through involvement in the daily care of the animals."[49]

Matsuzawa considered daily feeding for at least six months an essential step toward building a "good relationship" with animals they considered "research partners." He could enter an experimental booth with Ai and even with the group's alpha male, Akira, and stick needles into their arms. They would patiently watch Matsuzawa as he injected them with vaccines or anesthetics. No handling cage restrained the animals and no food rewards conditioned their compliance. Without such mutual trust between primates and primatologists they would not be able to safely go into a room together.

Participant observation required a leap of faith. After months of daily feeding and doing overtime to groom a potential research partner through the mesh, the moment of truth arrived. "One day, you're suddenly convinced that you're safe," Matsuzawa explained. "It's not the chimpanzee, but you who decides. If you're afraid of him, he will be afraid of you, too. No authority can tell you that you may now go into the booth. It's not an objective judgment. When you feel you will be okay, you open the door, and—no problem."

Most zoos prohibited such direct interactions, Matsuzawa noted, because they could be potentially dangerous. "'Potentially' means 'in general,'" he added. "But this is not about generalities. It's about the specific case. You can become a unique person who builds a special relationship. Once you're determined to make the effort, I'm 100% sure that you can become a friend of chimpanzees and you can be with them."

Once participant observers took up face-to-face interactions, they constantly had to assert their dominance. "When I play with Ai," Matsuzawa explained, "you will notice that I'm quite rough. I slap her back and push her. I also softly groom her. It's an intimate relationship in which I consciously show 'I'm dominant to you.' If you can do it, the chimpanzee will easily follow. That is chimpanzee nature."

Subordinating a high-ranking individual also improved his status vis-à-vis those below her. "When he was young, Ayumu always saw Matsuzawa treating his mother like this," Matsuzawa recounted. "He learned: 'This is a close friend and he is dominant to my mother. So I better behave myself in front of this guy.'"

Cultivating such relationships served as the basis for what Matsuzawa called "participant observation" or "participation observation."[50] While he often used these terms in an extended sense to describe any direct physical interaction between humans and chimpanzees that wasn't constrained by bars,

FIGURE 13. Participant observation: Matsuzawa playing with Ai. Photo by author.

he reserved a narrow definition of participation observation for studies of the sociocognitive development of mother-infant pairs in their social context. Matsuzawa was especially interested in "the cultural transmission of chimpanzee knowledge and skills from one generation to the next. When, what, from whom to whom, and how are knowledge and skills passed on?"[51]

While many ape language projects had focused on isolated individuals and many laboratories housed motley collections of chimpanzees of different origins, Matsuzawa hoped to grow a multigenerational community of related animals analogous to those he observed in the wild. Because many captive chimpanzees did not procreate spontaneously even when females were taken off contraceptives, Matsuzawa reverted to artificial insemination to re-create a primate society based on blood relations. He had no qualms about increasing the number of chimpanzees that had to live in captivity: "We are not ultra-radical animal rights extremists. We believe that having a baby is the best enrichment for female chimpanzees." Since the purchase of Ai, Akira, and Mari as the three original subjects of the Japanese ape language project in the late 1970s, a second generation of chimpanzees had been born in 1982 and 1983, and a third in 2000.

This human-made group opened a social space for the new paradigm of participation observation, which operated in "a triadic relationship between a mother chimpanzee, an infant chimpanzee, and a human tester."[52] As the Ai Project had entered its third decade, Matsuzawa announced, it had "progressed from the study of a single individual to a simulation of the chimpanzee community as a whole."[53]

Participation observation enabled both knowledge and care. "In 1982, we naively believed that mothering was an instinct," Matsuzawa said in his last lecture at Kyoto University in 2015. "But we had to realize that maternal behavior was learned." Human-reared chimpanzees often did not know what to do with the crying creature that had emerged from their body and even appeared to expect their human caretakers to also take care of their babies. That the primatologists could enter a room with a chimpanzee mother and her newborn allowed humans to show her how to hold and nurse the infant rather than having to take it away for human rearing (which would only have perpetuated the incapacity of captive apes to raise their own offspring).

Teaching expectant and new mothers maternal behavior comprised a variety of pedagogic approaches. For example, researchers gave them stuffed animals; they showed them video footage of how their wild conspecifics nursed; they demonstrated how humans took care of a gibbon baby; or they facilitated the right posture for breastfeeding by offering the inexperienced mother honey in a way that made her crouch and thus move her nipples closer to the baby's mouth.[54] Participant observation—maybe more aptly called "participant intervention"—entailed forms of social learning across species boundaries.[55]

As chimpanzee mothers learned to raise their own young, this nongenetic transmission of behavior transformed social relations in this captive community of chimpanzees. But it might have fallen short of Tomasello's demanding criteria for culture as the product of imitation, teaching, ratcheting, and so on.[56] Matsuzawa defined *culture* more liberally as "a set of behaviors that are shared by members of a community and are transmitted from one generation to the next through nongenetic channels."[57] He readily applied the term to describe differences between communities of wild chimpanzees. But when it came to the chimpanzees raised in his own laboratory, he refrained from presenting it as the kind of *Pan/Homo* culture that Savage-Rumbaugh and colleagues claimed to have created in her Primate Learning Sanctuary in Des Moines, Iowa.[58]

The bond between the participating mothers and the researchers allowed the experiments with chimpanzee children to be conducted under conditions very similar to those in human laboratories: in a face-to-face situation, in the presence of the mother, and with what Matsuzawa described as her cooperation.[59] Developmental psychology provided most of the paradigms to test object manipulation, tool use, drawing, imitation, or the cultural transmission of knowledge and skills from one generation to the next. The laboratory environment allowed the primatologists to determine when the captive chimpanzees got to use which objects. For example, they had never had an opportunity to crack nuts until Matsuzawa gave them stone hammers and anvils previously

FIGURE 14. Face-to-face interaction: Matsuzawa in an experimental booth with Ai as she paints. Photo by author.

used by the Bossou chimpanzees to explore the cognitive capabilities necessary for the emergence of nut cracking under more controlled conditions. In a participant observation session, he had first shown the mothers-to-be how to crack a macadamia nut. After the birth of their infants in 2000, he again provided them with nuts and tools, creating an opportunity for the infants to learn from their own mothers as well as from human participant observers who, if necessary, actively taught the mother by molding her hands to crack open the nuts.[60]

By contrast, some of the touchscreen experiments were set up in such a way that infants could not see what their mothers were entering because researchers were trying to get at the animals' individual responses to certain cognitive tasks, and social learning would only have confounded the results. The Ai Project intercalated participant observation, field observation, and controlled experiments without eradicating differences between these approaches.

While Matsuzawa and his team sought to make the experimental setting as naturalistic as possible without gambling away the benefits of being able to control and observe every step in the acquisition and transmission of the new behavior, they also acknowledged the effects of "enculturation": "All three chimpanzees exhibited behaviors that could not occur in the wild. They 'gave' nuts to the tester after failing to crack them open. Some authors have reported that apes in captivity attempted to have the tester solve a problem that was beyond their capabilities, in what is referred to as 'social tool use.'"[61] If stones didn't do the job, maybe humans would—an idea that could not have been more alien to the captives' conspecifics in African forests.

Although participation observation created its own artificiality, it protected the KUPRI laboratory against some of the allegations that Tomasello's group faced. By evading many of Boesch's critiques of unfair differences in experimental conditions privileging our species, Matsuzawa's approximation of human and chimpanzee testing also provided greater validity to cross-species comparisons in the eyes of fieldworkers otherwise distrustful of supposedly controlled experiments.[62] But it did not solve the problem that the different life histories and environments of laboratory populations posed for the generalizability and external validity of experimental findings.

Hayashibara and the Cultural Correction Device

Matsuzawa emphasized how careful he was in his face-to-face interactions with chimpanzees. He met them one-on-one, or a mother with an infant, but never two adults. "If a fight breaks out between the two, both would ask me for help. That would be dangerous."

In 2010, when Ayumu was about nine years old, Matsuzawa stopped going inside when Ai's son accompanied her. As the now-adolescent male began to work his way through the pecking order, he occasionally charged Matsuzawa. "I wasn't scared," Matsuzawa said. "I have been saying that chimpanzees aren't dangerous. And I really prefer a chimpanzee to the strange guy on a Manhattan subway at night. But it was time to stop. Everybody knows Ai, Ayumu, and me. Suppose I would just be scratched, people's perceptions would change."

This concern about the political ramifications of accidents curbed Matsuzawa's love of experimentation. "I never jumped into the outdoor compound," he told me. "That would be the extreme case and my dream. In an extended sense of participant observation, Matsuzawa would become a member of the chimpanzee community." It was the next generation of this new lineage of Japanese primatology that actually lived Matsuzawa's dream.

At about the same time as the Primate Research Institute was preparing for the second series of births that would give rise to participation observation not with isolated individuals but with mother-infant dyads living in a whole community, the Japanese sugar manufacturer and biotech company Hayashibara began to support an even more radical offshoot of the project. The owner was a natural history enthusiast who had built a dinosaur museum, exhibiting mostly Mongolian and North American specimens, near the company's headquarters in Okayama. He told Matsuzawa that he would fund a small chimpanzee research station—although the facility would turn out to be more spacious than the already extensive chimpanzee enclosure at KUPRI, including a 7,400-square-meter partly forested outdoor area, a psychology and an ethology laboratory, and other indoor spaces. In 1998, ecological anthropologist and primatologist Genichi Idani, his colleague Kohki Fuwa, and animal welfare researcher Naruki Morimura opened the Great Ape Research Institute

(GARI) of Hayashibara Biochemical Laboratories, located on a peninsula owned by the company in Setonaikai National Park.[63]

The scientific mission of GARI was "to study the evolution, life, behavior, society, intelligence, and culture of apes and humans" to answer the questions "What is a human being?" and "What should the present and future of humans be?"[64] The institute subscribed to the standard evolutionist rationale for comparative cognitive science: since neither the minds nor the behaviors of early hominoids had been preserved in fossils, we could learn about our ancestors only by comparing the minds and behaviors of extant humans and other animals, especially chimpanzees and bonobos as our closest living relatives. Idani, director of the institute, along with Matsuzawa's former student Hirata, who joined GARI upon completion of his dissertation in 2002, tied this research program to a naturalist ontology, assuming that "humans are a species of animal," that the assumed gap between apes and humans was "artificial," and that "most of the observed differences now appear quantitative rather than qualitative in nature."[65] At a time when the other ape species were on the verge of extinction, they argued, understanding the place of humans in nature was "crucial to our continuation as a species."

Hayashibara bought six three-year-old chimpanzees from the pharmaceutical company Sanwa Kagaku Kenkyusho, which had acquired a large number of chimpanzees for hepatitis research. But the individuals transferred to GARI had not been subjected to any invasive biomedical trials, and in their new home they would become the subjects and objects of participation observation.

Morimura and Hirata knew the practice from their time in Inuyama. Morimura had visited KUPRI as a graduate student from the University of Tokyo while Hirata was conducting his doctoral research. They had often watched Matsuzawa go in with Ai and other chimpanzees. After the births in 2000, Hirata had even been able to enter the booth and play with little Ayumu—"just for fun"—during time slots reserved for students and staff. Otherwise, participation observation, especially with adult chimpanzees, remained the preserve of professors. Consequently, Hirata and Morimura could really learn this practice only after leaving KUPRI.

Matsuzawa compared his pedagogy to that of chimpanzees who—*pace* Boesch—did not actively teach their young. "There is no schooling in chimpanzees.[66] However, they have their own way of education, known as 'education by master-apprenticeship,'" Matsuzawa contended. "Each chimpanzee community develops its own set of cultural traditions based on observational learning."[67] While Tomasello considered teaching to be a pinnacle of human social learning, Matsuzawa modeled his training of doctoral students on chimpanzee culture.[68] "I can be the master, they are the apprentices," Matsuzawa told me. "They can follow my way of taking care of chimpanzees, feeding them, observing their behavior, designing cognitive tasks, conducting experiments, analyzing data, and writing up papers." Time and again, he emphasized that he would *not* tell students what to do. Ultimately, they all had to become "a pioneer," finding their own way of doing something that nobody else had done before.

His Dutch colleague Frans de Waal pursued the analogy between the transmission of Japanese and chimpanzee culture further, comparing social learning in apes to the making of a sushi master: the apprentice's "education seems a matter of passive observation. The young man cleans the dishes, mops the kitchen floor, bows to the clients, fetches ingredients, and in the meantime follows from the corners of his eyes, without ever asking a question, everything that the sushi masters are doing. . . . He is waiting for the day on which he will be invited to make his first sushi, which he will do with remarkable dexterity."[69] At the Great Ape Research Institute, that day had come for Morimura and Hirata. Maybe they did not pioneer an entirely new approach (who ever does?). But they developed their own style and pushed the envelope of participation observation.

Hirata and Morimura did not just go into a booth with mother and infants but entered a whole group of chimpanzees. It certainly helped that they had begun to interact with the animals when they were infants and still weaker than their adult human caretakers. They could still behave aggressively but were "totally non-lethal," as Hirata put it.[70] For the first six years, they would even spend nights among the chimpanzees. This extreme experiment of cohabitation required that humans maintain good relations but also remain in charge of the situation. For example, Hirata and Morimura determined where each individual had to sleep. "It was not based on operant conditioning but there were certain rules they had to follow," Hirata explained, "because if they had not been sure what we expected them to do, then it would have been like in a chimpanzee group. They would have behaved too naturally and that's dangerous. Humans are too weak. We had to create an artificial situation, so that we could coexist comfortably." Teaching them rules such as where to spend the night or not to seize the fruits they would receive as food rewards in the course of an experiment was key to establishing a shared form of life.

As an onlooker rather than a practitioner of this more daring form of participation observation, Matsuzawa drew my attention to the fact that the GARI team entered the enclosure as a group. "Humans are involved in the chimpanzee community and can use the hierarchical structure of this male-dominated society. As you know, all adult males are dominant to all females in chimpanzee society. If three people go in with seven chimpanzees and the 'human chimpanzees' dominate on an everyday basis, this means that three top-ranking males are present. In this situation, the fourth- and fifth-ranking males—the actual chimpanzees—won't start a fight."

Participant observers took high risks. While Matsuzawa had never had any problems with Ai, the other two faculty members who conducted participation observation, Masaki Tomonaga and Misato Hayashi, had been attacked by their most trusted research subjects.[71] Tomonaga had been bitten by Chloé, a female that KUPRI had acquired at age four from the Ménagerie du Jardin des Plantes in Paris.[72] In 2011, Akira displayed while Hayashi was with him and bit her in the hand and forehead. It took three surgeries to fix her hand bones. Both Hayashi and Matsuzawa were sure that the brawny alpha male had no beef with her but had acted out tensions that arose from Ayumu challenging his position at the time. They also pointed to the presence of a visitor outside the booth who made Akira feel uneasy. For chimpanzee standards, the attack had not been severe. Although Hayashi believed that her species identity as a dominant human eclipsed her gender identity as a subordinate female, she noted that he treated females of his own kind more harshly and they usually suffered no injuries. But human skin was thin and human bones fragile. When she later showed Akira the bandage, his play face—an expression inviting another to have some fun together—immediately vanished and he ran off. Hayashi intended to resume participation observation when the already planned fourth generation of chimpanzees would be born. She was determined to continue her developmental psychology experiments with chimpanzee babies and infants and still wanted to test them under the same conditions as human subjects.

In the wake of these incidents, Matsuzawa emphasized that he would never tolerate the prohibition of participation observation. The institutional review of the accidents had not resulted in stricter rules, he claimed: "Telling people not to dry their cat in the microwave is the Western way. In Japan we don't regulate everything." Nevertheless, the chimpanzee research group grew more careful to preserve their scientific latitude. Everybody who had experienced the institute in the 1990s and early 2000s noted that human-chimpanzee interactions were not as casual anymore. At the time of my fieldwork, only Matsuzawa entered the experimental booth with Ai and occasionally with Akira.[73]

Many European and American primatologists worried that one day things would end badly for their Japanese colleagues. Hungarian chimpanzee management consultant Michael Seres, whom Matsuzawa had brought to Kumamoto Sanctuary, had noted tense moments in these interactions. William McGrew had asked Seres to talk the Japanese out of this "suicidal practice," but they were too proud of their tradition, Seres believed. Matsuzawa, on the other hand, felt perfectly at ease with Ai and Akira and was sure that nothing would ever happen to him.

Participant observers believed that their freedom from bodily harm was an epistemic problem. Morimura explained how their direct interactions provided a most immediate mechanism to falsify the researchers' hypotheses about chimpanzee behavior in daily life, with no need for experimental controls or statistical analyses: "If we misunderstand the chimpanzees, we will be in trouble. It's a very serious situation. We try to enjoy playing with them. But, if we fail to read their minds, they might attack us." From the point of view of a behaviorist experimenter, interpreting an animal's intentions might be pure speculation. Consequently, no "thick description," to use Clifford Geertz's term, could possibly reveal the meaning of what a chimpanzee did.[74] But in participation observation it was the chimpanzees who would clear things up in their own potentially violent ways should a misunderstanding occur. In an account of her work with bonobos, Savage-Rumbaugh and colleagues referred to this as the "cultural correction device" of *Pan/Homo* ethnography, "which acts as an automatic constraint preventing incorrect perceptions from compounding themselves."[75]

At GARI, one important way in which Hirata and Morimura had learned about chimpanzee social interactions was by interacting with the chimpanzees socially. But when I asked Hirata what they had gained from such participation observation, much to my surprise he responded: "We don't know the merit." He believed that participation observation enabled them to do many more things with the chimpanzees, but he had no way of proving his conviction scientifically. After all, they had conducted a social experiment with no control condition. It endowed them with a "feeling for the organism" that remained largely a form of tacit knowledge.[76]

Participation observation did not generate the kind of data peer-reviewed journals would accept. "Our stories don't make a scientific paper," said Hirata. Or the Japanese researchers would have had to revert to philosophy journals like the Spanish *Theoria*, where Savage-Rumbaugh and colleagues had found an outlet for the lessons they had learned from their own brand of participant observation of bonobos.[77] Instead Hirata planned to write a book, which would allow him to share their uncontrolled laboratory experiences. "What is published in scientific papers is true," he explained in an interview with me, "but it is only the tip of the iceberg." He hoped for books rather than articles to make public what participation observation had taught them about "chimpanzee nature."

As the juveniles at the Great Ape Research Institute grew older they became more rambunctious and began to sound out their relationships with the scientists. More and more minor transgressions with stronger and stronger animals ensued. When the chimpanzees were about six years of age, Morimura and Hirata decided that they would stay with the group only until the animals had fallen asleep and would spend the rest of the night in their own beds again. It seemed as if they had pushed participation observation to its very limits. Hirata explained to me that it was not humanly possible to spend whole days with subadult or adult chimpanzees. The apes would always get into fights and human bystanders were bound to get involved, possibly losing fingers or worse.

Research at Hayashibara came to an end in 2013 after GARI's patron declared bankruptcy. In 2008, however, Kyoto University took over a facility near Kumamoto where the pharmaceutical company Sanwa had housed its chimpanzee test subjects. Matsuzawa was entrusted with the task of converting the place into a sanctuary. He invited Hirata, Morimura, and their chimpanzees to this new location on Japan's southernmost island, Kyushu, some eight hundred kilometers from Kyoto and Inuyama. There they developed further the synthesis of participation observation and high-tech experiments that led Matsuzawa to declare Kumamoto Sanctuary "the spiritual twin" of his KUPRI laboratory.

Kumamoto: One Sanctuary, Different Human-Chimpanzee Relations

Surrounded by rice paddies and subtropical forest, the sanctuary sat on top of a foggy hill so steep that Satoshi Hirata's car struggled to get us to the gates of the extensive compound. Giant cage structures reverberated with chimpanzee screams betraying rising excitation, a ferocious energy contained only by metal bars and Plexiglas walls so strong that they would, the researchers hoped, withstand the powerful earthquakes that always threatened to rock the region. From a high-rise climbing structure reminiscent of the one at the Primate Research Institute, the chimpanzees could overlook the sea. The place appeared both halcyon and eerie.

Since Sanwa had passed the authority over the station to Kyoto University, this public institution had taken responsibility for apes the private company had previously used in invasive biomedical research. Thus Kumamoto Sanctuary became part of Matsuzawa's scientific empire and morphed into a large-scale experiment on rehabilitating damaged chimpanzee lives.

While Matsuzawa remained in charge of the steering committee, he appointed Hirata as director of the sanctuary. The latter's job was to transform a biomedical research facility into an old-age home for chimpanzees that would explore new approaches to animal welfare and double as a comparative cognitive science laboratory.

As a student, Hirata had been introduced to primatology by Toshisada Nishida, head of the Laboratory of Human Evolution Studies at Kyoto

University. But he was wary of the Kyoto School, which applied Lévi-Strauss's structural anthropology to chimpanzee groups. Hirata decided not to pursue primatology as a humanities discipline (*jinbungaku*). For doctoral research, he moved from this remaining stronghold of Imanishiism to Matsuzawa's laboratory. He maintained the Kyoto School's interest in primate sociality, though. While Matsuzawa had found his own way by turning from primate societies to primate minds, Hirata reconnected these different foci in his work on chimpanzee social cognition.

However, bringing Sanwa's retired test subjects into this scheme proved to be difficult. When I visited Building 1, where the biomedical experiments had taken place, a friendly veterinarian stood in front of the cages and tickled the belly of one inmate. The man had turned gray over the thirty years that he had worked for Sanwa, which still paid his salary. The University of Kyoto had taken over some of the pharmaceutical company's staff, who had often known the chimpanzees since birth. First they had infected them with hepatitis C viruses to test a gene therapy. Now they treated their incurable but only slowly progressing cirrhoses of the liver. The animals appeared restless. One kept going in circles, while another rocked back and forth. Many displayed against us, and a chimpanzee called Mizuo approached the bars, his mouth filled with water, and spit at me. "They are not zoo chimpanzees," remarked a young professor. "They get excited when they see strangers. Those raised by humans can't stop caring about humans. Some are very friendly to us, others nasty."

Nobody was eager to conduct participation observation with these behaviorally abnormal virus-carrying animals. Morimura had tried cognitive testing that did not require direct interactions, but he soon recognized that the Sanwa chimpanzees were too old to learn how to use a touchscreen or to tolerate other high-tech equipment. For the time being, they remained pensioned off.

Many laboratories housed apes on which they could not experiment. Savage-Rumbaugh had considered those individuals she could not work with as a control group that would allow her to better understand the effects enculturation had on Kanzi and the other human-raised bonobos that did participate in her ape language project, Hirata told me. But Savage-Rumbaugh and colleagues also thought of them as members of their *Pan/Homo* culture who simply remained on the *Pan* side of this human-ape continuum.[78] "Researchers feel like parents of several children," Hirata said. "Some are good at school, others less so, but you have to take care of everybody." Although in Japan it was easier to obtain funding for scientific research than for animal welfare, Kumamoto was not primarily a laboratory but a sanctuary.

At Kumamoto, the apes closest to the *Homo* side of the human-ape continuum were the six chimpanzees that Hirata and Morimura had brought with them from GARI. The primatologists had devised an outdoor enclosure with six touchscreen workstations where these chimpanzees could take cognitive tests without having to be separated from the group. Hirata and Morimura also continued the practice of participation observation with individual animals or mother and child.

At Kumamoto, two groups of chimpanzees lived in the same anthropogenic environment, but only the group that had experienced the rich upbringing at GARI was open to cognitive experiments and face-to-face interactions. Participant observation had opened them to the human world in ways that had profoundly transformed their form of life. This situation appeared to vindicate Boesch's insistence that different laboratory populations behaved differently because of their particular histories.[79] Morimura agreed with Boesch rather than Tomasello that neither the GARI chimpanzees nor any other captive group could represent "the" chimpanzee.[80]

Face-to-Face Measurements and Their Limits

Although Hirata and Morimura maintained the practice of participation observation, its place in their methodological tool kit had changed. While they had recorded some twenty terabytes of video footage at GARI, they no longer filmed their face-to-face interactions. At Kumamoto, the approach hardly served as a method of its own but had become a means for applying cutting-edge measuring devices. Its value for producing quantitative data that could easily be published in scientific journals became apparent when I watched Hirata and Morimura record electrocardiograms from the GARI chimpanzees. In captivity, and more so in the United States than in Japan, many males died of heart attacks. The Japanese primatologists suspected that the difference in mortality had to do with the unhealthy food pellets that American caretakers fed the animals. As part of their animal welfare research, they monitored the chimpanzees' cardiac condition.

Morimura and Hirata took me into a small chamber almost entirely filled by an experimental booth. We waited in the narrow gallery surrounding it until the first chimpanzee rushed into this fishbowl through a meshwork passageway above our heads. A thunderous display during which Zamba drummed the transparent walls with his massive fists shook the whole booth and culminated in the ape jumping with his feet against the bulging Plexiglas right in front of us. "Okay, I won't go in today," Hirata immediately decided. The high level of excitation would have made the encounter too risky.

Right after this show of force, however, Zamba sat down calmly at the barred door to his sanctum. Morimura and Hirata walked over to him with their ECG, and when asked, the chimpanzee presented his chest. They stuck two self-made electrodes through the bars and began to record his electrocardiogram while maintaining his compliance with talk and a steady supply of grapes. However, the measurement was frequently interrupted as he moved and the electrodes lost contact with his skin.

By contrast, when Hirata and Morimura subsequently went inside with Mizuki and her daughter Iroha they managed to keep them still: one researcher would entertain the unoccupied animal while the other could record the ECG continuously. Participation observation facilitated measurements with a superior signal-to-noise ratio, even when the primatologists had to deal with two subjects simultaneously.

This unhindered physical contact with the chimpanzees enabled the application of measurement technologies that required a direct and trustful management of behavior not possible through wire mesh and Plexiglas walls. ECG recordings were only the beginning. Ultrasound exams of pregnant females provided novel insights into fetal development.[81] The neural processes underlying chimpanzee thought could be recorded in a fully awake individual who allowed a trusted scientist to attach EEG electrodes to her head and to keep her from moving around during the measurement.[82]

Fumihiro Kano, a postdoctoral researcher at Kumamoto, had profited from Hirata and Morimura's ability to put a head-mounted eye tracker on his test subjects. As a PhD student in Inuyama, Kano had discovered an unused table-mounted device in Tomonaga's lab. Gaze behavior could serve as a window on social cognition—an interest that Kano, like Hirata, had brought from the Laboratory of Human Evolution Studies at Kyoto University to Matsuzawa's Section of Language and Intelligence at KUPRI.

The eye tracker had allowed Kano to develop not just a research object but also a style of experimentation that set him apart from Matsuzawa's touch-panel designs. While the latter required extensive training—even if Matsuzawa would eventually test how the animals generalized from what they had learned—looking around was as spontaneous a behavior as any. In this respect, Kano felt closer to what he called the Tomasello/Call approach, which sought to test innate cognitive and behavioral dispositions that had not been polluted by learning.

What continued to align Kano with Matsuzawa and Tomonaga was his propensity toward high-tech gadgets like the eye tracker. "The Max-Planck people are low-tech," Kano had noticed as a visiting researcher in Leipzig. "Their experiments have very sophisticated designs and control conditions, but the technology could be from Darwin's times." While the table-mounted eye tracker that he introduced to Leipzig required some way of restraining head movements and allowed only study of how apes watched two-dimensional images or movies on a computer screen, wearable devices captured how gaze behavior in freely moving subjects responded to real-life environments.[83]

Such instrumental recordings would not have been feasible without direct interaction in a face-to-face setting. In itself an approach devoid of technological mediation, participation observation enabled the application of high-tech devices that had previously been impossible to use on animals as willful as chimpanzees. But these combinations of direct physical contact and technical measurements came with their own limitations.

Although only participation observation allowed equipping a chimpanzee with a head-mounted eye tracker, Kano was not interested in cultivating the relations necessary for going in with the animals. "I'm a scientist. I just want data," he told me. For this purpose, he preferred so-called no-touch experiments. The problem with measurements based on participation observation was that only the very few primates and primatologists socialized to interact with each other in this demanding way could do research together. Being more art than science, this practice did not travel well and had become a Japanese peculiarity.

It would have been impossible to test the apes of Pongoland in this fashion. The research apparatus of the Max Planck Institute was designed for visiting scientists who lacked familiarity with the animals, Kano noted. During his time in Tomasello's laboratory, he developed an alternative apparatus: subjects kept their head still as they drank from a fixed straw while a table-mounted eye tracker recorded their gaze behavior in response to a short movie.[84] "Now I don't need any training," Kano explained. "I just need grape juice." And no researchers had to risk their life to coax the apes into compliance.

Kano admitted that participation observation generated individual measurements of higher quality. But at the Primate Research Institute, only Pan would wear the eye tracker; at Kumamoto Sanctuary it was Mizuki. Although technically impressive, these were basically single-subject studies. By contrast, Kano's no-touch approach enabled a much larger sample size, which translated into increased statistical power. Quantity became quality as chimpanzee eye-tracking data could be freely exchanged and pooled between Kumamoto and Leipzig. It became the empirical foundation of more than local knowledge.

Japanese-Chimpanzee Cultures?

When Lindeman coined the methodological distinction between *observer* and *participant observer*, he emphasized both complementarity and difference. Instead of looking at exactly the same phenomena as the observer, just from the inside, "the participant observer should be free to see many things which the outside observer can never see."[85] Instead of relying exclusively on behavioral observations or self-representations, Lindeman proposed to examine the harmony or disharmony between these perspectives. Conflicting accounts of what people do and what they say they are doing might either draw attention to ulterior motives behind stated purposes or call into question the analytic categories that the observer initially used to describe what people are doing.

As an observer I have pointed to geographical differences in scientific practices between the research groups described in this book. Matsuzawa, Hirata, and their colleagues did quite a few things with their chimpanzees that neither the comparative psychologists in Leipzig nor the field primatologists in Loango or Taï did—and vice versa.

As a participant observer I have noted that Matsuzawa thought their synthesis of laboratory and field methodologies was very much a product of "the Japanese way of thinking." As far as participation observation was concerned, Hirata offered a different interpretation. He acknowledged that since the end of the ape language projects, including Savage-Rumbaugh's so-called *Pan/Homo* ethnography, and the closing of Sarah Boysen's laboratory at Ohio State University in 2006, only Japanese primatologists continued the kind of face-to-face interactions described in this chapter. But he pointed to the many European and American researchers who, historically, had also interacted directly with their apes in laboratories and zoo nurseries. Sociologist Lawrence Wieder had studied such encounters between US chimpanzee researchers and their chimpanzees.[86] In an interview, Hirata pointed out: "Faced with chimpanzees, we independently developed a similar style based on close relations and friendship."

Akira Takada concluded from his behavioral analysis of interactions between the researchers at GARI and their captive apes that "the chimpanzees also domesticated the humans."[87] Both in Japan and in the West, there was no shortage of counterexamples where neither amity nor participant observation had grown out of the encounter between *Pan* and *Homo*. But where the bonds of mutual trust did emerge, Hirata and Takada suggested, chimpanzee behavior shaped the resulting research practices as much as any human culture did.

Despite Pamela Asquith's portrayal of Japanese primatology as more open to anthropomorphism than Western primatology, Matsuzawa and most of his students sided with Tomasello rather than Boesch when it came to debates over imitation and active teaching.[88] They also refrained from adopting the humanlike conception of culture that primatologists Sue Savage-Rumbaugh and William Fields proposed in collaboration with Swedish philosopher Pär Segerdahl. Segerdahl and colleagues had rejected the reduction of culture to a mechanism for the nongenetic transmission of information and instead thought of it as "a shared way of living containing characteristic activities, tools, environments, communication means, social relations, personalities, games, gestures, and so on."[89] Matsuzawa and other post-Imanishi primatologists felt that some of their European and American colleagues had taken the continuity between *Pan* and *Homo* too far.

Hirata and colleagues even expressed moderate skepticism regarding the existence of chimpanzee culture: "It seems inappropriate to answer yes or no to the question of whether nonhuman primates have culture."[90] While Idani and Hirata generally insisted on the artificiality of the gap between apes and humans and did not categorically object to anthropomorphic terminology, Hirata elsewhere postulated a "big gap between human culture and monkey culture." [91] Monkeys neither taught each other nor did they enforce norms: "In the case of humans, those who do not follow the culture of their own community will be treated as unorthodox and will be socially punished; human culture has the power to restrict or constrain the behavior of a person who belongs to that community. . . . On the other hand, a monkey who does not follow a culture will not be blamed and no social restriction works against a violation."[92] In the article, coauthored with Imanishi's student Masao Kawai, Hirata accommodated this difference by reverting to the early Kyoto School's distinction between culture and preculture.

Like their European and American colleagues, Japanese primatologists knew that genetically apes were more closely related to humans than to any monkey species. But whether and to what extent chimpanzees had the cognitive ability to teach and enforce norms remained a matter of empirical research and conceptual altercation. In an interview with me, Hirata interpreted the chimpanzee culture controversy as a war over words: "The problem is that we might never reach consensus about what *culture* means. What we can establish empirically is that different groups of chimpanzees behave differently. Calling that culture is a matter of subjective judgment."

Although Matsuzawa also emphasized differences between *Pan* and *Homo*, he had no qualms talking about chimpanzee cultures in the wild.[93] Moreover, he described the role of Hirata and Morimura in the chimpanzee community they had fostered at GARI as that of top-ranking "human chimpanzees." And yet Matsuzawa would speak neither of their research object as a "*Pan/Homo* culture" nor of participation observation as a form of "*Pan/Homo* ethnography." The terminology of Savage-Rumbaugh and colleagues annoyed him for two reasons.[94]

First, Matsuzawa remained committed to a conception of ethnography as long-term participant observation of people in naturalistic settings rather than laboratories. The goal of such ethnography was ethnological comparison. "A *Pan/Homo* ethnographer should be a cultural or ecological anthropologist studying the way of life of hunter-gatherers for many years," Matsuzawa told me. "And they should become a member of a community of chimpanzees or bonobos to get to know their way of life. Then you can talk about *Pan/Homo* ethnography."

Second, Matsuzawa denied that even the bonobos in Savage-Rumbaugh's Primate Learning Sanctuary had formed a real community, not to speak of bonobos and humans. "For me, it's not a society, not a community, and Sue had no intention to make such a thing," he charged. "It's just an aggregation of random individuals without social structure." His critique echoed Imanishi's distinction between an undeveloped "aggregation of like individuals" and the "wholeness" of "human society" based on division of labor—a variation and transvaluation of Durkheim's narrative of a social evolution from mechanical to organic societies.[95] At KUPRI and its spiritual twin institutions, by contrast, the goal was to re-create in the laboratory the social cohesion of small undifferentiated societies, which the French sociologist had attributed to primitive humans.

If participants and observers called such forms of life a chimpanzee culture within a human culture or a *Pan/Homo* culture, their decision was a matter of grounded judgments that reflected their social values, political orientations, and cosmological schemes. There was no scientific answer to the question of whether the so-called friendship between primates and primatologists warranted speaking of a hybrid community, even though the researchers could no longer immerse themselves in their subjects' social life for extended periods. Although my Japanese interlocutors shunned the polemic positioning of their more outspoken European and American peers, they were hardly neutral in the chimpanzee culture wars. Matsuzawa's synthetic primatology could not pacify the battle over definitions.

Whether or not Savage-Rumbaugh had intended to create an integral community of humans and apes, both she and Matsuzawa assumed that it was not simply a question of knowing nature but of engineering synthetic forms of sociality.[96] At the Primate Research Institute and Kumamoto Sanctuary, these unique forms of life had been constructed in institutions that fostered a deeper understanding of chimpanzees through a methodological synthesis of experimentation, field observations in the laboratory, and participation observation.

Within the bounds of captivity, these scientific practices simultaneously created opportunities for and knowledge of chimpanzee self-determination—self-determination not as metaphysical freedom from determination tout court but as freedom from *human* determination. Paradoxically, it took humans to carefully design research settings where chimpanzees could be themselves and volunteer in captive studies.

The problem of enabling and maintaining such epistemically fecund forms of self-determination required distinct responses in the laboratory and in the field. In Bossou, the integration of experimentation, observation, and interaction presented itself very differently from the knowledge cultures of Inuyama and Kumamoto. Could a chain of translations between these approaches put an end to at least the *methodological* battles over how to study chimpanzee cultures?

6

Field Experiments with a Totem Animal

IN BOSSOU, a small Manon village in a remote corner of Guinée Forestière, Tetsuro Matsuzawa extended his synthetic primatology from the laboratory to the field. Could a Japanese outdoor lab in the West African countryside pacify the methodological controversy over controlled experiments and naturalistic observations?

Rated the most heavily human-impacted chimpanzee research site, Bossou manifested a vision of chimpanzee culture that was radically different from that of both Taï and Leipzig.[1] In 1976, the very year when Boesch first visited his future field site in Ivory Coast, Yukimaru Sugiyama from the Kyoto University Primate Research Institute initiated the other long-term study of western chimpanzees (*Pan troglodytes verus*). A decade earlier, the population that lived around the village had already been studied by Adriaan Kortlandt. The Dutch ethologist believed that great apes were capable of culture.[2] But he did not imagine wild cultures. Instead he speculated that the Bossou community had learned nut cracking from local humans rather than from their fellow apes.[3] When Matsuzawa inherited the field site, he also inherited this cosmology, which set him apart from both Michael Tomasello and Christophe Boesch.

Long before Dutch and Japanese scientists had turned Bossou into one of the most renowned sites for primatological field experiments, the Manon had initiated an experiment of human-chimpanzee cohabitation utterly different from those at Kumamoto Sanctuary and the Primate Research Institute in Inuyama. The Zogbila clan, descendants of Bossou's founder, considered chimpanzees their totem animal. As other clans, often of different ethnic origins, joined and eventually came to identify as Manon as well, people born and raised in the village tolerated and even protected the free-roaming apes that regularly raided their plantations.[4] This seemingly peaceful coexistence of *Homo sapiens* and *Pan troglodytes* made French American primatologist

Tatyana Humle hopeful: "The Bossou community of chimpanzees importantly reveals that chimpanzees may thrive in human-impacted habitats," she concluded from two decades of research at Matsuzawa's Guinean research station.[5] Maybe more important than any findings about the cultural transmission of nut-cracking skills, this Afro-Japanese experiment in multispecies living appeared to provide a model for chimpanzee conservation in the Anthropocene. In such an arcadian world, human interventions, both scientific and agricultural, could do no harm, could they?

This chapter examines the peculiar brand of fieldwork that Japanese, European, and American researchers around Matsuzawa had devised in Bossou—well integrated with his laboratory work in Inuyama but in stark contrast to the detached field observations in pristine nature (or at least something a little more like it) that Boesch and his group conducted in Taï Forest. Human-chimpanzee cohabitation on farmland and the apes' voluntary participation in open-air experiments created the mirage of a more than human society that included both *Pan* and *Homo*. A paradise for multispecies ethnographers eager to foster "biocultural hope"—if only the chimpanzees would survive.[6]

Nonnatural Chimpanzees

In December 2015, a few days before the World Health Organization would prematurely declare the end of human-to-human transmission of the Ebola virus in Guinea, Tetsuro Matsuzawa and Naruki Morimura returned to Bossou. In the Guinean capital, Conakry, one still could not buy a local SIM card in a mobile phone store without passing an employee at the entrance who held an infrared thermometer gun to the valued customer's head. Yet, as a stopover between a cognitive neuroscience conference in Bangalore, visiting Baka pygmies in Cameroon with a Japanese TV crew, and giving a talk at the Sorbonne in Paris, Matsuzawa wanted to show his colors as soon as the hemorrhagic fever began to peter out. After withdrawing all students and researchers from his field sites in Bossou, Nimba, and Diécke when the outbreak had started two years earlier, he needed to assure the Guinean authorities and his field assistants that the Japanese would be back. All too easily could these international relations unravel.

After a two-day, one-thousand-kilometer taxi ride along dirt roads and past mud huts and baobab trees from Conakry to Bossou, Matsuzawa, Morimura, our chauffeur Omar Bah, and I drove through the marketplace. Behind the stalls rose Mont Gban. The top of this one-hundred-meter hill was covered by the last remaining patch of primary forest. The crowns of the emergent trees served as observation platforms from which the chimpanzees could follow the

FIGURE 15. Bossou, Guinea: from the sacred hill the chimpanzees can see the market. Photo by author.

FIGURE 16. The Manon protect their totem animal: "In Guinea, it is prohibited to sell or buy chimpanzee meat." Photo by author.

hustle and bustle of market day. Matsuzawa said that they often used this time when nobody worked on the fields to raid pineapple trees.

A hand-painted sign showed a dismembered chimpanzee for sale. On a second picture, a policeman led away the handcuffed market woman and her customers. "In Guinea, it is prohibited to sell or buy chimpanzee meat," the sign warned in French. The apes were widely hunted in this far-flung corner of the country. In Bossou, however, they not only enjoyed the state's protecting hand but also took pride of place in the villagers' cosmology. Concern for chimpanzees was among the few interests that the Manon minority and the central government in Conakry shared. People in the village abided by a food taboo against the consumption of the apes' flesh. A field assistant would later explain to me that "eating a chimpanzee would be like eating a human."

Legend had it that Zogbila, the founder of the village first described in sixteenth-century Portuguese documents, chose the family's totem animals.[7] According to one myth that circulated in Bossou, his mother had given birth to a chimpanzee and a dog, which led Zogbila to bar his descendants from harming these species.[8] Another myth claimed that he had killed a chimpanzee on the hunt, but not before the creature had severely wounded him, too. Convalescing, he announced to his family that unlike the other forest animals, his adversary had not been a wild beast but resembled a human being and might even have been a sorcerer from the village. Henceforth, the Zogbilas revered chimpanzees as their totem and refrained from killing and eating them.[9] In any case, the apes gave the clan its power, its honor, its sense, and its good fortune, one of Zogbila's descendants assured me.

In Bossou, other clans such as the Goumy and the Doré defined their social identity through different totems such as the lizard or the snail. The corresponding food taboos determined clan membership and prevented endogamy: taking a wife from his own totem group would be "like marrying my sister," an assistant belonging to the Doré clan explained.[10] Each family worshipped two or three totem animals. But no Manon from Bossou would eat chimpanzee meat. As the founders of the village, the Zogbilas had persuaded the other families to protect the local chimpanzee community, independent of whether or not the apes were one of their own totems. "They are like our ancestors, like humans," the Doré assistant maintained. Consequently, the only confirmed case of poaching around Bossou dated back to 1976 and had been committed by a soldier deployed from elsewhere who was chased out of the village.

While their totemic status protected all chimpanzees, the community living around Bossou occupied a special place in the villagers' spiritual geography.

For thirty thousand Guinean francs, roughly four US dollars, and somewhat upset about such miserliness (my Japanese hosts had asked me not to give in to his higher demands), one of Bossou's notables, Mato Topin Zogbila, told me about the inhabitants of the sacred forest of Mont Gban, one of four hills surrounding the village. The apes in Bossou were "no natural animals," he claimed. This set them apart from the wild chimpanzees of the nearby Nimba Mountains. "There is a very big difference between the Bossou chimpanzees and the chimpanzees in the rest of the world because here the chimpanzees crack nuts," Mato Topin claimed. When I objected that many West African chimpanzee populations had been observed cracking different kinds of nuts and that I had seen with my own eyes how the Taï chimpanzees used stone hammers and anvils to open *Panda* nuts, he seemed incredulous and ultimately undeterred in his local ontological patriotism.

If one looked no farther than the Nimba Mountains, located behind a five-kilometer stretch of anthropogenic savanna, Mato Topin's ontology might even have appeared empirically grounded. In her study of the Nimba chimpanzees at neighboring Seringbara, Dutch KUPRI researcher Kathelijne Koops had found that members of this population indeed did not use hammer and anvil to crack open oil palm nuts, nor did they exploit other nut-bearing species such as *Detarium senegalense*, *Parinari excelsa*, or *Parinari glabra*.[11] No genetic or ecological explanation suggested itself: at other sites, *Pan troglodytes verus* cracked these very nuts, and the Seringbara community could have found them in its home range. Instead the inhabitants of Seringbara employed a new percussive technology, breaking open large *Treculia africana* fruits with cleavers and anvils made of wood or stone. They also used sticks for ant dipping and digging. Thus, it was true that the Bossou chimpanzees were different from their neighbors.

The Bossou community also differed from many other nut-cracking communities in West Africa in that they used stone hammers not only with fixed anvils, usually roots or rocks, but also with movable stone anvils. Matsuzawa had even discovered that they occasionally stabilized these anvils with one and sometimes even two wedge stones.[12] None of his colleagues had observed the combination of three or four tools for the purpose of nut cracking. Matsuzawa interpreted "this advanced percussive technology as a case of progressive problem-solving."[13] In line with his nonconfrontational approach to the chimpanzee culture controversy, Matsuzawa refrained from using the loaded term *cumulative culture* in his publications—a term that Boesch, Tomasello, Whiten, and McGrew had often fought over. While no chimpanzee ethnographer had been able to document cultural change in the wild, he entertained the possibility of a modest ratchet effect in chimpanzee cultural history and took pride in the fact that it had enabled the Bossou community to develop the most complex nut-cracking tool kit documented in the ethnographic archive of chimpanzee cultures.

This observation of cultural particularity did not support Mato Topin Zogbila's ontology, though. In light of the comparative research conducted by primatologists at Bossou, Nimba, Taï, and various other West African field sites, the great metaphysical divide between the Bossou chimpanzees and all other communities shrank to a multitude of small differences between chimpanzee material cultures.

Yet the Zogbilas were not the only people who considered the nut-cracking skills of the Bossou chimpanzees a sign of their special relationship to the local human population. When Adriaan Kortlandt first observed this group, he already believed that great apes "might be characterized as 'cultural Primates', while gibbons and monkeys would be classified as 'instinctual Primates', provided we may use the term 'culture' for a non-verbalized system of social traditions."[14] He also considered tool use the paradigm of such prelinguistic culture. Jane Goodall had just reported that the Gombe chimpanzees used sticks to fish for termites and collect honey from beehives.[15] At the time, however, long-term observations of chimpanzees had only just begun. Kortlandt's bibliography suggests that he did not yet know about Harry Beatty's observation of nut cracking in Liberia, and during Kortlandt's brief visits to Bossou in 1960 and 1964 the chimpanzees' use of stone tools escaped him.[16] He had noticed that "chimpanzees in the wild often spy on human activities" and wondered "why they have not adopted some of our cultural achievements."[17] When Sugiyama and Koman first described how the Bossou chimpanzees used stones to crack open the nuts of oil palms, which did not grow well in dense forests but had been cultivated by West African farmers for about five thousand years, Kortlandt quickly connected the dots.[18] Since he had observed local people cracking nuts in identical fashion, he speculated that the Bossou chimpanzees had seen this behavior in Bossou humans and began to copy it as they found fewer and fewer wild foods in the wake of the massive transformation of their habitat by *Homo sapiens*. In Kortlandt's eyes, "this community and Beatty's may represent the first identifiable cases of direct transmission of technology from man to animal in the wild."[19]

In 2015, Matsuzawa held on to a modified version of Kortlandt's controversial thesis. He believed that Bossou chimpanzees had first acquired the skill not through observation learning but by playing around with the stones humans had left at nut-cracking sites. Primatologists categorized such facilitation learning as one form of social learning. Matsuzawa surmised the area around Bossou to be the cradle of the West African nut-cracking culture. A place with plenty of stones—and stones were indeed abundant in the surrounding forests—and a strong human presence would have provided ideal conditions. Subsequently, migrating females might have spread the new craft up to the shores of the N'zo-Sassandra River—and all the way to Taï Forest. If Matsuzawa's speculative naturecultural history was true, the seemingly authentic chimpanzee culture of Taï represented a derivative of the encounter between *Pan* and *Homo* in the agricultural landscape surrounding Bossou.

Bossou and Taï are not far from each other: when Kortlandt visited Christophe and Hedwige Boesch in 1984 to observe nut cracking at their site, he took a bush taxi. But the 350 kilometers of potholed roads did not enable much intellectual traffic. These primatological field sites represented radically different ontological and methodological paradigms. Boesch had developed his notion of wild cultures in opposition to Kortlandt's proposal of social learning between humans and apes, and he doubted whether chimpanzees that had not been raised in captivity would ever copy human behavior. While Boesch's variety of naturalist metaphysics inspired a research practice of naturalistic observation to be as hands-off as possible, Kortlandt and Matsuzawa favored field experiments and had no qualms about intervening in the chimpanzees' lives.

Methodologically, the Dutch ethologist seemed closer to the Japanese psychologist than to his French Swiss colleague. Maybe an ecological explanation could better account for the pronounced differences between the knowledge cultures of Bossou and Taï than any reference to disciplinary divides or the cultural gap between European and Japanese primatology.

The underlying ontological differences might appear to be a secondary theater of the chimpanzee culture wars. They had never made it into the scientific literature, presumably because they were speculative in nature. But although expressed only in personal conversations, this contemporary metaphysics of apes informed important research decisions—choice of field sites, allocation of funds, preferred methods, and interpretation of data—and thus fueled the public controversy.

The Outdoor Laboratory

Upon our arrival in the courtyard of the Bossou research station, a spacious grassy compound at the foot of Mont Guein right behind the village, seven field assistants lined up in rank and file. But no roll call followed. Our bush taxi driver, Morimura, and I stood back as Matsuzawa hugged each man for a long time, inquiring about his family and health. I had never seen such intimate physical contact at the Primate Research Institute in Inuyama. This was a different world.

How different became even clearer as we gathered in the main building's kitchen and common room where Matsuzawa showed the Manon staff video clips from his laboratory in Japan. On the small screen of a laptop computer, we watched Ayumu's fingers flying over the touchscreen in his test booth, tapping the now-masked numbers, which he had seen for only milliseconds, in correct ascending order. After this elaborate experimental demonstration of the chimpanzee's superhuman powers, one of the assistants eventually broke the ensuing silence: "Professor, how does one come up with such an idea?"

In Bossou, the only touchscreens were those of the villagers' cheap Chinese smartphones, and no keepers summoned the chimpanzees to test booths in a basement. Here, the scientists were the ones called in by an assistant when the apes happened to pass by the lab—and that was open-air and in the forest. From the kitchen table we could reach Le Salon, the newer of two outdoor laboratories established in 2009, in two and a half minutes, following a well-maintained footpath up to a small clearing.

When, two days later, the crackling of the walkie-talkie first announced the imminent arrival of the chimpanzees, we took our positions approximately twenty meters behind a large two-meter-high screen freshly constructed from palm fronds that minimized the impact of our presence. The apes found a large oil palm bunch containing the fleshy red fruits, which chimpanzees all over Africa like to eat, as well as seven piles of dried oil palm nuts, which only West African communities crack. Moreover, forty-nine stones, arranged in a seven-by-seven grid, could serve as hammers and anvils. We installed our photo and video cameras on tripods and watched this artificial nut-cracking workshop through peepholes cut out of the screen. "We invented this fixed stage," explained Matsuzawa. "It's a theater. The audience is waiting for the chimpanzees."

Enter the apes. First the alpha male, Jeje, emerged from the forest, eighteen years young and powerfully built, accompanied by his four-decades-older mother, Jire. They pant-hooted and Fana called back from the forest. Minutes later this senior showed up as well. In the new year, she would turn sixty—although still hale and hearty enough to raid papaya trees in the village, she was a true Methuselah among wild chimpanzees.[20] Soon her son Foaf arrived and then her infant grandson Fanwa rode into the arena on the back of his mother, Fanle. Now the so-called F family was complete.

Matsuzawa had first met Ai's free-roaming relatives in 1986. During his sabbatical in David Premack's laboratory, he had come to realize: "I cannot become David Premack. He is very, very smart. I would not be able to catch up with him. My intuition told me: Don't follow him. Go elsewhere. And that was Africa because David doesn't go to Africa. No one had ever tried to conduct

FIGURE 17. Outdoor lab: Matsuzawa and his team record a nut-cracking session from behind a screen. Photo by author.

FIGURE 18. Le Salon provides an unobstructed view of the chimpanzee group's use of manipulated objects. Photo by author.

lab and fieldwork in parallel. My question was very simple: 'How do chimpanzees use their intelligence in a natural habitat?'"[21]

Matsuzawa took a special interest in the cognitive foundations of lithic technologies. On the very first day of Matsuzawa's two-month visit to Bossou, his KUPRI colleague and host Sugiyama suffered such a severe bout of malaria that Matsuzawa got him onto the next Aeroflot plane home (the Soviet Union eagerly supported the Guinean experiment in African socialism). Assisted only by local guides, Matsuzawa had to develop his own way of conducting fieldwork. Initially, he adopted the kind of hands-off approach that Sugiyama practiced in the face of Bossou's anthropogenic environment. To develop a feeling for the animals' lives in the outdoors, he first followed the chimpanzees without intervening.

As an alpinist, Matsuzawa had acquired outdoor skills. Organizing Himalayan expeditions had prepared him for the demanding logistics of field research. But he came to Guinea as an experimental psychologist who had never been trained to conduct naturalistic observations in the wild. Moreover, he had a laboratory to take care of in Japan. This job did not allow him to take the time necessary to patiently collect behavioral data at the chimpanzees' pace. Since he would be able to come to Bossou for only one month every winter, he made a virtue of necessity. In 1988, he built an outdoor laboratory.

This research site soon made him "the foremost practitioner" of chimpanzee field experiments, as his British colleague William McGrew attested.[22] "Suppose that you follow wild chimpanzees at Bossou," Matsuzawa said, explaining the advantages of his approach.[23] "If you are lucky enough, you may be able to witness stone-tool use. However, you may also fail to observe stone-tool use even if you follow the chimpanzees from dawn to dusk, every day for weeks." In the field laboratory, however, "3 days suffice to observe stone-tool use in all able nut-crackers of the community, which is in clear contrast with the more traditional approach that relies solely on natural observations."

The objects and objectives of Matsuzawa's outdoor laboratory also diverged from those of a classical indoor laboratory like the one he ran in Inuyama, which sought to transform its objects for the purpose of experiment so as to be quite unlike anything in nature. However pronounced the behavioral differences between wild chimpanzee populations might be, nowhere in Africa did one come across chimpanzees like Ai or Ayumu describing the colors of the rainforest in Japanese kanji or using Arabic numerals to count the number of nuts they had collected. By contrast, field experiments, at least as Matsuzawa understood them, did not aim to create such artificial behaviors but sought "to stimulate the performance of the chimpanzees' natural behavior."[24]

Hybrids over hybrids, the outdoor laboratory did not just mix experimental and field research but was also located at a site defined by both scientific and religious practices. The first laboratory sat on top of the sacred Mont Gban. In precolonial times, this hill had presumably accommodated a secret initiation camp, which Ivorian cultural anthropologist Bohumil Holas had not dared to climb in the 1940s for fear of committing a potentially lethal sacrilege. In 1967 and 1969, Kortlandt brought the ethological tradition of combining fieldwork and experiment to this hilltop, where his team observed the chimpanzees' reaction to an electric-powered stuffed leopard that could shake its head.[25] In the late 1980s, Matsuzawa turned the scene of these ad hoc experiments into a more permanent research site. He furnished the space with wooden chairs, which allowed the primatologists to get much academic paperwork done, as they sometimes waited all day for the chimpanzees to stop by. Since French served as the lingua franca between Japanese researchers and their Manon assistants, the profaned place came to be known as Le Bureau.

The site was surrounded by the last patch of primary forest. Neither hunting nor farming was allowed on the hilltop. The Manon maintained it as a reserve for their spirits and devils. Until the early 1960s, the villagers had made offerings of rice and fruits to these beings, but eventually it was the chimpanzees and other forest animals that ate the food. Then, in 1961, the socialist government of the newly independent Republic of Guinea launched its violent demystification campaign, suppressing such backward pagan practices for more than twenty years.[26] By the mid-1980s, however, the ceremonial site, forbidden to all but the cult's initiates, had been reopened on a plot of secondary forest at the bottom of the hill right next to the village.[27] At about the same time, the feeding of the chimpanzees also recommenced on Mont Gban, this time for the sake of science.

When Matsuzawa provided oil palm and other nuts to study how chimpanzees learned to crack them, this experimental intervention seemed to fit into the tradition of food provisioning that has often been described as a hallmark of Japanese field primatology.[28] Yet European primatologists like Jane Goodall had also lured chimpanzees with sugarcane and bananas, while some of their Japanese colleagues did not appreciate such human interference in the observed animal behavior. Sugiyama, for instance, refrained from facilitating the habituation of the Bossou chimpanzees with edibles. He wanted them to stay natural.[29]

Sugiyama's doctoral student Osamu Sakura, who had helped Matsuzawa set up Le Bureau while doing his own sociobiologically inspired fieldwork on chimpanzee sociality, expressed his bewilderment: "I was quite surprised when I first saw Bossou. The chimpanzees lived just behind the village, which was home to 2000 people. Yes, they were wild, but their environment was quite artificial. It was true that the researchers did not feed them, but the chimpanzees always ate human-cultivated crops and stole pineapples. So I was really confused about what was natural." (At the time of our interview, Sakura had left primatology for a field that had made the confusion of nature and culture its ontological paradigm: he taught science and technology studies at Tokyo University.)

Field experiments had provoked tensions among primatologists similar to those arising from provisioning. For a long time most researchers working with great apes had shunned such interventions for much the same reasons, worrying that they might lead to unwanted long-term effects on the animals' behavior.[30] And if one did conduct a field experiment, not altering their habitat was usually considered an essential precondition. But while provisioning had dropped out of favor, field experiments were quickly gaining in popularity. Matsuzawa had no ethical or epistemological qualms when he not only reintroduced one-off trials in the forest but established a permanent outdoor laboratory, supplying oil palm nuts year after year during research season. "In Bossou, humans are providing nuts everywhere," Matsuzawa said in justification of their interference with the apes' feeding ecology. With or without field experiments, this remote corner of Guinea was anything but a pristine habitat.

Because of the West African Ebola outbreak in late 2013, which had begun in a nearby part of Guinée Forestière with a two-year-old boy playing in a hollow tree full of infected bats, the Japanese primatologists had missed the last two years of Fanwa's development. As Matsuzawa resumed his research in Bossou, he was eager to learn whether the four-year-old had started nut cracking—a skill that chimpanzees acquire between the ages of three and four. Initially, it did not seem so. Fanwa's mother broke open the shells with hammer and anvil and tolerated her son's theft of kernels. Fanwa keenly watched her every move. Eventually, he began to copy her actions. At first by hitting the ground with his bare hands. Then with a stone but no anvil. Soon with stone and anvil but no nut. And eventually he also placed nuts on an anvil, adroitly hit them with a hammer, and ate their contents. In the course of each session we observed over the next days, the four-year-old seemed to recapitulate the entire ontogenesis of this behavior for us. Although Matsuzawa was willing to conjecture about the phylogenetic and historical origins of nut cracking in informal conversations, his scientific publications had shifted the focus away from Kortlandt's speculation about cultural transmission between species to the observable development of nut cracking in individual chimpanzees.[31]

In the forest, lush vegetation frustrated the detailed observations necessary to reconstruct the fiercely contested details of how chimpanzees learned nut cracking. This problem haunted the naturalistic observations favored by Boesch and Sugiyama. Taï consisted mostly of primary forest where the crowns of emergent trees kept the floor dimly lit. Although the darkness repressed the uncontrolled proliferation of undergrowth, visibility was still limited to about twenty meters. By contrast, most of Bossou's giant trees had fallen victim to large-scale industrial forest exploitation in the mid-twentieth century. But the thickness of the remaining secondary forests often made it even more difficult to see. Even many well-habituated chimpanzees experienced an observation distance of five meters as too intrusive. Moreover, the researchers' self-imposed rules required them to stay at least seven meters away from the animals at both Bossou and Taï to minimize the risk of disease transmission. In a low-visibility environment, the outdoor laboratory provided a space that was kept free from vegetation to provide an unobstructed view of all activities, even of the most timid individuals. Allowing both watching and intervention, this new research site was in equal parts laboratory and observatory.

But sometimes it was not just low visibility but also the singularity of events that undermined the truth claims of fieldworkers. Matsuzawa designed the outdoor laboratory to stay clear of the kind of controversies—or rather "anti-controversies," since they were characterized mostly by a roaring silence—that haunted naturalistic observations that nobody else would ever make again. Sugiyama's report on the first immigration of a male who had supposedly joined the Bossou community between 1980 and 1982 and Boesch's insistence on having seen active teaching among the Taï chimpanzees were cases in point.[32] "Fieldwork is filled with anecdotes in which a researcher claims that he or she witnessed a behavior," Matsuzawa wrote. "No one can deny it: the observer claims that he/she saw it."[33] But no one had to accept it either. To avoid such awkward situations, Matsuzawa had introduced camcorders in 1987, shortly after Sony had commercialized the first battery-powered models in Japan. As in the Inuyama lab, these cameras filmed the chimpanzees' every move. The video recordings allowed colleagues not present at the time and place of their occurrence to carefully examine even singular behaviors again and again, if necessary in slow motion. The resulting footage enabled fieldworkers to go beyond anecdotal evidence and amounted to a video archive spanning thirty years. Thus the outdoor laboratory gave researchers many more opportunities than the forest to systematically observe rare behaviors, generating more robust data in less time. It almost seemed like the best of all primatological worlds.

At nightfall, the nut-cracking session in the outdoor laboratory ended with one of the old females trudging off into the forest. One by one the others followed, giving the human observers behind the palm-leaf screen a wide berth.

Only alpha male Jeje swaggered across the clearing straight toward our small peepholes, around the partition, past our tripods, cameras, and research equipment, and right through the auditorium. "That's how he shows his braveness," whispered Matsuzawa. Then Jeje, too, disappeared between the trees. Exit the apes.

Chimpanzee Cultural Zones

Once the cogs of this knowledge-producing machine started turning amid rustling leaves and singing birds, it spat out pieces of information that fueled the culture wars within the discipline and challenged local understandings of the Bossou chimpanzees. In the winter months between 1992 and 1995, Noriko Inoue-Nakamura and Tetsuro Matsuzawa provided the chimpanzees visiting Le Bureau with oil palm nuts and stones to observe and videotape how three individuals under the age of four learned nut cracking.[34]

At the time, one of the few other studies of how wild chimpanzees acquired tool-using skills was Boesch's much disputed report of two cases of active teaching among the Taï chimpanzees.[35] Confronted with the problem of fieldworker's regress—the fact that in the field sciences differences between scientific observations can always be attributed to differences between research sites—Inoue-Nakamura and Matsuzawa remained doubtful but polite, describing active teaching among chimpanzees as "rare": "So far, there have been no instances of active teaching or guidance except two reported episodes in which mothers influenced their infants' attempt to crack nuts."[36] The analysis of their footage, filmed in the course of ninety-four days, provided no hint at any form of schooling. At least in Bossou, they reported, "the chimpanzee mothers' attitude toward their infants was characterized by the lack of any feedback to infant chimpanzees' attempts at nut-cracking behaviors."[37]

Although Inoue-Nakamura and Matsuzawa did not openly challenge Boesch, they also sided with Tomasello and colleagues' denial of true imitation learning in chimpanzees because the infants had not exactly copied the demonstrator's methods of tool use.[38] Instead they had learned by way of emulation, understanding the demonstrator's goal as well as the general functional relation between stones and nuts while figuring out the details of the process through trial and error—a result that Matsuzawa further corroborated in laboratory experiments based on participation observation. Sitting face to face with a chimpanzee, he demonstrated how to open a box with a tool and found that his subjects hardly imitated his exact body movements (which, of course, were the movements of a different species of animal) but paid close attention to where on the box he applied the tool. Like their wild cousins, these captive apes did not focus on the path but on the goal.[39] This form of silent observational learning without direct instruction was what Matsuzawa and colleagues had called "education by master-apprenticeship," comparing it to how one learns the Japanese art of sushi making.[40]

Frans de Waal popularized Matsuzawa's theory in his book *The Ape and the Sushi Master*.[41] By now, my skepticism toward the cultural essentialism informing orientalist accounts of "Japanese primatology" should have become apparent. A similar case could be made for the people of Bossou: although most natives identified as Manon, their clans, defined by shared totems or food taboos, cut across ethnic boundaries between Manon, Konon, and Guerzé—a historical consequence of marital alliances, invasions, and migrations.[42] In the 1980s, the insight that human cultures do not form bounded and homogeneous wholes became the antiessentialist rallying cry of cultural anthropologists.[43] As it ossified into a hackneyed truism, cultural primatologists found that the same held true for chimpanzee cultures. In Bossou, observers noted that two adult females, Nina and Pama, who had disappeared years before my arrival, did not know how to crack nuts. Matsuzawa took them to be immigrants from the neighboring Seringbara community, which lacked a tradition of stone-tool use.[44] Yet primatologists did not take the diffusion of socially learned or not-learned behaviors to indicate a predicament of the just adopted culture concept.

Whatever political agendas might have led some contemporary anthropologists, most prominently Eduardo Viveiros de Castro, to advocate an ontological self-determination of the world's peoples, long-term observations of female migration between chimpanzee communities appeared to bridge the metaphysical gap between the chimpanzees of Mont Gban and those of other communities.[45] Primatologists assumed that chimpanzee groups exchanged their pubescent females, which spread not only their genes but also the behaviors they had acquired from their native groups. In all likelihood, the older females of the Bossou community had been born as the kind of "natural animals" from which Mato Topin Zogbila wanted to distinguish the Bossou chimpanzees.

The reason this traffic in individuals and their knowledge and ignorance had come to an end in Bossou was that human agricultural activities had isolated the local chimpanzees from their neighbors. In less anthropogenic environments, however, every generation of migrants introduced into their host communities new cultural differences, soon to be ironed out by the following generation. Matsuzawa and his team looked at migration as a natural experiment analogous to Hans Kummer's translocation experiments, which had transferred individual baboons from their native group to an unfamiliar one. While such interventions would have allowed study of how a chimpanzee adapted to and transformed a new social environment, there was widespread consensus among cultural primatologists that it would be highly unethical to mess with an ape's life in such a reckless way. Instead they observed the fate of migrants and their offspring.

Despite their unskilled mothers, Nina's and Pama's children turned into skillful nut crackers without any developmental retardation. Like all chimpanzee infants they also observed other adult females and were quickly enculturated into the Bossou community. It was these inquisitive children who restored some conformity in the group's social traditions—until the next immigrants arrived.

Such diffusion of cultural competencies and incompetencies became the key to interpreting the results of the first truly manipulative experiment in the outdoor laboratory. Widespread human cultivation made oil palm nuts available to the Bossou chimpanzees. No other nut grew in the population's territory. Matsuzawa sought to elucidate the sources of cultural diversity across different wild chimpanzee communities as well as the mechanisms behind the maintenance of this variation. In dry seasons between 1993 and 2006, his team presented the Bossou chimpanzees with two species of nut, coula and panda, that were found and cracked by neighboring groups (and also in Taï), but not in Bossou. If the Bossou chimpanzees turned out to be unfamiliar with these nuts but found a way of opening them, the field experimenters could study chimpanzee innovation in action and observe its subsequent transmission in the community.

In the outdoor laboratory, most chimpanzees first handled, sniffed, and mouthed the coula nuts before trying to open them like oil palm nuts. The exception was Yo, who, seemingly acquainted with coula nuts, unhesitatingly picked out and cracked the ripe ones. When it came to the unusually hard and less nutritious panda nuts, however, all members of the group gave up after only a few tries. Dora Biro, an animal behavior researcher from Oxford who had collaborated with Matsuzawa on this experiment, reasoned that Yo had probably immigrated from a neighboring community such as Yealé on the Ivorian side of the Nimba Mountains that habitually cracked the coula nuts growing in its home range. Yo could serve as a model, especially for the juveniles, and appeared responsible for the spread of the behavior in subsequent years. By 2002, almost all nut crackers also cracked coula and often preferred them over oil palm nuts. However, the less attractive panda nuts failed to catch on in an environment that offered tastier alternatives and no reliable model for how to open them. Biro and colleagues concluded: "Through immigration, social transmission, and subsequent 'education' through generations, sets of neighbouring chimpanzee communities can come to comprise 'cultural zones' characterised by the possession of certain shared behavioural traditions."[46]

Neither Biro nor Matsuzawa and colleagues provided a bibliographic reference for the notion of cultural zones, but their usage suggests that they defined these zones in terms of a correspondence between a geographical area and the cultural features shared by neighboring groups inhabiting this territory.[47] Social scientists from Max Weber to Samuel Huntington had previously employed the term *cultural zones* in similar fashion.[48] Anthropologist Alfred Kroeber conceptualized the confluence of territories and traditions as *culture areas*, which overlapped so much that he would have preferred to map their differences through shading rather than boundary lines.[49] Ernest Gellner opposed cultural holism by emphasizing that cultural zones like the Habsburg Empire were internally so diverse and conflicted that "they do not in any sense have a 'single' culture."[50]

Harvard primatologist Richard Wrangham accepted the idea of local chimpanzee cultures but questioned whether they hung together as what he called culture zones. Considering that Bethan Morgan and Ekwoge Abwe had observed nut cracking in Cameroon's Ebo Forest, about 1,700 kilometers east of the N'zo-Sassandra River, which supposedly demarcated the West African chimpanzee culture zone, he argued that the culture zone concept had become untidy: "If invention is the chief constraint on distribution, a series of discrete culture-zones should be evident within which neighboring populations show similar traditions. But of the dozens of chimpanzee behaviors that appear to be social traditions, from ant-dipping and palm-pounding to leaf-clipping and hand-clasp-grooming, many have distributions at least as quirky as nut-smashing. . . . Disjunct distributions are the norm, not the exception."[51]

Maybe Gellner's remark about discord within the Habsburg Empire indicates where the notions of chimpanzee and human cultural zones diverged: as collections of largely isolated cultural traits that adapted primarily to an environment but not to each other, chimpanzee cultural zones lacked not only the coherence that holists expected of cultures but also the tensions resulting from the antagonistic relatedness of discordant values or antithetical beliefs. Although the social life of chimpanzees was full of conflict and drama, primatologists did not describe their cultural life in such terms. Neither field experiments nor observations had so far revealed any clash of civilizations or a *Kulturkampf* arising from incommensurable traditions and innovations (although Shirley Strum described how a "disagreement" about whether to maintain their old way of foraging in the savanna or take on the easier but risky lifestyle of crop raiding split a group of baboons).[52]

Considering the importance of conflict in storytelling, this lack of tension strictly limited the narrative potential of chimpanzee ethnography as a genre. As I write this book, the reignited culture wars in the United States and Europe keep newspaper readers in suspense. Although more civil in tone, the chimpanzee culture wars chronicled here unfold the antagonisms between proponents and opponents of an extension of multiculturalism beyond the human, between European, American, and Japanese primatologists, and between laboratory workers and fieldworkers. No such commotion animated the monographs of chimpanzee ethnographers because their subjects were not divided over cultural issues.[53]

Since most primatologists saw ethnographic storytelling at best as a byproduct of their research, another limitation of their approach to cultural zones weighed heavier. If cultural zones were indeed the outcome of both social learning and ecological affordances, the outdoor laboratory could not control for whether it was the absence of a model or access to more attractive food sources, or both, that prevented the adoption of panda-nut cracking. And yet the more controlled setting of an indoor laboratory would not have allowed even asking that question in the first place. In a captive setting where chimpanzees were not free to leave and join other communities, the discovery that an immigrant facilitated the spread of coula-nut cracking would have been most improbable (unless she had learned it in another institution before being translocated). In Bossou, the very lack of human control had provided the conditions of possibility for an unexpected finding and a new insight. Moreover, by providing field experimental evidence for a cultural zone, Matsuzawa and his colleagues extended the significance of their knowledge claims from the field site of Bossou to a higher regional, although not global scale.[54]

Synthetic Primatology

Historian of science Gregory Radick contrasted the lab-based ape language projects, "which had bent the wills of the animals to the experimenters' interests," to field experiments that "had let the animals be free."[55] Instead of teaching them in the scientists' world, the scientists "had taken themselves to the animals' world in order to learn." This combination of relatively natural life and relatively controlled experiment led Radick to claim that "the conducting of a field experiment brought with it an unbeatable combination of moral and epistemic authority: the authority of experiment over observation or speculation, and the authority of nature over artifice."[56] In his career, Matsuzawa banked on the epistemic power of such hybridization:

> My research approach has focused on synthesizing these two different approaches (observation and experiment) and these two different study environments (natural habitat and laboratory setting). I thus developed two paradigms: (1) field experiments in the wild and (2) participation observation in the laboratory. In my view, this holistic approach is the most

suited in providing us with a truer and deeper understanding of chimpanzees as a whole.[57]

In a conversation about the underlying philosophy, Matsuzawa explained that there was something Hegelian about his methodology: "When I spoke of synthesis I thought of dialectics: thesis, antithesis, synthesis. For me, the thesis is laboratory work, the antithesis is fieldwork, and the synthesis is my work." Matsuzawa's assemblage of laboratory experiments and field observations created a space for chimpanzee research in the borderland between bench and field that European and American biologists had begun to carve out in the late nineteenth and early twentieth centuries.[58] In the 1910s, Robert Yerkes, one of the founders of primatology in the United States, had already explored alternatives to the laboratory such as his farm and other private estates, barns, or circus quarters to establish a continuum between highly controlled observations in the lab and naturalistic observations of wild primates.[59] Yet Matsuzawa attributed the holistic spirit behind his methodology to the "Japanese way of thinking," which preferred assembling rather than segregating things.

If one could have the experimental cake and eat it naturally, primatology would need no other approaches. But Matsuzawa did not mean that either field experiments or participant observation in the laboratory represented the ultimate method. None of these practices amounted to a Hegelian form of *Aufhebung* that simultaneously abolished, preserved, and transcended thesis and antithesis.

Field experiments always compromise and come with their own drawbacks. As Matsuzawa's longtime collaborator William McGrew remarked about Bossou: "Their 'outdoor laboratory' is limited by natural constraints: trials cannot be scheduled nor subjects assigned randomly to treatment."[60] Field experiments trade control for ecological validity. For instance, Osamu Sakura noted that laboratory research usually isolated test subjects: "But in the field, they form their natural social groups. If the dominant animal wants to use this stone, the other animals have to wait till the dominant one finishes. So we can observe a lot of the social parameters." However, if researchers wanted to know what physical qualities chimpanzees were looking for in a good stone hammer, but the group's dominance rank hierarchy constrained its members' choices, a more human-controlled situation excluding dominant individuals could provide less noisy data by giving subordinate individuals the freedom to pick their favorite hammer. Here, an indoor laboratory would produce superior data because confounding variables could be managed more efficiently. If, on the other hand, the question was how much of their caloric intake chimpanzees met by eating nuts during the nut-cracking season, then human provisioning of those nuts in the outdoor lab would distort the results. Consequently, naturalistic observation would produce superior data. Field experiments served as a bridge between field observations and laboratory experiments. They did not sublate but complemented and connected classical fieldwork and benchwork.

Matsuzawa explained that field experiments helped "improve the ecological and social validity of cognitive experiments in the laboratory. This kind of inverse translational approach gave us the idea of 'participation observation', which relies on the daily-life, direct face-to-face observation of captive chimpanzees."[61] This chain of translations between rigidly controlled experiments, field observations and participant observation in the indoor laboratory, field experiments in the outdoor laboratory, and field observations in the forest had become the hallmark of Matsuzawa's synthetic primatology.

Since it is virtually impossible to witness the entire process by which infant chimpanzees acquire nut-cracking skills in the wild, the field experiments at Bossou led to participant observation and laboratory experiments at KUPRI. As discussed in the previous chapter, Misato Hayashi and colleagues devised an experiment in which Matsuzawa provided three female chimpanzees, Ai, Chloé, and Pan, with macadamia nuts (oil palm or coula nuts were too difficult to obtain in Japan) and stone hammers and anvils from Bossou.[62] This first session in 1997 was followed by a second session four years later. In the meantime, all three subjects had given birth, and in 2001 their infants accompanied them. Every single opportunity to observe and practice nut cracking was on record.

While the original demonstration had been based on observational learning of the sort practiced by wild chimpanzees, the second session involved active teaching, which Matsuzawa took to be a unique feature of human culture. Like some of the ape language tutors teaching their students the gestures of American Sign Language, he even molded the chimpanzees' hands to crack open the nuts. The model was human. In Matsuzawa's eyes, interspecies social learning was not necessarily an ecologically invalid laboratory situation but possibly the historical origin of this form of tool use in West Africa.

Pan and Chloé quickly learned to open their nuts. Chloé even began to perform all the right actions after a single demonstration. Despite her acumen in the symbolic realm, Ai, however, never learned to crack any nuts. Ai's failure pointed to the same question as Nina's and Pama's inability to crack nuts: "Why do some individuals fail to acquire the skill?" Hayashi and colleagues asked.[63]

Whereas culture was the answer provided for Nina and Pama—they were immigrants from a chimpanzee population that had no tradition of nut cracking—Ai had spent less than a year in her mother's group in Guinea before she was abducted and raised in the same *Pan/Homo* community as Chloé and Pan. The laboratory observations of Ai's behavior zoomed in on the details of her object manipulations: she placed her nuts on an anvil, but instead of hitting them with another stone she pressed them with her feet or knuckle, a behavior previously observed in Bossou infants.[64] Yet no developmental psychology experiment could explain why the entire Yealé community, where Nina and Pama had supposedly grown up, did not know how to crack nuts. That question had to be taken back to the field, where an ecological analysis of chimpanzee culture suggested that "the density and distribution of nut trees contribute to the likelihood of nut cracking being invented, as well as later transmitted, in a population of chimpanzees."[65] And so it went, back and forth, back and forth between outdoor observations and indoor experiments with stepping-stones in between.

In a conversation with me, Andrew Whiten recognized Matsuzawa as "the shining star" in the borderland between laboratory and field research. However, Whiten situated his Japanese colleague in an otherwise Euro-American lineage of "catholic yet integrated methodological programmes" that complemented systematic and quantitative observation with experimentation in both field and laboratory, ranging from early ethologists like Konrad Lorenz and Nikolaas Tinbergen to primatologists like Hans Kummer, Dorothy Cheney, and Robert Seyfarth.[66] Whiten's own research on social learning in chimpanzees had also taken him "from the field to the laboratory and back again."[67] Like Matsuzawa, he had added links to this chain of translations. He not only conducted and advocated field experiments to address a "laboratory/field mismatch" but also designed more naturalistic captive experiments.[68] Experimenting on whole groups of animals rather than isolated individuals, he examined the controlled transmission and open diffusion of newly acquired behaviors through a population of chimpanzees.[69] While Matsuzawa had become the most prominent bridge builder between lab and field, he was not the only one, and the endeavor was just as German, Dutch, Swiss, American, and British as it was Japanese.

In contrast to Hegel's linear philosophy of history and early twentieth-century expectations that biological research would progress from the field to the laboratory, Matsuzawa's—but also Whiten's—synthetic primatology moved in circles, slowly spiraling toward an entwinement, not a fusion, of fieldwork and benchwork.[70] Different methods continued to confront researchers with trade-offs between control and ecological validity, but a growing number of intermediate approaches mitigated the antagonism between controlled experiments and field observations that fueled the chimpanzee culture wars.

Fieldwork with Garden Scissors

Naturalistic observation was slowly fading out at Bossou, despite its complementary role in Matsuzawa's holistic approach. Sugiyama had still accompanied the chimpanzees from dawn until dusk, picking them up at their nests every morning—just as the students and field assistants did at Taï. But Matsuzawa's group felt that this kind of fieldwork had run out of steam. "Fifty years ago, you could go to Africa with a notebook and a pen and you could publish your results," Morimura told me. "That's what Goodall and Imanishi did. Today, that's no longer possible."

While Matsuzawa's data came from the outdoor laboratory, Morimura's was recorded by approximately ten camera traps set up at different points of interest. The field assistants had maintained the devices during the Ebola years and would continue to service them while the Japanese returned to their home institutions. Morimura preferred this hands-off approach. Four decades of direct observation might have brought chimpanzees and humans all too close.

During the three weeks of my stay, however, we followed the chimpanzees every day. Matsuzawa honored my interest in how his team dealt with their research subject in this slightly folkloristic way. By 2015, the practice of naturalistic observation served mostly educational purposes. About half of the students and researchers visited Bossou to get a better feeling for what chimpanzee life was like if not confined to a laboratory, sanctuary, or zoo. Morimura, for example, had first come in 1998 to find inspiration on how to design the Hayashibara Great Ape Research Institute. And he continued to take a guilty pleasure in spending half days with the chimpanzees, taking notes on their activities, although for no specific study.

On a chilly December morning, the first after our arrival, Morimura and two field assistants took me to the sacred Mont Gban, where we listened for chimpanzee pant-hoots to locate the group's whereabouts. In Taï, nest-to-nest following made this practice superfluous unless one had lost the chimpanzees. In Loango, where the chimpanzees were poorly habituated and forest elephants made it too dangerous to hike back in the dark, such intense listening carved out moments of highly concentrated, almost reverent silence. We would sit on the forest floor, facing different directions on guard against elephants, absorbing bird songs, the shrill droning noise of cicadas, and the creaking of trees. In Bossou, however, it was the sound of a motorbike that cut through the murmuring of leaves. Looking up from his cell phone, one of the Manon asked: "Today, it's the 24th. Isn't there this festival? What's the name again?" Morimura explained to the assistants who worshipped the god Wala and sacrificed to the devils inhabiting the hilltop that this was an ordinary workday for him because he came from a Buddhist rather than a Christian culture. The birth of Jesus and the deep time in which Darwinians experience something of our primordial past in the encounter with wild apes seemed equally removed from contemporary Bossou, where the signal of a nearby mobile phone mast extended early twenty-first-century life well into the forest.

Text messages and emails were received and sent, phone calls from the village taken. No dress code required us to wear forest-green clothing—presumably blue jeans and bright red sweatshirts did not perturb chimpanzees as used to encounters with humans as the Bossou community. And no *barrière* separated human life at the field station from animal life in the forest. Boots were occasionally cleaned at the well and facial masks put on when in direct contact with the chimpanzees, but we could wear the same clothes in the forest as in the village.

The one significant hygienic measure introduced during my stay prohibited fetching takeaway food from the village. Instead the assistants now had to return to the station for their lunch break. Although the Ebola epidemic had come to a halt six kilometers from Bossou, Matsuzawa worried about putting the chimpanzees at risk by bringing potentially contaminated food items into the forest. Note the contrast to the established Ebola cosmologies: Matsuzawa feared less the zoonosis lurking in the wilderness, threatening to spill over from its natural reservoir into the human population, than a virus that had now found its place in human settlements from where it endangered wildlife. With meals relocated from the forest to the kitchen table, even the aftermath of the Ebola crisis lent fieldwork at Bossou a distinct sense of everydayness.

When we finally heard the chimpanzees screaming at some distance, we first hurried along footpaths that the apes also liked to walk. With no elephants or other megafauna left, only *Homo* kept open these tracks leading through the otherwise almost impenetrable secondary forest. One field assistant blazed a trail with a pair of garden scissors. How alien such a pruning of nature would have been to the philosophy of Taï. Not to mention that there the chimpanzees often moved so fast through the more open primary forest that neither guides nor researchers had time for trimming branches and vines—they just stumbled onward. Matsuzawa later explained to me that he had introduced the garden scissors to primatology. “I don't like the sound of the machete,” he said. “This is the Japanese way of observing chimpanzees.”[71]

Since no poacher roamed the forest with pruners, the gentle clicking of the blades also announced to the apes that the harmless researchers and their assistants were coming. Still the chimpanzees uttered a soft *ouh* as we approached, and a field assistant responded in kind. “It means: ‘Who is there?’” translated Boniface Zogbila, one of the more senior guides. “And we assure them that it's just us.”

The fact that people like Matsuzawa and the older assistants whom many of the chimpanzees had known for their entire lives frequently brought new doctoral students or visiting researchers like me did not pose a problem. The apes trusted us by association while avoiding close encounters with the other villagers, the assistants told me. Boniface's relative Mato Topin had claimed that the chimpanzees would hide if they heard human languages other than Manon. But they certainly did not mind my broken French, nor English, nor Morimura and Matsuzawa's Japanese.

These forms of interspecies communication and eavesdropping served as another reminder that in great apes, habituation gave rise to a more complex relationship than the blending of human observers with the scenery. As with the South Group in Taï, not all members of the Bossou community experienced us "like a breeze, like the air, like a rock," as Matsuzawa described habituation. Fanwa, the youngest, could not take his eyes off the strange animals following them around. At age four, he was becoming rambunctious. In the outdoor laboratory, he often threw stones in the direction of one of the senile females, who usually took this as an incentive to play. But the challenging nature of the gesture was unmistakable. Should the old ladies still be alive a few years from now, Fanwa would beat every one of them into subordination. He had already extended his provocations to us: sometimes he threw a stone, more often he dropped branches from above—a prank we pretended to ignore, but it always reminded us that in the summer, one of the doctoral students at KUPRI's bonobo project in Congo had been hit by falling wood and ended up quadriplegic. Whether the apes had anything to do with it nobody knew. Morimura expressed concern about Fanwa's lack of distance, which could get the ape into serious trouble with humans.

Probably Peley had fallen victim to his own transgressions. Born in 1998, he belonged to a generation that might have learned from its already habituated parents that the local specimens of *Homo* were not dangerous predators. At age fifteen, he had fought his way up to the alpha position in the group and displayed his dominance, among other things, through increasingly disrespectful behavior toward researchers and assistants. He walked up to them from behind, touched them, and tested the limits. Whereas in Taï the rule was to remain absolutely passive in such situations, the Manon guides reacted like the local farmers when conflicts with the chimpanzees arose: they threw branches or waved a tripod at this powerful teenager in an attempt to regain their standing. Morimura grew convinced that within the next months there would be a serious incident.

Before this could happen, however, Peley and two other chimpanzees disappeared in 2013. Japanese efforts to regrow patches of forest in the savanna between Bossou and Seringbara had enabled this courageous party to visit the Nimba Mountains, raiding the local farmers' crops along the way. The fact that none of them returned from their expedition pointed to a coordinated act of killing, possibly by shooting them at night in their nests. Nobody knew for sure.

If Loango and Taï—despite the incursions by poachers, militias, and Chinese oil companies—still maintained some semblance of fierce and primeval wilderness, Bossou could easily be mistaken for a bucolic idyll: nature as a garden. But such gardens were treacherous. Hearing cheerful voices from the village and texting with one's friends while doing fieldwork in the forest with a pair of pruners gave a false sense of security. It was no coincidence that as devastating a disease as Ebola had emerged from the recently torn apart ecosystem of Guinée Forestière, where human plantations had cut deeply into the forests while the forest's former inhabitants found new homes and food sources in fields around human settlements. These emerging contact zones with species that *Homo sapiens* had previously had little to do with could turn into hotbeds for zoonoses.[72] Such sites could also spawn intense human-animal conflict, even in a village as tolerant toward chimpanzees as Bossou.

A "Naturecultural" Experiment

At 9:00 a.m. the chimpanzees descended from Mont Gban to visit the outdoor laboratory at the foot of Mont Guein. They crossed the first of two dirt roads leading to the nearby Liberian border without incident. The field assistants adhered to the "observation rules" pinned to the wall of the field station's kitchen: "If chimpanzees approach a road or path, it is important to ensure that at least one guide reaches the path or road rapidly in order to avoid any problems between chimpanzees and people." This required sufficient familiarity with the apes and their territory to predict where they would be heading. When the animals approached one of the traffic arterioles of rural Guinea, one of the observers ran ahead, occasionally waving down a motorcycle or car until the chimpanzees had passed. As much as the apes appreciated the well-pruned trails through the forest, they treated the much broader and more open roads with utmost caution.

While we waited at the second road, a woman walked by toward the Liberian border, carrying a big plastic bucket full of goods on her head. When she had just passed the point where we expected the chimpanzees to appear, her dog also ran past us. Although dogs figured as the Zogbilas' second totem animal, they were not revered by everyone in Bossou. Other families ate them. I caught my breath as the first chimpanzees traversed the road right in front of the pet. The dog briefly stopped and then dashed forward toward his owner just as the other chimpanzees emerged from the shrubs. Having already passed through the group, he turned around and started barking. Jeje, Foaf, and one of the females ganged up and moved in his direction, first slowly, then at a run. The woman started screaming at the top of her panicked voice. My heart pounded as I expected to see her canine companion torn to pieces while the field assistants broke out in laughter. A moment later the dog got away and the chimpanzees let him be. They had never killed a dog, the assistants assured me.

After fifty years of long-term studies, field observations of wild chimpanzees no longer stirred up a swirl of surprise findings of new "natural" behaviors, although Boesch and other contemporary naturalists continued to produce important insights as they revamped chimpanzee ethnography with twenty-first-century technologies. But primates all over the world changed their behavior in response to drastic transformations of their habitats brought about by humans. In pristine forests no roads had to be crossed, but by 2015 traffic cut through even national parks. In a report to Japanese car manufacturer Toyota, which had donated a jeep and some money to the Guinean field station, KUPRI-trained primatologist and Africanist Gen Yamakoshi wrote: "Research findings from Bossou were sometimes dismissed as biased or unreliable on the grounds that the area has been the subject of considerable human disturbance."[73] But Yamakoshi and his British colleague Kimberley Hockings made a virtue out of necessity by casting a naturalistic eye on those "disturbances" as value-neutral interactions between two primate species—for that is all they were, if one gave up what Yamakoshi dismissed as the "romantic and mystical desire" of ecologists "to place themselves in primitive environments as far as possible from the modern world, . . . closely linked to attitudes from the colonial era."[74]

Road crossing, for instance, could serve as a natural—or "naturecultural"?—quasi experiment to study how chimpanzees organized socially and spatially in response to fearful events.[75] Crossings provided rich opportunities for observation because researchers could easily anticipate when they would occur. So could the apes, of course, which might make a significant difference to being ambushed by a predator. Thus roads were certainly no perfect model for encountering something frightening. But leopards rarely preyed on chimpanzees in the presence of humans. Even a forest housing as many big cats as Taï provided scientists with too few chances to observe such attacks directly, and around Bossou leopards had been driven extinct altogether. By contrast, within only four months, Hockings, Anderson, and Matsuzawa witnessed almost thirty road crossings.[76] They found that the animals' hierarchical social structure enabled the chimpanzees to coordinate their responses to such potentially risky situations. Often, the alpha or beta male would go first, standing guard on the road as the alpha female led the others across, while another high-ranking male covered the group from behind. This spatio-social organization indicated to Hockings a "division of roles" similar to the one Boesch had described for the Taï chimpanzees' cooperative hunting.[77]

FIGURE 19. Road crossing: a natural experiment to observe how chimpanzee groups respond to threatening situations. Photo by author.

If Hockings was right that education by master-apprenticeship allowed the Bossou community to pass down its recently invented approach to road crossing, did other primatologists observe any chimpanzee populations that had traditionalized different ways of dealing with human traffic?[78] At Sebitoli, Uganda, Marie Cibot and colleagues also found that it was usually healthy adult males who led their parties across the roads of Kibale National Park.[79] Yet Hockings explained to the *New Scientist* that the Sebitoli community "tended to split into smaller subgroups when crossing, whereas chimpanzees at Bossou often, but not always, crossed in progression lines. This might be down to a higher intensity and speed of traffic at Sebitoli, forcing chimpanzees to split up."[80] Another media report compared this geographical difference in chimpanzee behavior to "how a local's opinion of J-walking can change from city to city."[81]

At Wamba, a field site in the Democratic Republic of Congo directed by KUPRI primatologist Takeshi Furuichi, Shinya Yamamoto studied how bonobos dealt with human traffic. He discovered that the "hippies of the primate world" ran across roads without the chimpanzees' hierarchically organized cooperation, Matsuzawa noted in summarizing the study of his former student.[82] Whether these behavioral differences within and between *Pan troglodytes* and *Pan paniscus* were genetic, ecological, cultural, or all at once, such hardly controlled naturecultural experiments could not tell.

Ethnoprimatology: Naturalism after Nature

Foaf and his elderly mother scurried up the *sous-préfet*'s papaya tree into a human world both luscious and menacing. Without her last surviving son, a brawny but shy thirty-five-year old, Fana might have preferred to doze next to her daughter and grandson in the plantation where we had left the group of chimpanzees. When they entered a field accompanied by the Japanese researchers or their local guides, farmers frequently demanded compensation, claiming that the ape watchers had not just followed but driven the animals onto their property. Chimpanzee mother and son must have bided their time between the crops, waiting for the right moment. Only minutes before, dozens of men with machetes and dogs had passed by to look for another villager who had vanished into the forest—a victim of either his own hand or sorcery, they feared. And two human infants had frolicked around in the withered grass. Now they watched their mother, Madame la Sous-préfète, making a fire in a small thatched hut. Inside, a radio host chitchatted with his guests as the Manon family prepared lunch. For a few minutes, field assistant Vincent and I were the only humans left outside, sitting in the shade of a small concrete house as we waited for the chimpanzees to resurface from the field. Foaf and Fana knew the two of us to be harmless and grasped the opportunity to dart out of the shrubs to the papaya tree at the edge of Bossou village. Seconds later, Fana and Foaf slid down again, big ripe loot in their mouths, and disappeared into the field.

Today, nonhuman primates, including chimpanzees all over Africa, inhabit increasingly anthropogenic worlds. In response to the loss of untouched habitats that endowed ecologically oriented research such as Boesch's with an aura of authenticity, a new field of inquiry had emerged since the 1990s. When ecologist Leslie Sponsel introduced the term *ethnoprimatology*, he sought to resolve the paradox that "biologists tend to consider *Homo sapiens* as part of nature in an evolutionary sense, yet apart from nature in an ecological sense."[83] In the spirit of "dark green religions," which since the nineteenth century had established new forms of nature worship in Europe and North America, Sponsel sought to revise the place of humans in nature through a metaphysical integration of naturalism and nature mysticism.[84] In contrast to the intellectual projects of cultural primatology and chimpanzee ethnography, however, ethnoprimatology did not study monkey and ape *cultures*, as the "ethno-" prefix might suggest. Sponsel even listed "the adaptive importance of technology and other aspects of culture" among the discontinuities that distinguished humans from other animals.[85] This original ethnoprimatology did not study the co-construction of cultures by human and nonhuman primates, either. Instead it continued to operate in a particular biological rather than ethnological register, taking as its research object the interface between human and primate ecology.

Epistemically, ethnoprimatology aspired to a disciplinary reintegration of biological and cultural anthropology. Sponsel explained the relevance of cultural anthropology to this endeavor by the fact that "the primatologist cannot know what is 'undisturbed' natural behavior and ecology until anthropogenic influences on the monkey species are known, if for no other reason than to factor them out or to establish their absence in areas where monkey populations are not 'disturbed.'"[86]

While Sponsel had devised ethnoprimatology to better understand how Amazonia's "original peoples of the forest" lived in "relative harmony with their environment" as an integral part of the ecosystem's faunal community, subsequent rearticulations of the project abandoned this primitivist imaginary, turning its gaze from "traditional indigenous societies" that hunted monkeys to human-primate encounters in modern agricultural and even urban settings.[87] These interactions included many novel phenomena, from Balinese macaques attracting tourist business around a Hindu temple, which in return provided them with food, to the increased cross-species exchange of pathogens like respiratory infections, AIDS, and Ebola viruses in the wake of research or logging activities in the forests of Central and West Africa.[88]

As human activities accelerated environmental change, primates were catapulted from the deep and slow temporality of so-called natural history into the staccato previously reserved for the history of *Homo sapiens*—a history that had originally excluded even human communities imagined to live in an unchanging state of nature, whose lives salvage anthropologists sought to capture in a quickly fading ethnographic present. Just as late twentieth-century cultural anthropologists like Johannes Fabian had pressed for the acknowledgment of supposed Stone Age people as our contemporaries, twenty-first-century ethnoprimatologists like Agustín Fuentes and Kimberley Hockings admitted their subjects into historical time.[89]

In ethnoprimatology nobody has conversed as much with current cultural anthropology as Fuentes. He has proved to be open even to the project of recent posthumanistic scholarship—extending agency, the supposed ability of humans to determine their own lives and make history, to other animals. For Fuentes, the object of ethnoprimatology is the "multispecies interface . . . in which the two species are simultaneously actors and participants in sharing and shaping mutual ecologies."[90] Taking his cues from both niche construction theory in evolutionary biology and animal studies literature in the humanities, Fuentes suggested that the "traditional language of evolutionary adaptation" was insufficient to understand how other primates not just passively adjusted to an environment created by humans but actively contributed to creating the ecological niche in which they coexisted with *Homo sapiens*.[91] Although Fuentes's own account had grown out of European and American scholarship, he noted that "the perception that organisms assimilate their environment through their body and that their bodies 'environmentalize' their local ecologies" converged with Kinji Imanishi's critique of the classical Darwinian paradigm of natural selection, in which organisms either adapted or were eliminated.[92]

That Fuentes situated ethnoprimatology in proximity to both the human sciences and the Kyoto School was no coincidence. Speaking of animal culture, society, and history, Imanishi and his heirs understood primatology as a field of knowledge closer to cultural anthropology, sociology, and philosophy than to the natural sciences.[93] The school's theoretical framework, Imanishi's account of evolution, was marked by the dialectic of agency and determinism, which Amanda Anderson and Joseph Valente identified as characteristic of the human sciences.[94] Imanishi rejected Darwinism because its conception of natural selection had been modeled on animal husbandry and "simply replaces human agency with nature."[95] In opposition to this view of nonhuman animals as fully determined by external forces, Imanishi maintained that "living things control themselves and their environment by their integrating nature," which is "in fact the same as humans' autonomy and independence."[96] The idea that organisms were "not passive objects of natural selection, but creative agents of evolution" did not imply that they were unaffected by their habitats, but it highlighted their vital role in shaping and interacting with their surroundings: "Only when living things react, does the environment make them live."[97]

Hans Kummer claimed that "most primates do not shape their environment in any adaptive way."[98] But even in a largely agricultural environment like Bossou, chimpanzees transformed the habitat they shared with humans, as both Imanishi and contemporary ethnoprimatologists claimed. In an email exchange, Hockings gave me an example: chimpanzees fed on and dispersed cacao seeds across their home range. But the cacao plants failed to reach maturity within the forest. They produced fruit only when deposited by chimpanzees on human farms where humans took care of the saplings by restricting weed growth and providing the right amount of shade. Overall, however, this *Pan/Homo* coconstruction of an ecological niche was highly asymmetrical. As far as the ability to remake environments for one's own purposes was concerned, modern humans trumped all other ape species by far.

At Bossou, the chimpanzees had lost their traditional livelihoods as most forest had given way to fields. But they had also gained access to an anthropogenic land of plenty, which bore almost nothing but edible and particularly nutritious fruit. Chimpanzee researcher Richard Wrangham emphasized the evolutionary gap between the digestive systems of *Pan* and *Homo*: having evolved into a species of cook, humans would be hard pressed to live on crudités.[99] But what nourishes us also nourishes chimpanzees, so they turned to the plantations surrounding Bossou and even raided papaya trees in the village.

Supervised by James Anderson at the University of Stirling in Scotland and hosted by Tetsuro Matsuzawa, ethnoprimatologist Kimberley Hockings showed in her doctoral research that the Bossou chimpanzees ate roughly 20 percent of their fruit and all their tubers in abandoned and guarded fields where they spent almost 10 percent of their feeding time.[100] Crop raiding had become part of their everyday life and became more and more frequent as deforestation proceeded.

Much rarer—it took a collective of researchers seventeen years to record twenty instances—was the pillaging of plastic containers, which people set up to collect the fermented sap of raffia palms.[101] Such field observations were the least controlled form of knowledge production in primatology. They left open why certain individuals abstained, whether the chimpanzees drank the juice because of its alcohol or sugar content, and so on. Only the behavioral consequences were unmistakable: after drinking the palm wine, the male chimpanzees often beat up the females, while human observers also had to beware. Eventually the drunkards would spend the afternoon sleeping off their intoxication, field assistant Bonifas Zogbila told me as we stood in front of his family's palm trees. From the point of view of the Manon farmers concerned, at least those still committed to their traditional cosmology, allowing the chimpanzees to feast on their cultivars amounted to an offering to the villagers' ancestors.

Not surprisingly, in a region as poor as Guinée Forestière, ancestor worship had its limits. While papayas, for instance, were considered children's food and could easily be left to the apes, a large pineapple sold for half a day's salary. Thus the owners of pineapple trees often used dogs to protect their crops against the chimpanzees.[102] Others, especially immigrants and natives of Bossou alienated from the old ways, tried to deter crop raiding more generally—at best by making noise, but sometimes with stones and worse. Although the Japanese researchers always emphasized that it was the Manon's own will to conserve the chimpanzees, they did use their resources to strengthen this engagement, as Hockings pointed out: "Much money from KUPRI has been invested in improving the health and education facilities at Bossou, which probably contributes to a 'social pressure' to accept a certain degree of crop-raiding by the chimpanzees."[103]

As ethnoprimatology sought to integrate primate ecology and sociocultural anthropology, its interest in human perspectives was more strategically than intellectually motivated. The primary goal was not to understand and take seriously what the Manon or other ethnic groups knew or believed about nonhuman monkeys and apes and to rethink the cosmological premises of Japanese, European, or North American primatology from a Manon point of view, but to enroll local traditions in the protection of primates and their habitats. As French anthropologist Vincent Leblan put it: "Ethnoprimatology is less governed by anthropological issues than by adaptive approaches to biological conservation."[104]

Maybe that was the reason Matsuzawa's holism extended to the interface of human and chimpanzee ecology but stopped short of assimilating human ethnography and ethnology into his methodological synthesis. It took a major conflict for the villagers' understanding of their coexistence with the apes to transform the Japanese primatologists' conservationist philosophy.

No Pastoral

Again and again the totemic bonds between human and chimpanzee inhabitants of Bossou were put to the test. Chimpanzee visits to village and fields did not bespeak an arcadian multispecies community. The reason almost everybody but the *sous-préfet* had cut down the papaya trees at the village edges was that the chimpanzee raids occasionally entailed attacks on humans, especially unaccompanied children, who could suffer severe injuries. In contrast to reports from other study sites, the Bossou chimpanzees did not seem to target human infants as prey. Instead Hockings, Matsuzawa, and their colleagues assumed that the children had become a political plaything in power struggles within the chimpanzee community. The attacks were exacerbated between 2003 and 2007, right when Yolo, the male suspected as the main perpetrator, was fighting his way up in the dominance hierarchy.[105] Encounters with human infants might have served as welcome opportunities to demonstrate his boldness in front of the group.

Historically, such attacks were not entirely new in West Africa. In the 1920s, British colonial officers had already compiled evidence relating to such incidents in neighboring Sierra Leone.[106] But if the numbers collected by KUPRI researchers and their assistants were correct, their frequency had increased in Bossou. While village elders could remember only three attacks before long-term observation began in 1976, the intensified habituation efforts under Matsuzawa's direction were accompanied by eleven attacks between 1995 and 2010 alone. And just as Hockings and colleagues published their study of these interspecies conflicts, the chimpanzees found three children whose family was visiting from another village playing in a field and killed a three-year-old boy.[107]

These incidents divided the villagers. Some elders excused the chimpanzees, who had probably drunk too much palm wine from a nearby grove, for otherwise they would never have harmed the people of Bossou.[108] Convinced of the ontological singularity of this ape community, others claimed that it must have been chimpanzees from elsewhere who had gone after the children.

On a third account, people from Liberia or from Bossou itself had transmuted into chimpanzees to commit these attacks.[109] Such belief in shape-shifting clearly had a sociological dimension, argued Paul Richards in his ethnography of the Mende, another West African people culturally related to the Manon.[110] The Mende also attributed chimpanzee attacks to shape-shifting witches who sought to obtain ingredients from the bodies of their human victims to concoct a magic potion believed to bring wealth and power. No wonder that the chimpanzees often tore off the children's digits, ears, and genitals—the very parts that the "human chimpanzees" needed for their charms. Following Edward Evans-Pritchard's social explanation of Azande witchcraft, Richards acknowledged that such accusations of shape-shifting were frequently occasioned by threats to the moral order such as power struggles within and between human communities.[111] However, Richards's work with Mende hunters also suggested an epistemic dimension. In light of their own knowledge of chimpanzee behavior, the fact that these usually most evasive animals approached and even attacked humans appeared out of character. This biological puzzle was resolved by assuming that the chimpanzees weren't really chimpanzees but humans in disguise. At Bossou, the exoneration of the local ape community must also have been fueled by the totemic identification and the extension of the villagers' kinship network to the chimpanzees.

A fourth account blamed the intensifying human-chimpanzee conflict on "the Japanese." The foreigners' presence had made the apes behave more aggressively toward the locals. All Manon families who had been subject to attacks expected monetary compensation from the researchers, whom many now considered to be responsible for "their"—the researchers'—chimpanzees. The afflicted demanded to be recompensed because they had given up cultivating the forest to protect the chimpanzees and felt that they should be supported when the animals caused damage. During his ethnographic research in Bossou, Vincent Verroux learned that the villagers held the scientists responsible for dissolving the traditional bonds between the local Manon and the chimpanzees.[112] They felt expropriated of their—the Manon's—sacred animal.

The day the apes killed the child, the forces that considered them a liability rather than a source of spiritual power and social identity gained the upper hand. An incensed crowd knew exactly who the culprits were and went into the forest with machetes and guns to slaughter the chimpanzees. Their status as totem animals might have increased many farmers' tolerance toward crop raiding, but not toward beating human children to death.

This put the field assistants in a difficult spot vis-à-vis other people in the village. The assistants took the side of the animals they knew so well and on whom they depended for their own livelihood. At the time of the incident, Bonifas Zogbila explained to me, local farmers had turned a lot of forest into plantations and chased the chimpanzees away with stones. Eventually, the apes vented their spleen on the child, he believed. But in Zogbila's eyes, the chimpanzees were innocent. The little boy's sister was to blame for his death: at age twelve, she should have picked him up instead of abandoning her brother in the field. So the assistants informed the gendarmerie to save the apes. In the ensuing conflict between the human inhabitants of Bossou, the guides of the Japanese were bewitched. Morimura had to pay a sorcerer in a neighboring village to cast a counterspell—an expense that presumably never made it into KUPRI's research budget report.

In the decades following Guinea's independence from French colonial rule in 1958, such magical practices, the totemic identification with different animal species, and the resulting particularism marked the ethnic minorities of the forest region as backward—an obstacle to realizing the vision of African socialism in a modern nation-state. The ruling Susu and Malinke elites imagined themselves as belonging to a sphere larger than the local and even the national. Their cosmopolitanism took the forms of Islam and Marxism. They envisaged the citizens of the new republic adhering to a monotheistic faith and conceiving of themselves as part of an internationalist *communitas* (although, lacking industry, Guinea hardly had a proletariat to unite with the "workers of the world").

In 1961, the state launched an iconoclastic social engineering project to root out the Forestiers' retrograde cultural practices: representatives of the Conakry government exposed men's masks to women and children to demonstrate that such sacrileges wouldn't kill anyone; they publicly burned fetishes and humiliated and beat ritual specialists in front of the village community.[113]

For the revered Bossou chimpanzees such Enlightenment zeal bore its risks, which made ethnoprimatologists like Hockings weigh the advantages of scientific disenchantment against the conservation of myth: "Education may increase knowledge and decrease fear, and consequently apes may become de-mythologized, making them more vulnerable to direct exploitation."[114]

Ultimately, the government's so-called demystification campaign failed. Since the 1980s, the suppressed rites had come back into the open. But the collective memories of these violent interventions into local cultures explained why members of ethnic minorities still considered the state as an instrument of the Susu and the Malinke majorities to extend their rule from the capital to remote Guinée Forestière.

Suspicious of the information provided by government officials, the young men tracking the chimpanzees had just helped to burn down an Ebola clinic near Bossou before it could even open. Rumor from Liberia had it that medical staff had poisoned patients and misrepresented them as Ebola cases to receive international aid payments. The field assistants wanted to protect their community against this threat. Needless to say, nobody believed those lies about bats being the origin of the ongoing epidemic: people in Bossou continued to eat them and not a single villager had fallen sick, one assistant enlightened me.

Forestiers also perceived the establishment of nature reserves as projections of external powers. Yamakoshi pointed out that "Western countries used their colonies to realize ideals that they could not put into effect at home," where they did not dare to forcibly seize their citizens' estates.[115] Around Bossou, the French first protected the Nimba Mountains as a Strict Nature Reserve in 1944. Piggybacking on the remnants of colonial rule, the Republic of Guinea continued to confiscate local people's land to establish conservation areas. Its officials had adopted the Euro-American idea of an untouched nature, which had no roots in West African cosmologies.[116]

In 1981, the socialist Guinean government designated Nimba as a UNESCO Biosphere Reserve, and ten years later a postsocialist administration added the Bossou forest to the reserve's core area. Traditionally resolving problems regarding the forest and the chimpanzees through discussions in the village's decision-making bodies, the inhabitants of Bossou did not welcome the imposition of national and international regulations on what they considered local affairs.[117] Without compensation they were now prohibited from extending their fields into the reserve. Caught between the nature reserve and Guinea's border with Liberia, they were scrambling for space to live.

Although the Japanese researchers greatly valued the Manon's homegrown protection of the chimpanzees, they worried that it might not suffice to maintain the population and involved the Guinean government and UNESCO. They initially believed that the apes would profit from enlarging the forest area as much as possible. Yamakoshi remembered his frustration with the Manon: "I often longed to carry out research in a wider natural environment, and sometimes I resented the actions of the villagers who cut down or burned the forest that had become the habitat for the chimpanzees."[118] As foreigners dependent on the hospitality of the local population, however, the scientists were in no position to enforce or tighten regulations.

With the help of the Japanese embassy, Matsuzawa lobbied for the establishment of a Guinean counterpart to KUPRI that would relieve him of the complex political negotiations with different factions of the village community. In October 2001, the national government founded the Institut de Recherche Environnementale de Bossou, which was meant to conduct primatological, genetic, meteorological, and sociological research. The Direction Nationale de la Recherche Scientifique granted the director of IREB, a Malinke, the political authority to manage conflicting chimpanzee-related interests in the area. From now on, a representative of the national government rather than the village controlled access to the chimpanzees for research and ecotourism. Moreover, IREB supported UNESCO in declaring the very animals that the village's founding lineage regarded as their family's ancestors a "world heritage."

In my interview with him, Mato Topin Zogbila asserted that the chimpanzees "are there for us and above all for my family," rather than for foreign scientists or humanity at large. In February 2002, the Manon responded to what they perceived as an intrusion of state power into their own way of coexisting with the chimpanzees by a defiant swidden campaign, cutting down trees in the core area of the reserve, which they had inherited from their forebears. Then the objectors spread a curse powder around the entrances of the forest. Despite orders by the IREB director to ignore the powder, the field assistants, now officially employees of IREB, refused to follow the chimpanzees, and research activities came to a temporary halt. Focusing on stakeholder conflicts between humans rather than coconstructed ecologies they shared with apes, African area studies professor, primatologist, and incoming director of KUPRI's field station Gen Yamakoshi and his West Africanist colleague Vincent Leblan provided a detailed account of the confrontation, which was eventually resolved as some families accepted lump-sum payments while others went to jail.[119]

Surprisingly, this cosmopolitical conflict largely converted the Japanese researchers to the defiant villagers' ecological and ethnoprimatological perspective. Old photos of Mont Gban show that by the 1960s, almost all tall trees on the lower part of the hill had been cut for commercial timber exploitation and had been supplanted by fields. As the state put more and more environmental regulations in place, however, the villagers had to abandon this farmland, and secondary forest took over in the late twentieth century.[120] An unidentified elder cited by Yamakoshi and Leblan and other Bossou notables argued that the "traditional" agricultural use of this area had served both humans and chimpanzees better than the new conservationism. The fields functioned as a buffer zone between the village and the apes, preventing assaults on children, which this spokesperson of the revolt blamed on the scientists' habituation efforts. Plantations also provided a much richer source of nutrition to the chimpanzees than any forest, as long as their crop theft was tolerated. Following this logic, the problem was not that humans had destroyed the animals' natural habitat but that reforestation had forced the chimpanzees to come closer to the village because they could not find enough to eat in the forest. The anonymous elder stood in for this belief and continued the cultivation of protected land, for which he was repeatedly imprisoned after 2002.[121] Verroux identified this rebel as Mato Topin Zogbila.[122]

That very year, Yamakoshi reported to Toyota that, setting aside his preoccupation with unspoiled nature, "what I had previously regarded as a village where people were a threat to the survival of chimpanzees I now saw as a place where people were successfully living in harmony with them."[123] Although both the Japanese primatologists and the Guinean IREB employees held on to the conviction that to reconnect the Bossou chimpanzees to neighboring communities they needed more forest, not fields, the Japanese came to consider the Manon model of chimpanzee conservation as the most efficient way of protecting humans against chimpanzee attacks and chimpanzees against human infectious diseases.

The conservation and restoration of nature gave way to the embrace of an anthropogenic landscape because its "transformation through human activities may imply habitat gain rather than loss for chimpanzees."[124] Matsuzawa attributed the group's extraordinary longevity and the fact that Bossou females had given birth at an earlier age than the members of other study populations to their superior nutritional condition.[125] He told me that ten or fifteen years ago, at about the time of the conflict with the villagers, he came to realize that, being confined to such a small territory, the Bossou chimpanzees profited more from their agricultural environment, which provided a high density of extremely nutritious foods, than they would have from a primeval forest that offered a diversity of wild but frequently inedible plants.[126]

The chimpanzees' rustic life would have seemed idyllic, if the community's overall number had not been in free fall. Mato Topin blamed this drastic decrease not on the quadrupling of Bossou's human population, but on the researchers: "When the scientists came, the chimpanzees died!" Matsuzawa and Morimura took this accusation to heart. But their own contribution to the chimpanzees' demise remained a matter of speculation and compunction.

Just as the relationship between experimental psychologists and their ape subjects had been barred from scientific study, the relationship between field primatologists and free-roaming primates constituted the blind spot of ethnoprimatology. Although ethnoprimatologists studied the interface of human and nonhuman primates, they focused almost exclusively on interactions between monkeys or apes and indigenous populations or tourists.[127]

There might well be political reasons for this omission. Neither ethnoprimatologists nor most other breeds of primatological fieldworker saw themselves as participant observers. They awkwardly recognized that their own presence affected what the observed animals did and possibly even whether they survived. But the consequences of the primatologists' being there were adverse effects to be minimized, not cultivated and highlighted in the manner of reflexive cultural anthropologists. Unnecessarily drawing attention to the anthropogenic nature of observed phenomena would have undermined the researchers' epistemic authority as fly-on-the-wall witnesses of spontaneous behavior. Even worse, in light of debates over whether human presence was detrimental to observed primate populations, it could call into question the moral legitimacy of their scientific work.

Assessing the researchers' impact also came up against a self-evident but profound epistemological problem: only researched populations were researched. At least before this dawning age of camera traps and genetic sampling, there was little reliable data on primate groups that did not interact with primatologists.

Concerned about what would become of little Fanwa, who once again threw twigs at us, Morimura pulled out his smartphone one afternoon and calculated how many chimpanzees had disappeared each decade since Sugiyama had set up shop in 1976: it had been 5 during the remaining 1970s; 15 in the 1980s, although Sugiyama came mostly alone and not even every year; 14 in the 1990s when Matsuzawa intensified the habituation efforts, bringing in significant numbers of visiting scientists and students; 10 in the 2000s, including the 5 victims of a flu-like respiratory disease probably transmitted by humans, and 5 in the first half of the 2010s. Considering that losses had peaked *before* chimpanzees and researchers saw each other more often and at shorter distances, the reasons for Morimura's and Matsuzawa's pangs of conscience were not statistically obvious.

Yet Morimura decided to curtail the time they would follow the animals after my departure. He vowed to rely more heavily on camera traps for the collection of scientific data and to dedicate his field trips to Bossou more fully to building what appeared to be an escape route from the sacred hill. Since 1997, Matsuzawa's team had collaborated with the Guinean government on a reforestation program, which sought to plant a so-called Green Corridor of trees across the savanna that separated the totem animals of Mont Gban from their wild relatives in the Nimba Mountains. In the face of the rapid decline of both biodiversity and cultural diversity within primate species, such forms of salvage primatology moved center stage in cultural primatology.

7

Salvage Primatology

CULTURAL PRIMATOLOGY EMERGED not just as an epistemic but also as an ethical and political enterprise, which sought to preserve a new form of biocultural diversity. Chimpanzee ethnographers fought for the survival of the species in its cultural multifariousness.

At the same time, primatology had itself been conceived as a multicultural field. While Europeans, Americans, Japanese, and Africans constantly learned from each other, they also fought over how to relate to other primates. The chimpanzee culture wars extended far into the realm of conservationism. The interests of foreign scientists, field assistants, local farmers, state authorities, and apes rarely aligned.

These political conflicts corresponded with different scientific interpretations and cosmological visions of the place of humans in the world. Who or what was to blame for the unabated overkill of other species that threatened to eradicate the great apes as well?

Amid this harrowing turmoil, chimpanzee ethnographers had quietly begun to archive the quickly evaporating chimpanzee cultures for future generations. This chapter examines how cultural primatologists modeled their enterprise on the salvage anthropology that had emerged since the nineteenth century, desperately trying to document and preserve human cultures that the colonial encounter was about to disfigure or eradicate. In the process, they created an archive in the spirit of a mournful positivism.

The Rebirth of Catastrophism

Diversity has become one of the supreme values of late modernity, appreciated across very different domains: conservationists fight for the conservation of biodiversity; people living with autism diagnoses demand a recognition of their neurodiversity; the European Commission and students of endangered languages advocate linguistic diversity; European and especially American institutions foster cultural and racial diversity. Historian of science David

Sepkoski associates the valuation of biodiversity in recent decades with the rebirth of catastrophism in evolutionary theory.[1] His narrative helps us understand how primatology came to model itself on salvage anthropology in its attempt to archive as much information and material as possible before wild apes would forever disappear.

Extinction is a modern idea. Before the eighteenth century, animals did not enter and exit the stage of history. In revolutionary France, natural historian Georges Cuvier proposed that entire species like the mammoth could die out—and not just single species but large numbers of them and all at a time. This catastrophism made paleontological discoveries of species that no longer existed commensurable with the Christian doctrine of creationism. In recurrent geneses, God re-created life on Earth differently each time flora and fauna had been wiped out by disaster.[2]

Charles Darwin and other nineteenth-century evolutionary theorists noted a distinct theological odor surrounding catastrophism. They replaced this belief in sudden and violent events transforming the whole planet by the philosophy of uniformitarianism, assuming that current natural laws and natural processes had always operated in the universe and that natural historical events occurred at the same rate now as they had in the past. Darwin's own theory of natural selection adopted Cuvier's insight that species could go extinct—but not in bulk. Instead they vanished one by one in the perpetual struggle for existence as new and better-adapted species showed up, competing with established residents for scarce resources.[3]

It had already dawned on some naturalists that the invasion of modern humans could speed up this process. Darwin (and even more so the Social Darwinists) extended this logic to colonial conquest: "Extinction follows chiefly from the competition of tribe with tribe, race with race. . . . When civilised nations come into contact with barbarians the struggle is short."[4]

Although many observers regretted witnessing such processes of gradual extinction, they were no source of cosmological angst. As long as speciation and extinction occurred at the same pace, they did not disrupt the dynamic equilibrium maintained by a fiercely competitive nature red in tooth and claw. In this Victorian worldview, biodiversity was a zero-sum game and did not require human appreciation and protection. And it wasn't considered a positive good. By contrast, extinction—although regrettable from a sentimental point of view—removed the unfit, including savage tribes that the inexorable logic of biology had doomed anyway, and thereby improved races and entire species. It made the world a better place.[5]

For almost a century, the uniformitarian paradigm reigned over evolutionary theory. After the 1960s, however, catastrophism made a comeback, although in secular form. The fossil record persuaded paleontologists that natural history had not always proceeded in the gradual manner that Darwin had imagined. Mass extinctions did occur, not as acts of God but in the wake of meteor strikes and climate change, and they followed a different dynamic than the everyday extinction processes between these big biotic crises. Instead of only driving extinct species that weren't well adapted, they introduced an element of chance: not genes but bad luck seemed responsible for the deaths of many species. This new catastrophism fit well into a Cold War era marked by political uncertainty and the threats of nuclear Armageddon and environmental disaster, noted Sepkoski.[6]

The renaissance of this eighteenth-century outlook corresponded with the growing realization that mass extinctions had happened not only in deep history. Researchers found themselves in the middle of such an ecological cataclysm. Unlike the previous five times that biodiversity had taken such a dramatic hit, this "sixth extinction," as Edward O. Wilson had first called it at a conference in 1986, turned out to be of our own making.[7] "Dominant as no other species has been in the history of life on Earth, *Homo sapiens* is in the throes of causing a major biological crisis, a mass extinction, the sixth such event to have occurred in the past half billion years," Louis Leakey's son Richard noted. "And we, *Homo sapiens,* may also be among the living dead."[8]

So far, however, humans had done exceedingly well while nonhuman primates were among the especially affected taxa. In 2017, mostly due to agriculture, but also because of logging and hunting, 75 percent of all primate populations were in decline, and 60 percent of primate species counted as threatened.[9] In 2016, the International Union for Conservation of Nature had listed four out of six great apes—both types of gorilla and both types of orangutan—as "critically endangered." Classified as simply "endangered," chimpanzees and bonobos were still a tad further from the brink of extinction. But their populations also trickled through the anthropocenic hourglass at a swift pace.

For cultural primatologists, this process felt especially painful: with each primate community disappeared a unique culture. One century later, chimpanzee ethnologists confronted the very dilemma that Bronislaw Malinowski had already experienced in his fieldwork on the Trobriand Islands in the 1910s: "Ethnology is in the sadly ludicrous, not to say tragic, position, that at the very moment when it begins to put its workshop in order, to forge its proper tools, to start ready for work on its appointed task, the material of its study melts away with hopeless rapidity."[10] Marshall Sahlins described the mood that had colored ethnographic writing from the start as "sentimental pessimism."[11] Considering that the meeting of ethnographers and their subjects had usually been part of a colonial encounter, Renato Rosaldo diagnosed the mourning for what the anthropologist's supposed colonialist accomplices had destroyed as a sense of "imperialist nostalgia."[12] Social and cultural anthropologists responded by lobbying on behalf of endangered cultures and by documenting them as extensively as possible.[13] Similarly, primatologists took action as conservationists while collecting all the data and material they could to preserve at least a record of the behavioral diversity of nonhuman primates for posterity.

The realization that biodiversity was no natural constant, homeostatically maintained by a balance of speciation and extinction, had led to a transvaluation of nineteenth-century values. Extinction no longer appeared as a force for good but was seen as an evil to be prevented at all costs, while diversity, both biotic and ethnic, had become an endangered end in itself. It had to be defended against the dominating forces unleashed by modern humans, especially those who had spread across the globe from Europe. At a 1996 conference co-organized by UNESCO, the World Wildlife Fund, and Terralingua (dedicated to the preservation of endangered languages), concern about the endangerment of cultural and biological diversity, although of different historical origins, converged as manifestations of the same phenomenon: colonizing cultures came to be seen as simultaneously replacing biological *and* cultural diversity—now amalgamated into "biocultural diversity."[14] Yet cultural primatologists sought to preserve a different kind of biocultural diversity: the plurality of chimpanzee cultures.

William McGrew put it succinctly: "Just as cultural anthropologists are active advocates on behalf of the traditional societies that they study, so must cultural primatologists do the same. Conservationists may seek to save the species *Pan troglodytes*, but cultural primatologists must seek to preserve cultural diversity. This means going beyond a few famous, long-term study sites like Gombe or Taï. It means safeguarding Tenkere, where the apes make cushions and sandals, and Tonga, where the apes dig up tubers for moisture. Both of these populations are unprotected and on the verge of extinction. What a pity it would be to lose them."[15]

"Man has never lived in harmony with nature"

Before visiting the Taï Chimpanzee Project, I had accompanied Christophe Boesch on his travels through Gabon and Côte d'Ivoire for almost a month in 2014. During this trip, he frequently changed his two hats, that of director of the Max Planck Institute for Evolutionary Anthropology and that of president of the Wild Chimpanzee Foundation. One day, he organized the logistics of a large-scale research enterprise that spanned West and Central Africa and involved over one hundred scientists, field assistants, technicians, and administrators. The next day, conservation politics took precedence over scientific research. After all, the ethnography of African apes would come to an end as soon as the last wild culture had vanished.

In Gabon's capital, Libreville, he met with Lee White, a Gabonese of British ancestry now directing the National Agency for National Parks, and famous American environmentalist Mike Fay, a consultant of Gabonese president Ali Bongo Ondimba, whose predecessor and father, Omar Bongo White and Fay had persuaded to protect Loango and eleven more national parks in a still largely forested country. Ensuring support in the highest echelons of national government prepared Boesch for a confrontation with Loango's park authorities.

Following a gorilla group, students of the Max Planck Institute had run into young men from Bonne Terre, a village within the park, who had thrown sticks at the apes. Fortunately, the gorillas did not retaliate, but the situation had been extremely dangerous for everybody involved. If anyone had gotten hurt, the locals would have blamed it on the gorillas and on him, Boesch contended. His field assistants had also found a man logging trees with a chain saw as well as sling traps, which often injured the apes. Boesch assumed that it wasn't the elderly villagers themselves but their nephews from the capital who drove over for a weekend of poaching. While Gabon's overall population continued to grow, the rural area around Loango National Park experienced an exodus of young people from the countryside to the cities.[16] In contrast to the situation around Taï National Park in Côte d'Ivoire, these new urbanites had no desire to extend their parents' and uncles' slash-and-burn fields into the protected forest. But they did connect the subsistence economy of this remote part of the country to the markets of Libreville, where a truckload of bushmeat would earn them a whole month's salary.

"Either the villagers are removed from the park, or the scientists will leave," Boesch threatened *le conservateur* and the park's director of tourism. The latter relied on the researchers to habituate a group of gorillas, which would generate not only scientific data but also revenue from a luxurious ecolodge that had already been built and would include a gorilla safari in its tour packages. Offering primary rainforest, savanna, and a beach frequented by elephants and what *National Geographic* called "surfing hippos," Loango would be able to attract a well-to-do clientele. A resettlement of the villagers was not possible, the officials objected. Gabon operated on the principle of *nature et culture,* they informed Boesch. He replied that since their habituation brought the gorillas closer to the village, the old arrangement no longer worked. But the local population had to profit from the park, the director of tourism insisted; otherwise there would be an uproar against protecting the forest. After more tactical back and forth of this kind, the warring parties settled on a compromise, which Boesch had already discussed with Lee White. The inhabitants of Bonne Terre would stay, but a representative of the National Agency for National Parks and two ecoguards would be installed in the village to keep an eye on human activities in the forest.

Boesch's agenda diverged from the conservation of biocultural diversity in that he did not trust indigenous people (*pace* their romanticizing colonial designation as *Naturvölker*) to preserve nature any more than European or Chinese neocolonialists did (in this respect, he followed in the footsteps of Dian Fossey, who had first introduced him to wild apes).[17] Providing automatic rifles to foragers certainly hadn't made local hunting practices any more sustainable, and now bushmeat markets provided a venue to sell all the game that hunters did not need for their own subsistence. "Man has never lived in harmony with nature," Boesch assured me.

This brand of environmentalist misanthropy was in line with a paleontological literature accumulating more and more evidence that the disappearance of megafauna, from mammoths to Neanderthals, since the Ice Age was not just due to natural climate change but usually coincided with the arrival of *Homo sapiens*.[18] As this highly invasive species migrated from Africa to colonize Europe, Asia, Australia, the Americas, and thousands of islands, its use of increasingly sophisticated weapons and tools promoted humankind from a scavenger that had to wait for the lions to finish their meal to the apex predator in all ecosystems.[19] Modern colonialism represented only the latest chapter in this naturecultural history of predation.

The Politics of Ivorian Nature and Culture

In Côte d'Ivoire, the second leg of our voyage, Boesch's relationship with local and national authorities had grown even more complicated over the decades. Since his arrival as a student looking for a dissertation project in 1976, he had built up his long-term field site in the country and had metamorphosed into a major player in Ivorian research and conservation. Former president Houphouët-Boigny, who had led the French colony to independence in 1960, had been enthusiastic about the crazy white man who lived in the forest with his wife and baby, Boesch recounted. This dual citizen of France and Switzerland dodged the widespread resentment against the French and their postcolonial meddling with Ivorian politics mostly by aligning himself with the Swiss institutions in Ivory Coast.

While Boesch understood science as war by other means, he practiced conservationism as a form of muscular diplomacy, which had preserved access to the president's office through several coups d'état. He had recently met Alassane Ouattara, who came to power after the so-called crisis of 2010. This armed conflict between the supporters of former president Laurent Gbagbo and Ouattara's militias had eventually been decided in favor of the latter as UN and French forces intervened on Ouattara's behalf. An economist by training, the new Ivorian head of state recognized how the conservation of chimpanzees and their dwindling habitats could serve ecotourism. Pragmatically, Boesch embraced a pact with the travel industry. Not only would the pauperized population of Taï profit economically, but research had identified the presence of scientists and tourists as the best predictors for the survival of great apes.[20]

Our journey to Boesch's field site in Taï National Park was simultaneously a promotion trip for the ecotourist project of Djouroutou, about sixty kilometers from Taï. Accompanied by an Ivorian tourist operator, five international journalists, and one of the main sponsors of the Wild Chimpanzee Foundation, a Swiss botanist and old friend of the Boesch family, we drove in a white SUV, which the Swiss Embassy had lent the Wild Chimpanzee Foundation, from the WCF headquarters in Abidjan to Côte d'Ivoire's politically volatile southwest. The goal was to generate publicity for tours to a chimpanzee community, which the WCF had habituated under the auspices of the Office Ivoirien des Parcs et Réserves (OIPR). Along the way we crossed the enormous N'zo-Sassandra River, which separated the nut-cracking chimpanzee cultures to the west of this natural geographical barrier from those to the east that supposedly did not know how to crack nuts. In this remote part of the country, people struggled to make ends meet. Over more than three decades, Boesch had made many friends in the area and wanted to help. Before we entered the rainforest, a night of folkloristic dances allowed an Oubi village to profit from the allure chimpanzees and indigenous cultures held for Westerners. Here it was the local WCF representative who advocated the principle of *nature et culture* as he explained their vision of ecotourism to the journalists.

Yet not everybody in the region appreciated the WCF's efforts. Stuck between the Liberian border and the national park, Taï's fast-growing population, further increased by migrants from Burkina Faso and Mali, could not develop new farmland unless people cleared new fields in the protected forest. Poaching provided an additional source of income, which conservationists sought to cut off. In the 1990s, there had also been conflicts between the locals and the federal government over the town's power supply, which involved regulations concerning the national park. In the course of these negotiations, Boesch had briefly been taken hostage to strengthen the locals' bargaining hand. Of course, the TCP and the WCF also provided jobs. In that respect, protecting the apes was in the villagers' own best interest. Over the years, however, as major employers, the TCP and WCF had also fired many people who had not regularly reported for work, violated camp rules such as the prohibition of bushmeat consumption, or simply not learned to correctly identify individual chimpanzees.

Resentment between the European conservationists and Ivorian officials charged with protection and use of the environment came to the fore when we arrived in the town of Taï. The WCF shared an estate with the branch office of the Ministère des Eaux et Forêts. The ministry's loquacious representative, Captain Alain, invited me over. He had already heard of my arrival and wanted to know what "a philosopher" did in Taï. People here suffered from hunger and were not appreciative of philosophy, he told me (rather, they sought moral orientation and spiritual edification in numerous churches, as I came to realize the following Sunday). Nor did they value that the WCF cared more for chimpanzees than for humans. The NGO did not sufficiently align its activities with the needs of the local population, he claimed. For example, to reduce bushmeat consumption the conservationists had built a fish farm without realizing that many people in the region considered fish their totem animal, which they weren't allowed to eat. On the contrary, the local WCF manager explained to me, the reason their pisciculture project had failed was not that people didn't eat fish but that all the fish had been stolen from the pond. Now they trained people to breed agoutis instead, which could be safely kept in a locked building. As far as the problem of hunger was concerned, he pointed out that the Eaux et Forêts officers used every opportunity to extort a few fish from local fishermen—"pour le capitaine." Alain, in any case, came from a better-off part of the country, did not look malnourished, and appeared interested in chitchat with passing philosophers.

Not just friction but breakdowns of scientific and conservation activities resulted from the political crises Côte d'Ivoire had experienced since its first military coup d'état in 1999. When Boesch had begun to work in Taï in the late 1970s, the country was prospering economically and appeared to be a bastion of calm and relative prosperity among the violent conflicts haunting the other newly independent West African states. Its biggest source of revenue became the world's appetite for chocolate. Ivorians sold 40 percent of all cocoa to global markets. The seeds of later discord were sown as a Baoulé-dominated planter bourgeoisie initiated an internal colonization of the densely forested and sparsely populated southwest—the region where Taï was located and where cocoa grew best. The quickly spreading plantation agriculture required a labor force, which the ethnic groups native to the region could not provide. Soon the original inhabitants, whom the Baoulé regarded as backward, became a minority in their own villages as migrants from increasingly desertified Burkina Faso and Mali moved in, being granted ownership of the land they cultivated.

By the 1990s, the population had exploded and most forest had given way to cocoa plantations. With the exhaustion of the forest frontier, political conflicts broke out. In 2000, Marxist professor of history Laurent Gbagbo, a fierce critic of the cash crop economy, won the election with the support of the so-called autochthons of the country's western regions. After an armed conflict, he managed to wring power from the military junta. His populist discourse of *ivorité* and the spread of antiforeigner rhetoric served to keep his half-Burkinabé rival Ouattara from power and successfully mobilized xenophobic resentments against African immigrants, European neocolonialists, and UN peacekeepers. This resulted in the reclaiming of "patrimonial lands" from non-native planters, killings of and by French troops in 2004, and attacks on UN offices in Abidjan and the southwest in 2006. In 2002, Gbagbo successfully put down a rebellion against his government but was removed from power in 2010 after losing both elections and a violent confrontation with Ouattara's militias as well as French and UN troops, leaving more than one thousand people dead.[21]

Each of these political crises resulted in the evacuation of researchers and conservationists from Côte d'Ivoire. In the ensuing ethnic strife, the ecoguards and field assistants returned from the national parks to their villages to protect their families and join local militias, which in the area around Taï were usually pro-Gbagbo forces. As Ouattara's troops gained the upper hand, many field assistants had to flee to refugee camps across the Liberian border. The interruption of cross-country traffic led to local food shortages and increased demand for bushmeat.

Poachers used the opportunity to go after the otherwise relatively well-protected apes. The Guéré and Oubi traditionally living around Taï refrained from eating chimpanzees, whom they considered their ancestors. But the growing immigrant population had no such food taboos and cherished chimpanzee meat not only for its tenderness but also for its magical powers. During the coup d'état of 2002, seven of the Taï chimpanzees disappeared. Probably more would have been killed if one of Boesch's doctoral students, Tobias Deschner, had not volunteered to go back into the forest, signaling to the villages around the park that the Taï Chimpanzee Project would be a reliable partner in the face of political turmoil. Apart from a six-week interruption, field assistants continued to follow the TCP study populations throughout the crisis of 2010. Yet one-third of all Taï chimpanzees lost their lives during this period.

Even though the resolution of the political conflict had made the forest a safer place again, we still heard shots at night during my stay at the research camps deep inside the forest. A few months earlier, the authorities had confiscated a dead chimpanzee at a local bushmeat market. The hunting pressure gradually decimated the nonhabituated communities, many of which had already gone extinct.

The field assistants, who were either Oubi or Guéré, assured me that immigrants were to blame for the poaching. The Baoulé and the Dozo, traditional hunters from northern Côte d'Ivoire and southern Burkina Faso and Mali, violated park regulations. The government agencies entrusted with the protection of wildlife did not stop them, supposedly because the Dozo had supported Ouattara against the pro-Gbagbo militias in the area. For their services, former fighters were even allowed to cultivate parcels of protected forest, one indignant Oubi field assistant told me. A WCF employee in Abidjan frowned when I told him about this finger-pointing: even if the Oubi and Guéré did not participate, the immigrants would have no incentive to hunt illegally in the national park if the surrounding villages did not buy their bushmeat.

Demographic Fatalities

In the end, it was not European, American, or Japanese scientists and activists but African politicians who determined conservation policies. As politicians, they represented very different interests than those of the foreign representatives of the great apes. Hjalmar Kühl, one of Boesch's coworkers at the Max Planck Institute, remembered how a local politician visiting the research camp in Taï Forest explained to the scientists that from his point of view, the national park had been created for white people while his constituents did not have enough agricultural land to grow food for everybody. Being confined by the national border with Liberia to the west and the protected forest to the east, Taï and the surrounding villages were in dire straits. Of course, the Taï Chimpanzee Project and the Wild Chimpanzee Foundation spent money and offered jobs in the region. Moreover, the WCF initiated development projects such as the fish and agouti farms and habituated the chimpanzee group in nearby Djouroutou to bring ecotourists to the region. But far from everybody profited from these activities.

While the WCF did its best to reconcile human and chimpanzee interests, no NGO could stop the tectonic population growth in sub-Saharan Africa. At the time of my fieldwork, a study predicted that the number of human beings living in this region would double or maybe even triple by 2050.[22]

From a global perspective, the environmental consequences of such a proliferation might not have been unprecedented. British reverend and political economist Thomas Malthus had already worried that the linear increase of food production could not keep up with the exponential population growth he anticipated in Europe.[23] By 1850 almost twice as many Europeans had to share the continent's resources as in 1750.[24] Yet advances in farming technology and an extension of land available for agriculture suspended all Malthusian limits that could have checked the demographic explosion of humans after the Neolithic.[25] To increase their agricultural production,

Europeans had cut down most of their primeval forest centuries if not millennia ago.

Hjalmar Kühl identified the human population explosion as the main problem. Unless the process could be reversed as in Germany or Japan, providing more agricultural land could provide only temporary relief. After a few years the additional fields would no longer suffice, and more forest would have to be cut down to prevent human beings from starving.

Yet environmentalists shrank from telling Ivorians to make fewer babies. "To put it nastily, only the Pope can do something about this," said Kühl. "No conservationist organization can change how many children people have."

NGOs had not always been that reluctant. From the 1960s to the 1990s, international foundations and aid agencies had regularly pressed African governments to invest in family-planning programs. Since the mid-1990s, however, they had fallen silent. "There are always a few who say this should be addressed, but if an organization demanded steps in this direction they would immediately lose their financial support," Kühl reckoned. "Calling population growth a problem was seen as culturally insensitive and politically controversial. International donors shifted their focus to promoting general health care reform—including fighting HIV/AIDS and other deadly diseases," noted Robert Engelman.[26] In the eyes of the former president of the Worldwatch Institute, Africa's population explosion was a political problem: if women were empowered to take reproductive decisions into their own hands, the fertility rate would go down. In private conversations, Boesch expressed the same conviction. While Engelman hoped for a growing number of African politicians who realized that they could not increase prosperity and keep peace without curtailing family sizes, Kühl doubted that many chimpanzees would live to profit from such cultural change: "The problem is that the transition from a traditional society where people have many children to a modern society with significantly fewer children happens at a much slower pace than the environmental destruction."

Kühl's biosocial evolutionism had a fatalist bent. But he did not expect a total annihilation of wild chimpanzees. He replaced Darwin's dictum that "rarity . . . is the precursor of extinction" with a slightly more optimistic economic theory, which predicted that the chimpanzees' demographic free fall would come to a halt before the last communities died out: "Linear projections into the future are always inaccurate. If a resource becomes scarce, its value increases. When only a very few chimpanzees are left, there will be some tourist operator who would like to show people the last individuals. Some very small populations will survive, but they won't be representative. And, of course, their cultural diversity will have been lost."[27]

Anthropocene Overkill

British philosopher John Gray expressed an even crasser fatalism. He dismissed the humanist faith in political action as a secular surrogate of the Christian hope for salvation. The idea that humanity could shape its own future amounted to exempting it from Darwin's discovery that species were mere currents in the drift of genes. Echoing the misanthropic impulses many environmentalists harbored in the dark night of their soul, Gray argued that the mass extinction we are currently witnessing was neither an economic nor a political question: "The destruction of the natural world is not the result of global capitalism, industrialisation, 'Western civilisation' or any flaw in human institutions. It is a consequence of the evolutionary success of an exceptionally rapacious primate. Throughout all of history and prehistory, human advance has coincided with ecological devastation."[28]

Gray took sides in a controversy between paleontologists, archaeologists, and geologists over what had caused the eradication of large parts of the megafauna in different parts of the world. Why had much of Europe's native wildlife, from prehistoric cave lions to woolly mammoths and possibly the Neanderthals, disappeared, and why had Carpathian wisents, Caucasian moose, and Portuguese ibexes followed in their footsteps more recently? Why did biodiversity in the Americas take a hard hit around the time when the Clovis people populated the continents, and again when the New World's European colonization began? And why did many species of large animals fare better in Africa until quite recently?

In the 1960s, US geoscientist Paul Martin proposed that neither prehistoric climate change nor disease but early humans had driven large animals extinct wherever they migrated.[29] As Martin's theory of this "Pleistocene overkill" soon became engulfed in the culture wars, he fought back: "Talk of climate change has the advantage of not distressing those concerned with cultural sensitivity—some racial groups could suffer bad press if the word gets out that their ancestors might have helped to exterminate moas, *Megalania*, mammoths, or megatheriums."[30] At a time when everybody was talking about the Anthropocene, Martin's narrative of anthropogenic mass extinction seemed to be gaining the upper hand.[31] However, in contrast to Paul Crutzen's geology of humankind, which focused on greenhouse gas emissions, the story of the sixth extinction moved the drama of anthropogenic environmental change back from the late eighteenth century to the Ice Age.[32]

The reason more European and American species had initially fallen victim to *Homo sapiens* was that the African fauna had coevolved with hominids, while many animals abroad lacked even the simplest defenses against human predators, argued Martin.[33] But at the beginning of the twenty-first century, a similar process threatened to steamroller the African great apes just as it had done to many of the remaining primate species in South America and Asia. It seemed as if neither natural selection nor individual and social learning enabled them to keep up with the accelerating increase of capacities enabled by cumulative culture in humans. Yet culture was not the only decisive factor that primatologists considered in thinking about how *Homo sapiens* had managed to become the predominant primate species all over the planet.

Cooperative Breeders and Chimpanzee Grandmothers

Shortly before I visited Bossou, Fana's second grandson, Flanle, had gone missing—probably killed by humans. The assistants suspected a Manon farmer who had moved from another village—not all Manon revered chimpanzees the way people in Bossou did. The man had repeatedly attacked the chimpanzees with a sling and even with a machete when they raided his illegal plantation.

Amid this drama of human-animal conflict, nonprimatologist readers might have missed the truly surprising piece of information in the previous paragraph. It is not the killing of Fana's grandson, totem animal or not, but that she could experience the loss of a grandson in the first place. According to the "grandmother hypothesis," most prominently promoted by Kristen Hawkes and colleagues, grannies are a very human phenomenon.[34] Women frequently outlive their reproductive age by decades and then help to raise their children's children.[35]

Evolutionary anthropologists argued that so-called cooperative breeding enabled women to have the next child after two to three years, sometimes even after only one year. While chimpanzees, on the other hand, take four to seven years, gorillas four to six years, and orangutans seven to nine years before they can have another child, the involvement of grandmothers, fathers, other kin, and even nonrelatives in child rearing enabled forager women to give birth every three to four years, almost twice as fast as the apes, noted American primatologist Sarah Hrdy.[36] With the help of baby formula and other substitutes for breast milk invented to facilitate weaning, *Homo sapiens* even managed to bring down the interbirth interval to one or two years. "Humans are unique in terms of their capacity to rear multiple children at the same time," wrote Matsuzawa.[37] No ape could keep up with that.

Normally, chimpanzee females had no opportunity to serve as grandmothers. They avoided inbreeding by moving to another group when they reached sexual maturity. Thus the emigrants' mothers would never meet their daughters' children unless in a violent intercommunity encounter. Of course, infants fathered by their sons might well surround chimpanzee females. But since *Pan* enjoyed intercourse promiscuously and did not pair-bond, primatologists assumed that neither a mother nor her son would know who his children were. And even if they miraculously did, the biological grandmothers would not have much time for their son's offspring since female chimpanzees usually remained fertile and preoccupied with their own children until they died.

Half of the Bossou population had already outlived their reproductive age. Of the remaining eight individuals, Fana, Jire, Yo, and Velu were in their late fifties. Velu had already been identified on 16 mm film footage that Kortlandt had recorded in the 1960s when Guinea had just escaped French colonial rule and began its experiment in African socialism. She had witnessed the subsequent arrival of the Japanese in the 1970s and the construction of the outdoor lab in the late 1980s. Matsuzawa used this observational space to notice that the elderly ladies took twice as long to crack a nut than their younger conspecifics. A geriatrician from Japan planned to examine the effects of old age in wild chimpanzees in the Bossou community. Considering that only 7 percent of chimpanzees survived to age forty, at which point they were considered old, and that in the wild they rarely lived past their fiftieth birthday, the Bossou population demonstrated a previously unknown longevity, which Matsuzawa related to the particularly rich nutrition obtained from human fields.[38] In an anthropogenic habitat, they lived long enough to become grandmothers.

Cut off from neighboring communities by fields and a human-made savanna, the Bossou community was isolated from neighboring chimpanzee groups. They no longer recruited young immigrants, nor could their own females leave. Consequently, daughters stayed with their mothers, who got to meet their grandchildren.

In an obituary for Velu, whose dead body was found on Mont Gban in March 2017, Matsuzawa described how she had often taken care of her granddaughter Veve while her daughter cracked nuts: "In this example, grandmothering really helped the mother to increase her energy intake."[39] But it could not reverse the somber demographics of her community. Matsuzawa sought solace in the Buddhist metempsychosis doctrine: "Velu, I really enjoyed watching you and sharing the events of your life. Sleep well. Let us meet again in the next life. On that alternate plane, you may become a human while I may turn into a chimpanzee."[40]

In Bossou, humans were outcompeting chimpanzees by sheer numbers. Since primatological observation had begun in the 1960s, the population of the village had grown from one thousand to over four thousand inhabitants, while the number of chimpanzees had dropped from twenty-two in 1976 to seven in 2017. And when the other three elderly females would die, the size of the already diminished group would be cut in half again. "Morimura-san, you, and me," Matsuzawa had told me upon our arrival in Conakry, "we will be the last to see the Bossou chimpanzees."

The Green Corridor: Conservationism as Multicultural Strife

The evolutionist recognition of the demise of *Pan troglodytes* did not throttle the primatologists' last effort to conserve the species and its cultures. At their African field sites, Japanese researchers experienced the plight of their chimpanzee subjects firsthand. In the 1980s, they had begun to take active steps to ensure, or at least prolong, the apes' survival. Since its inception in 1985, the Primate Society of Japan had promoted nature conservation. This was also the year when the Japan International Cooperation Agency supported the Tanzanian government in protecting the Mahale Mountains, the Kyoto School's first and most important chimpanzee field site, as a national park.[41] At a time when Imanishi's influence had begun to wane and traffic between oriental and occidental primatology was increasing, Japanese researchers took on conservationism, although in their own ways.

When Tetsuro Matsuzawa first visited Bossou in 1986, the passionate mountaineer could not resist the temptation to climb the highest summit of the nearby Nimba Mountains. In the lush forests covering the mountain slopes, he found another chimpanzee group. Five kilometers of razor-sharp elephant grass and withered herbaceous plants cut off the revered chimpanzees of Mont Gban from their unhabituated neighbors in the Nimba Mountains. This wide-open savanna, originally created by humans and now sustained by the relentless African sun, disquieted forest animals. In the few instances when particularly brave chimpanzee females had crossed over to Bossou, they had never stayed for good—presumably because they experienced life amid such pronounced human presence as too stressful. In forty years of field observations, not a single new immigrant had joined the Bossou community. In the long run, the group could not survive such an insular existence.

In 1997, the Guinean government permitted Matsuzawa to start a reforestation program, which sought to plant a so-called Green Corridor of trees across the savanna to facilitate genetic exchange between the chimpanzee populations on both sides of this geographical divide. But Japanese institutions offered few funding opportunities. Although Matsuzawa regularly received large grants for his research, unlike his Western colleagues he had not received support from public bodies for the long-term conservation of his research subjects.

While regretting this lack of resources, he also resented how Europeans and Americans had turned conservationism into big business. Every time he took a plane from Conakry to Paris, he said, it depressed him to see the conservationists flying business class. He preferred grassroots initiatives that did not waste their money on white-collar workers in New York and London who proposed measures that did not work on the ground.

Instead of founding an NGO like Boesch's Wild Chimpanzee Foundation, Matsuzawa imagined himself as the protagonist of Jean Giono's tale *The Man Who Planted Trees*.[42] First published by *Vogue* in 1954, it tells the story of a shepherd who single-handedly grew a whole forest in a deserted valley in the foothills of the Alps. Even though the Green Corridor had been Matsuzawa's idea, he emphasized that it was the Manon who implemented this Japanese initiative, employing local practices and materials. Eventually, the funding for this modest endeavor came from Matsuzawa's lecture honoraria and book royalties.

Matsuzawa relied on Manon traditions for both political and pragmatic reasons. Politically, including the locals and recognizing their experience increased their acceptance of the project. Pragmatically, his strategy tallied with one of the rationales for protecting biocultural diversity: the natives of any place knew best how to live in accordance with their environment. That it was best to draw from local knowledge Matsuzawa had learned the hard way. His attempt to protect young trees in the Green Corridor with polypropylene hexatubes, an approach that had worked well in development projects in Afghanistan and the Himalayas, came to naught in the West African savanna because these plastic structures hadn't been designed for bushfires.[43]

In a tree nursery, one worker grew saplings from chimpanzee dung to ensure that the species populating the passage would be attractive to chimpanzees rather than humans. When the young trees had grown enough, the other two regular staff members and up to ten temporary helpers from the village planted them in the savanna. Instead of immediately aiming at a continuous stretch of forest, they set out by creating patches five meters by ten meters about every fifty meters, so animals could hop from grove to grove. Instead of continuing to use hexatubes to create a microclimate that sustained plant growth, they now protected the delicate little saplings against the sun by falling back on one of the Manon's traditional agricultural techniques: they constructed wooden scaffolds from bamboo growing around the village, which they covered with palm branches. But these so-called *hangers* had become a bone of contention in Bossou cosmopolitics.

Reforesting land owned by the Guinean state, Matsuzawa, Morimura, and their Manon assistants conducted the Green Corridor Project under the auspices of the Institut de Recherche Environnementale de Bossou (IREB). IREB had been established in 2001 as a state-run research facility to serve as the Guinean counterpart to the field station of the Kyoto University Primate Research Institute (KUPRI), conducting primatological, genetic, meteorological, and sociological research. But many inhabitants of remote Guinée Forestière experienced the state as an oppressive and exploitative force, dominated by two ethnic groups, the Susu and the Malinke, who looked down on the supposedly backward Forestiers. Since many people in the area understood the Green Corridor as a projection of state power into their lives, it became a target of protests. Especially cattle farmers, who often burned down part of the savanna for their animals to graze the fresh grass, which then started growing, did not appreciate that the protection of the corridor undermined their subsistence. And so sparks occasionally flew, and not just verbally, setting ablaze the newly planted trees.

But opposition also came from IREB itself. In 2007, the government replaced the institute's second Malinke director with a Susu already familiar with both the Bossou chimpanzees and their Japanese researchers. Aly Gaspard Soumah seemed the perfect candidate to strengthen international scientific collaboration. He knew the Japanese well. A student at Gamal Abdel Nasser University of Conakry in the early 1980s, he had met Matsuzawa's predecessor Yukimaru Sugiyama as he initiated long-term field observations of the chimpanzee community at Bossou. Sugiyama had invited Soumah to study with him at KUPRI in Inuyama.

But soon the relationship between student and professor turned sour. By the time Soumah returned to Guinea with a doctoral degree from Japan's premier natural science faculty, anger had taken hold of him—anger about what he perceived as Sugiyama's attempts to foil his own student's dissertation research on the Japanese macaques of Takasakiyama. The monkeys' strict and impenetrable social hierarchies had allowed Soumah to understand Japanese human society, he told me.

After he had begun to run IREB, more resentment and frustration had built up. Soumah felt excluded from the scientific activities of the Japanese and decried their failure to train a single Guinean student. Instead Matsuzawa brought Asians, Europeans, and North Americans—including me—who collected data in Bossou that fueled their careers and landed them jobs at some of the best universities in the world while the natives got to play the role of assistants. Soumah felt disrespected—Matsuzawa had reduced him to the international students' chauffeur, he complained—and he had been cut out of communication within KUPRI. "We are not collaborating," he complained to me in an interview. "There is no black on the team of Matsuzawa. This is colonization!"

The Japanese, by contrast, considered IREB a separate institution with which they coordinated primarily logistical matters. Politely, they granted

Soumah occasional coauthorship on journal articles. But since his dissertation on Japanese macaques, he had not conducted any more primatological observations. Although IREB was officially an institute for environmental *research,* two of its three civil servant scientists had died and the third one, a friendly old man who had just lost his wife, mostly tended to his field. Thus, Soumah had little data to contribute to a research collaboration.

From Matsuzawa's point of view, the IREB director's most important function was not academic but political. His greatest achievement was the integration of Bossou into the Natural World Heritage Site of the Nimba Mountains against fierce local resistance. Stuck in a remote Ebola-ridden corner of the country, two bumpy days by car from his family in the capital, *le docteur,* as the villagers called Soumah, found himself less than popular among the Manon and ignored by the Japanese. By 2012 he had had enough and decided to block the reforestation efforts that Matsuzawa and his Forestier workers were conducting in the savanna.

Soumah prohibited the construction of the *hangers* on conservationist grounds. He considered cutting off palm fronds to provide shade to the saplings a form of environmental degradation. "As director of the environmental research institute, I have a responsibility," he said to me. "You grow forest by destroying forest? Ecologically, that's nonsense." After all, reforestation projects in Mali and Mauritania, where the dry season lasted much longer, had succeeded without these structures.

Inspecting some of the dead saplings planted under the new regime, Matsuzawa, Morimura, and the Green Corridor workers expressed their frustration about Soumah's blockade. Bamboo had to be cut back every year to curb its otherwise rampant growth, the local assistants said, and they regularly pruned their oil palms to increase the trees' fruitfulness. Matsuzawa encouraged them to explain their agricultural practice to Soumah. "That's pointless," they replied. "We are Manon and he is Susu. He won't listen to us."

With or without *hangers,* in Soumah's eyes the Green Corridor Project had already failed. In almost twenty years, less than 10 percent of the five-kilometer stretch had been reforested. If the work continued at this pace, he asked rhetorically, how long would it take to restore migration and genetic exchange between the Bossou and Nimba chimpanzees?

None of the Bossou apes had set foot in the savanna since 2013 when three of their companions had not returned from an excursion to the mountains. Even if a fully-grown corridor would one day provide them with a greater sense of security, it could not guarantee that immigrant females would stay in Bossou as long as the area remained so densely populated—and the human population was rising steadily. If anything, the corridor would provide the Bossou chimpanzees with a way out, threatening to put an end to the village's mythical coexistence with its apes. But even that grim prospect seemed overoptimistic.

Soumah planned his own rescue of the Bossou community. As soon as he found money, IREB's director-general wanted to introduce confiscated pet chimpanzees from a sanctuary. When I mentioned this idea to Morimura, he threw up his hands in despair. Chimpanzees raised in human households had never had a chance to learn from their wild relatives how to survive in a forest—or on fields. Developing the necessary skills was part of chimpanzee culture, not their nature. They wouldn't even know what to eat, Morimura explained, and since they had grown up among humans who had fed them, their search for food would inevitably lead them into the village, maybe even into houses. Current human-chimpanzee conflicts around pineapple and papaya stealing would pale by comparison.

On his final day before departing for a lecture at the Sorbonne in Paris, Matsuzawa discovered something new. After observing one last nut-cracking session in the forest, he examined the hammers the apes had left under the palm tree. There was a red tinge to the worn stones, and they appeared unusually heavy. Iron ore, Matsuzawa surmised. He had the tools wrapped up in big leaves to take them to Japan for mineralogical analysis. "This is chimpanzee archaeology," he beamed. "One day, a chimpanzee will take ethnic pride and request that I return the tools of his ancestors."

Yet the chimpanzees and chimpanzee researchers were not the only ones interested in these rocks. While KUPRI, IREB, and the villagers quarreled over how to conserve the Bossou community, irrespective of the region's World Heritage status, the Guinean government had granted Sable Mining Africa Ltd. a license to exploit a deposit of an estimated six hundred million tons of iron ore in the Nimba Mountains. For faster shipping of this mineral wealth to a port in Liberia they considered constructing a railway through the savanna, threatening to make the Green Corridor obsolete before it had even been completed. Amid these geographical upheavals and historical transformations, the Bossou community continued to feast on oil palm nuts and pineapples, enjoying the comforts the cohabitation of *Pan and Homo* had on offer, before they would go down in natural history.

Could an African Primatology Save African Primates?

In the United States, cultural diversity had ascended to a supreme value that different factions of left and right fought over in the culture wars. Committed to many of the cultural left's preferences, Donna Haraway described primatology as a multicultural field.[44] Epistemologically, she had radicalized the post-positivist position that scientific observation was theory and value laden and sought to demonstrate that the cultural, racial, and gender identities of scientists shaped the knowledge they produced. Politically, she wanted her readers to recognize the unique contributions of non-Western researchers to postwar science, which was not the exclusive domain of Europeans and North Americans. In primatology, she saw two biopolitical struggles intersect: the challenge of multiculturalism to the hegemony of a neoimperialist system of multinationalism, and conservationist efforts to preserve "multispecies-ism" amid this multicultural, power-charged "conversation."[45] Yet one voice was missing from this cacophony of European, American, Japanese, and Indian researchers. Were there no African primatologists?

Haraway cited Tanganyika's first president, Julius Nyerere, as saying that personally, he wasn't very interested in animals, but he wanted to protect them as a great source of income, considering that many Americans and Europeans felt "the strange urge" to see crocodiles and other wild animals.[46] Whether because of this indifference toward domestic wildlife or because foreign researchers had few incentives to develop scientific, collegial relationships with black Africans until well after independence, by 1985, Haraway had not been able to identify a single black African primatologist.[47] But this situation slowly began to change.

Historian of science Georgina Montgomery adopted primatologist Susan Alberts's account of an "Africanization" of the Amboseli Baboon Project after 1981.[48] Since 1997, this process had included a US-trained Kenyan PhD. But for the most part, Montgomery reported the delegation of basic behavioral observations to African field assistants (similar to what chapter 3 described for the Taï Chimpanzee Project). Although the involvement of these academically untrained researchers hardly translated into intellectual contributions to the primatological literature beyond the diligent collection of long-term data, it certainly provided significant impulses in local conversations over the nature, culture, and conservation of nonhuman primates. Let me give an example from my own fieldwork.

In his free time, Louis-Bernard Baly, a senior field assistant who had been working for the Taï Chimpanzee Project since 1999, collected the rural legends that his fellow Guéré told each other about the apes. His goal was not to preserve Guéré culture but to dispel the myths of the elders by contrasting them with what he had learned about the animals in fifteen years of daily observations.

One of these folktales turned on its head Rousseau's momentous anthropological project of finding in chimpanzees and gorillas humans in the state of nature: here modern humans did not figure as ex-apes, but the apes were believed to be ex-humans who had moved into the forest.[49] A very long time ago, an old Guéré man had told Baly, two brothers had camped in a big forest where a genie turned one of them into a chimpanzee. The chimpanzee stayed behind as his scared sibling ran back to the village. But humans and chimpanzees continued to be family. In fact, the ape kept calling for his relatives by sticking a magical leaf to his chest, which made the sound of his chest beating travel all the way to the village. Other people claimed that he used a tom-tom instead.

In an interview with me, Baly wove the history of the TCP into this legend—spinning a myth that proclaimed to end all myths: "After many, many years had passed, a group of people really wanted to see the one who had stayed behind in the forest. In the time of globalization, in the time of wealth, in the time of truth—in the springtime of truth—we, who have become researchers, have gone back into the forest to study the behavior of the one who stayed behind, who is called the chimpanzee."

Returning into this "world of truth," the field researchers found that the chimpanzees neither used magical leaves to beat their chests nor drummed on any percussion instruments, but they banged against tree buttresses. The account of some hunters that the chimpanzees cracked coula nuts by hitting them on each other's heads had also turned out to be a cock-and-bull story. Ethologist and Nobel Prize laureate Niko Tinbergen had famously said that "many farmers and hunters . . . were better zoologists than the armchair professionals who never got their hands dirty or their boots muddy."[50] Following the chimpanzees in his own muddy boots day after day, Baly did not think as highly of the farmers and hunters in Taï. He was as critical of their primate folklore as European primate scientists had been of the fanciful stories brought back by early travelers and adventurers.[51] If a poacher did not want the chimpanzees to escape, he could not look at them for more than ten seconds before pulling the trigger, he assured me. And yet poachers provided much of the local knowledge. Thus, during his work for TCP, Baly had discovered that "many of the things, which our parents in the village had told us, were lies."

Not everybody in Taï welcomed Baly's attempts at enlightening his fellow citizens. An elder doubted that the field assistants would ever reveal what they had really experienced in the forest because these were their "professional secrets." Assistants who insisted that they had seen the chimpanzees' nut-cracking workshops with their own eyes had been told that they had bought into the lies of the whites. "When we said, no, watch the films," which the Wild Chimpanzee Foundation screened regularly in the village as part of its sensitization campaign, Baly recounted, "the elders responded: 'These are all montages. We know the stories of the whites. They make us believe things, but in reality it's not like that.'"

Despite his zeal to replace Guéré folk primatology with empirical knowledge, Baly also recognized the problems arising from a disenchantment with the apes: "People who lived here fifty years ago preserved the chimpanzees because they considered them their ancestors, but my generation thinks that these people were fools. Science showed the chimpanzee is not a human, thus we can kill him."[52] Of course, not everybody in Taï was as empirically minded as Baly, but many others had converted from animism to monotheist religions, mostly Christianity, which had no more taboos against the consumption of chimpanzee meat than zoology. Moreover, more and more Muslim immigrants came from Burkina Faso and Mali. Holy scriptures from the Middle East began to supersede West African legends. For chimpanzees, the ensuing multiculturalism was bad news. "Africa is big," Baly explained. "Everybody has their culture. Maybe this community respects the chimpanzee because he is their brother, so they don't eat him. But then somebody else comes from another community and says to himself: 'He is *their* brother, not mine. So I will kill him and make some money out of the meat.'"

The knowledge that Baly brought up against the myths of the elders was primarily experiential. Sharing the everyday life of habituated chimpanzee groups, he knew their ways much better than any hunter and wanted his fellow Guéré to recognize this. He did not promote the Darwinian cosmology introduced by his European employers.

Yet the opposition of African legends and empirical knowledge produced in the practical framework of a Western scientific research project raises the question of why no Ivorian primatology had emerged that contributed to primatology at large, just as the Japanese tradition had lastingly transformed the international field. If Buddhism could shift the primatological gaze from the focus on human exceptionalism, which had dominated discussions in Christian countries, to the recognition that nonhuman primates had culture, too, then why could West African animism not enable an equally consequential refashioning of primate studies?

Ontology aside, the economic conditions of the countries where chimpanzees, bonobos, and gorillas lived stymied academic leisure, which had enabled European, American, Japanese, and Indian scholars to reconstruct their metaphysics of apes in light of empirical research. Thirty years after Haraway's failed search for an African primatologist, however, a small number of PhDs like Soumah had been trained and funded by European, American, and Japanese research institutions. Would they be able to "Africanize" primatology? And would the resulting African primatologies—for they would inevitably emerge in the plural—be better suited to preserve the biological and cultural diversity of primates?

From Monkey Linguistics to Koulango Primatology

By 2014, three decades of research at Taï had given rise to a small community of Ivorian primatologists about to take their first independent steps. I met one of a handful of midcareer researchers in Abidjan where he worked at the Centre Suisse des Recherches Scientifiques. Financed by Swiss taxpayers, Karim Ouattara (no relation to Côte d'Ivoire's head of state) had done his dissertation research not with the Taï Chimpanzee Project but with the neighboring Taï Monkey Project. Under the supervision of Swiss primatologist Klaus Zuberbühler, he had studied the vocalization of Campbell's monkeys.

There was nothing particularly Swiss or Ivorian about this Koulango's approach. But whereas most of his European colleagues used their Guéré field assistants primarily to collect long-term data, he emphasized that he respected them as research collaborators and listened carefully to what they knew about the cercopithecines. With their help, he discovered that Campbell's monkeys broadened the meaning of their highly specific leopard and eagle alarm calls to warn against threats on the ground or in the canopy more generally by adding a suffix—just as humans extend word stems by morphemes as, for instance, in the transition from "brother" to "brotherhood."[53] Ouattara's professed respect for his field assistants had translated into a by now common expression of gratitude in the acknowledgments section of his article, but it still did not translate into coauthorship, the usual way of giving credit to scientific collaborators.

Science for the sake of science did not satisfy Ouattara. After his discovery about the vocalization of Campbell's monkeys he had been invited to many conferences in Europe, he told me. But what difference did that make to people at home? Côte d'Ivoire had no research culture, he lamented, and his scientific success had received little recognition.

Ouattara's new project was meant to benefit local communities as well. Just as he had listened to what his assistants had to say about the monkeys in Taï Forest, he now wanted to listen to villagers who created their local cultures as they interacted with local monkey species in his own home district in eastern Côte d'Ivoire. This situated him in the budding field of ethnoprimatology, studying the relations between humans and nonhuman primates.

Together with two other Ivorian primatologists and an ethnologist, Ouattara examined ethnic groups that tolerated the presence of monkeys in their villages because they regarded them—just as the Guéré did with the apes—as descendants of humans and of their own ancestors. In Soko, for example, a small settlement close to the Ghanaian border, populated by Nafana, Koulango, and five other ethnicities, people believed that in the late nineteenth century, their sorcerer had used his magical powers to turn many villagers into monkeys. As the troops of Samory Touré, a Muslim cleric who resisted French colonialism and founded the Wassoulou Empire, approached, this therianthropy enabled the people of Soko to hide in the forest. Since the chief himself did not survive the raid, the bewitched had remained monkeys until this day.[54] Consequently, the macaques of Soko could return to their village as they pleased, even attending soccer matches, but only certain initiates were allowed to enter the animals' nearby forest habitat for rituals of ancestor worship.

The village elders did not want Ouattara or any other scientists to follow the monkeys into the sacred grove. The primatologist suspected that they feared losing their privileged status in the community if he revealed the secrets of their ancestors. He wanted to understand the function of the monkey secrets in village life, but he neither intended to expose the power structures they supported nor wanted to disenchant them with the monkeys. Such diplomatic blunders would certainly have deprived him of access to the macaques.

Ouattara stressed that he regarded local traditions and modern science as equally valuable. Scientific methods determined what counted as evidence, he explained to me, and traditional knowledge about monkeys just resulted from a different methodology. Conceiving of scientific knowledge as socially constructed, Ouattara—unlike Baly—did not want to tell people what was right or wrong. Instead he sought to understand how they had arrived at their perspective. In contrast to important tenets of cultural anthropology, his goal was not to deploy the native's point of view to criticize and transform his own culture, either.[55] Like many ethnoprimatologists, Ouattara sought to replace a postcolonial approach to conservation that imposed European ideas and practices on Africans by mobilizing local traditions to conserve nature.

Ouattara positioned his conservationist agenda in the politics of a multicultural state. In the wake of the postelectoral crisis of 2010, poachers had brought the macaque population of Soko to the brink of extinction. He told me that so far, people had tolerated their killing, merely thinking to themselves that such acts would draw misfortune on the hunters and their families. Or they assumed that only their own ethnic group would be punished for killing the monkeys while others would not be subject to this prohibition. Ouattara hoped that his research would make the inhabitants of Soko more conscious and appreciative of their mythology, so that they would stand up for their values when people belonging to other ethnicities, which revered, say, snakes rather than monkeys, came to hunt the villagers' simian relatives.

Considering the toll of human lives arising from ethnic strife in Côte d'Ivoire, I could not help wondering whether strengthening particularist cultural identities and their sacred values would be as good for humans as it might be for monkeys. But political violence was usually initiated by educated people from the city who had lost their cultural roots, Ouattara assured me. Presumably, he alluded to former president Laurent Gbagbo, who had earned a degree in history from the Université Paris Diderot and had served as the director of the Institute of History, Art, and African Archeology at the University of Abidjan. Then the scholar took power and incited a civil war, for which the International Criminal Court in The Hague now charged him with crimes against humanity. Ouattara did not want his country to adopt a European model of democracy but hoped that Côte d'Ivoire would find its own way. This advocacy of *political* self-determination mirrored his support of *ontological* self-determination, to use a term coined by Viveiros de Castro, granting the inhabitants of Soko the right to think of macaques as they pleased—as long as it helped to conserve them.[56]

Of course, the biocultural diversity promoted by Ouattara's brand of ethnoprimatology represented a late modern rather than a traditional Koulango value. Whatever the genealogy of its morals, the proof of the pudding would be in the eating: Would it keep the monkeys of Soko alive?

Saving Data: The PanAf Program

By the late 2010s, it was less the slowly growing input from African cultures, or from any other ethnic traditions for that matter, that was changing the face of the still-budding field of cultural primatology, and more the rapidly changing conditions of the Anthropocene. Like many of their colleagues in the humanities, primatologists did not wait for the International Commission on Stratigraphy and the International Union of Geological Sciences to reach consensus about whether the evidence sufficed to make the new natural historical epoch official.[57] All who had done fieldwork among African apes had seen the gigantic human footprint for themselves. When we drove down the highway from Abidjan to Taï, Boesch remembered that in the 1970s, he had still seen chimpanzees and elephants along the way. In 2014, fields and plantations lined the road all the way to the Liberian border and wildlife had disappeared. Between 1991 and 2008, the Ivorian chimpanzee population had been decimated by 90 percent.[58] Between 1990 and 2014, the overall population of West African chimpanzees had been reduced by 80 percent.[59] Chimpanzee ethnographers didn't have much time left to document the diversity of wild cultures. "We realized that, considering how fast people were cutting down the forest to grow oil palm plantations and mine the soil for minerals, great apes are disappearing so rapidly that, if we waited for another ten or twenty years, not many would be left," Hjalmar Kühl explained to me. In 2008, Boesch and Kühl decided to use the remaining time to develop a radically different approach to cultural primatology.

Cherishing the value of seeing for oneself, which had been at the heart of scientific observation and empiricism for half a millennium, Boesch emphasized that one had to experience a culture firsthand to understand it.[60] But he also realized that most of what primatologists knew about wild chimpanzees was based on only a dozen field sites. These were the twelve sites where habituated communities allowed the researchers to follow them around and closely observe what Malinowski had called the imponderabilia of actual life and typical behavior.[61] If the goal was to document what was left of chimpanzee cultural diversity, however, cultural primatologists needed many more sites to build an ethnographic archive for chimpanzee posterity. And that required less immersive ways of learning about the apes' lifeways.

Habituating a large number of communities was out of the question. Not only would it have taken up to five years or more as well as significant financial resources and labor to get a new group sufficiently used to human presence that chimpanzee ethnographers could collect data; this process would also have exposed the apes to potentially lethal human pathogens and made them even more vulnerable to poaching. While the continuous presence of researchers did provide protection against hunters, it made the unsuspecting animals easy prey as soon as the researchers left. And there was a very real risk that they would have to leave despite their best intentions, either because their grants ran out (no funding agency would commit itself to unconditional and open-ended support of a field site), or because the political situation made it too dangerous to continue the work. Considering how volatile postcolonial African states remained, the decision to habituate groups at new field sites, especially in Central Africa where data was most badly needed, confronted researchers with serious ethical problems. Even in the case of already habituated groups such as the one in Bossou, researchers increasingly relied on camera traps to minimize the risk of disease transmission.

Boesch and Kühl responded to the dilemma of wanting to collect as much information as possible about the quickly vanishing chimpanzee cultures without contributing to their plight by initiating the Pan African Great Ape Monitoring Programme: The Cultured Chimpanzee. This PanAf program, as everybody called it, contributed to both conservation biology and cultural primatology. It grew out of the Ape Populations, Environments and Surveys (APES) database, which the Max Planck Institute for Evolutionary Anthropology had established in 2005. Kühl wanted to salvage and archive the large amounts of data idling on his colleagues' hard drives and in their drawers to provide an accurate picture of where chimpanzees, bonobos, gorillas, and orangutans lived and how many were left. This great ape census served to inform long-term conservation strategies.[62]

The PanAf program deepened this research agenda by providing the basis for evidence-based conservation of chimpanzee populations all over Africa. The census informed researchers and political actors about the regions in which chimpanzees had survived and at what density. The presence of this one charismatic species served as a criterion for decisions about which areas required special protection and which areas could be opened to increased economic exploitation. Unless the apes enjoyed a particular status among the local human population, as in Bossou, where no other large animals had survived, their presence frequently indicated the persistence of a rich fauna and flora that could be saved alongside them. Once the authorities agreed to implement protective measures, often in collaboration with NGOs such as Boesch's Wild Chimpanzee Foundation or Matsuzawa's Green Corridor Project, the census and the accompanying ecological data also enabled researchers to evaluate how effective different measures would eventually prove to be.

PanAf also sought to build an ethnographic archive that would save chimpanzee cultures from oblivion. To minimize human pressure, it relied exclusively on noninvasive methods such as camera traps, genetic analysis of fecal and hair samples, or collection of material artifacts such as tools left behind at termite mounds, underground beehives, and nut-cracking workshops. Its epistemology decoupled empiricism from ethnographic experience.

Before the Taï chimpanzees had been habituated in the early 1980s, Boesch had already gathered most of the data for his dissertation on chimpanzee nut cracking by way of indirect observations. Twenty-five years later, however, a range of new technologies enabled a completely new form of primatological fieldwork. Not only had video cameras become lightweight enough to be taken to the forest, but rapid advances in storage media and battery life enabled researchers to set up camera traps at strategic points all over the forest, recording whenever an animal triggered an infrared motion detector. The internet got citizen scientists involved in the coding of 350,000 one-minute video clips generated by these cameras.[63] The performance of these amateur researchers could be compared to newly developed animal biometrics, which employed algorithms automatically recognizing species and, in some scenarios, even individual animals.[64] Solar-powered audio devices allowed continuous recording of animal communication and poachers' shots over long stretches of time. The invention of the polymerase chain reaction (PCR) and new genetic analyses provided the tools to determine the demographics of chimpanzee communities on the basis of their excrements. And isotope analyses of hair left behind in their night nests provided information about meat consumption and thus about how much hunting different chimpanzee groups did, without ever watching them eat or prey on other animals.[65]

The PanAf program collected these different kinds of data at thirty-five field sites, eventually aiming at forty. In comparison with the twelve sites following habituated chimpanzees, this amounted to an almost fourfold increase in known chimpanzee communities. Analogous to how long-term observations had extended the temporal scale of primatological research, this complementary form of collective observation boosted its geographical range. Like many other field sciences, field primatology no longer contented itself with producing local knowledge and developed its own strategy for scaling up the significance of its truth claims from particular field sites like Taï to the whole African continent.[66]

But even Boesch's large and exceptionally well-funded research group at the Max Planck Institute could not do it on their own. They had to collaborate with already established field sites. Thus they built a vast network all over Africa. Having departed from a form of chimpanzee ethnography that starred the primatologist as lonesome hero, PanAf and the Collaborative Chimpanzee Cultures Project (CCCP and CCCP-2; see chapter 2) transformed cultural primatology into Big Science.

Since primatologists tended to be controlling of the field sites they built, bringing together a large and dispersed collective of observers required carefully balancing many interests and allaying concerns that anyone's data could be scooped. Still, not everybody agreed to participate in the Max Planck Institute's large-scale collaborations. Matsuzawa, for instance, preferred to maintain his independence: "I have no reason to pant-grunt Christophe," he told me with reference to a chimpanzee vocalization that signals subordination.

Moreover, the field sites that did join the effort had conducted and recorded their observations in idiosyncratic ways. Now everybody had to receive the same training and follow the same protocol. The primatologists running the PanAf program worked with computer scientists to implement a shared standard for the envisaged digital database, which complemented a cold room in the basement of the Max Planck Institute that preserved chimpanzee tools and poop.

Collaborating across disciplines posed yet another challenge: not only did primatologists and computer scientists have to overcome disciplinary language barriers, but everybody preferred to publish in journals that their disciplinary peers held in high esteem. Contributing to the literature of other academic fields hardly advanced one's own career. This made it difficult to agree on publishing venues.

Computer scientists played a key role in building an ethnographic record that could still be studied long after the demise of the documented chimpanzee cultures—presupposing that their storage media and data formats would survive the fast-paced disruptions of the digital revolution. The archive was built for future generations of primatologists who would pose questions and apply analytic techniques their predecessors did not anticipate. While testing hypotheses continued to be the standard model of science, nobody knew which hypotheses researchers to come would test, if any. In the spirit of big data approaches, Kühl imagined that instead of testing hypotheses, they might also mine the data for patterns and correlations. In any case, the organizers of the PanAf program wanted to salvage what they could: "If we don't do it now, we will never again be able to get these samples," Kühl told me. This required collecting and storing much more data than currently needed.

In the natural sciences, however, many funding agencies balked at providing money for the collection of surplus data that neither tested a hypothesis nor served any other analytic purpose at present. So the Max Planck Society stepped in: "No other institution could have pulled off such a high-risk, high-gain project," Kühl said.

Of course, the PanAf research strategy was not exclusively altruistic, incurring costs solely for the benefit of future generations of primatologists. The data also served the short-term interests of the current generation. It revealed more behavioral variations between chimpanzee populations. For example, at four sites across West Africa, camera traps had recorded how chimpanzees habitually banged and hurled rocks against trees or tossed them into tree cavities, resulting in visible stone piles—a finding taken to have implications for how to interpret cairns and other stone assemblages at ritual sites of ancient humans.[67]

Beyond a qualitative account of the lifeways of different communities, Boesch and Kühl hoped that the quantitative analysis of the PanAf data would shed further light on the relationship between genetics, ecology, and social learning.

Tristes archives

Once again cultural primatology recapitulated the history of cultural anthropology. Its concern about the rapid extinction of primate cultures echoed earlier worries about the extinction of human cultures in the wake of the colonial encounter, which had shaped the epistemological anxieties and practices of anthropology from the early nineteenth century to this day. Around 1800, the Société des observateurs de l'homme still imagined the world of humans as stable in its variety. By the 1830s, however, a growing number of observers noted the demise of colonized people. As Victorian humanists developed a sense of compassion for the wretched of the Earth, the British Parliament issued a report in 1837 that called for remedial action. From the start the concern was both humanitarian and scientific. Four years later, a committee that included Charles Darwin emphasized the importance of ethnological research, considering that "the races in question are not only changing character, but rapidly disappearing."[68] Unless anthropologists documented the varieties of humans before their extermination, we would never fully understand the history of human nature. Jacob Gruber argued that the resulting ethnographic salvage operation profoundly shaped the practices and institutions of anthropology, as it began to build an archive of human cultures.[69]

No book captured the melancholic sentiment pervading this ethnographic and archival work better than Claude Lévi-Strauss's *Tristes tropiques*.[70] Alienated from his own culture but inspired by French Enlightenment thinkers, most prominently Rousseau, the French anthropologist traveled to Amazonia to find "a society reduced to its simplest expression."[71] What he found in an intolerably oppressive rainforest was not a better life, but abject misery: societies "enfeebled in body and mutilated in form," which "had been pulverized by the development of western civilization."[72] The ethnographic task of documenting human destruction and cultural loss appeared so central to Lévi-Strauss that he concluded that the discipline devoted to studying this process of disintegration should not be called anthropology but "entropology."[73]

In the mid-twentieth century, just as Lévi-Strauss was writing *Tristes tropiques*, the insight that supposedly traditional human cultures were disappearing or changing beyond recognition at breakneck speed fueled a transition from participant observation conducted by individual anthropologists to large-scale cross-disciplinary fieldwork projects such as the Six Cultures study of child rearing, conducted by whole teams of ethnographers in India, Kenya, the United States, Mexico, Okinawa, and the Philippines. Such Big Social Science also amassed big data for which researchers constructed extensive databases and archives, enabled partly by microfiche technology, which did not even outlive some of the waning cultures it helped to record. Anthropologists became as concerned about the potential loss of ethnographic data as they were about the irrevocable loss of cultures, argued historian of science Rebecca Lemov.[74]

Anthropologists raised the late-modern "endangerment sensibility" that worried about the loss of cultural and biological diversity to the second order.[75] Lorraine Daston described the spreading concern about the preservation of data as "a moment of archival anxiety."[76] At the beginning of the twenty-first century, the conjunction of the sixth extinction and the digital media revolution brought back questions of how to create an ethnographic archive that would preserve abundant data on chimpanzee cultures for future primatologists who might no longer be able to experience these cultures for themselves.

The PanAf project and the data collection and storage under the auspices of the Collaborative Chimpanzee Cultures Project amalgamated a sense of *tristes tropiques* with the elegiac positivism that had colored the creation of many scientific and humanist archives since the late nineteenth century. By the time Max Weber gave his speech "Science as a Vocation" in 1917, Auguste Comte's self-assured positivism of the 1830s had given way to a gloomy mood that science progressed only by amassing modest but stable facts, not eternal truths, noted Daston.[77] Theories seemed to change like fashions. Today's hard-won scientific achievement was tomorrow's fad or stepping-stone. Archives represented repositories for those largely decontextualized facts and artifacts, numbers and samples, measurements and recordings that provided a minimum of continuity to scientific and scholarly enterprises.

Subsequently, the underlying positivism had a rough century, challenged and abandoned time and again, most prominently in the German *Positivismusstreit* of the 1960s and the postpositivist philosophies of science that began to mushroom and spread from the United States at about the same time. Philosophers like Willard van Orman Quine, Paul Feyerabend, and Thomas Kuhn tied the fate of seemingly stable facts to perishable theories, which arguably informed all scientific observation.[78] Although the label *positivism* had become a term of abuse, which the abused rarely claimed for themselves, elements of the ethos and pathos of different positivist traditions lived on in many scientific practices—not least in the archiving of chimpanzee cultures. Broken down into observable cultural traits and preserved in the form of stone hammers and honey-fishing rods in a Leipzig cooling chamber, thousands of kilometers away from the tropical heat and moisture that their makers and users had cherished, these artifacts would give future primatologists potentially important research data, but an anemic impression of the diversity of chimpanzee ways of life.

The chilled melancholia of these ethnographic archives, documenting a biocultural diversity on the wane, sheds light on our late modern cosmology. Philip Fisher's analysis of the structure of grief suggests that feeling the pain of this loss reveals a bright line between those who are and those who are not part of the inner fabric of our world. Although few human beings have spent much time with monkeys or apes, these animals take up an important place in African, Japanese, and European cosmologies. Fisher noted that one component of all mourning was an "advance payment of grief for ourselves that will be unpayable once that death has actually occurred, because we will not be in existence to feel the sorrow of the greatest loss of all, that of ourselves."[79]

Such anticipatory gloom had already shaped the pessimistic view of natural history and the anthropologist's vocation, which Lévi-Strauss had laid out in *Tristes tropiques*: "The world began without the human race and it will end

FIGURE 20. Ethnographic archive for chimpanzee cultures in the basement of the Max Planck Institute for Evolutionary Anthropology, Leipzig, Germany. Photo by the author.

without it. The institutions, manners, and customs which I shall have my life in cataloguing and trying to understand are an ephemeral efflorescence of a creative process in relation to which they are meaningless, unless it be that they allow humanity to play its destined role." That role was to hasten "the disintegration of an initial order."[80]

At the beginning of the twenty-first century, this sentiment was amplified by daily news media reporting about the accelerating loss of biodiversity and failed climate change negotiations. Popular books such as *The World without*

Us reveled in the details of how surviving species of plants and animals would recolonize the ruins of New York City.[81] The sadness pervading the ethnography of the last wild chimpanzee cultures reinstated a humanist view of the world, centered on our own finite existence as a species.

It was a humanism brimming with misanthropy. Its anthropocentrism recognized the exceptional role of modern humans in natural history—exceptional in our cataclysmic impact on the rest of the planet. "Only we have the power to destroy it," Boesch wrote to highlight our responsibility for the chimpanzees' fate.[82] This negative narcissism repeats the admixture of power, guilt, and grief that has shaped the colonial self-consciousness of Europeans and Americans since the nineteenth century and that has tinged anthropological writings ever since. The Anthropocene has scaled up this spiritual malaise from the level of Western culture to that of the human species. My fieldwork in cultural primatology left me wondering: What would it take for *Homo sapiens* to become a happy ape again?

Conclusion

AS THIS BOOK goes to print, the controversy over chimpanzee cultures has not been settled. Before the epilogue zooms out again to the wide philosophical angle of the prologue, which situated the chimpanzee culture wars between Enlightenment and Anthropocene, I would like to briefly review what the preceding chapters have taught us about cultural primatology.

Controversies and Comparisons

I organized my account alongside two of the debate's most important front lines: the first between laboratory and field research, the second between Japanese and Euro-American traditions in primatology. The ethnographic core of *Chimpanzee Culture Wars*—chapters 3 to 6—mimics one of the 2 × 2 contingency tables that organized Matsuzawa's comparative cognitive science when he contrasted, say, knowledge and care of wild and captive animals. My book juxtaposes a Western field station, a Western laboratory, a Japanese laboratory, and a Japanese field station. While this helps to give the book a clear-cut structure, much of the chapters' content has called into question whether these domains are neatly separable and suggests that national cultures and research approaches were hardly the only factors that determined scientists' positions in the controversy. Each of the four sites was too peculiar to represent the general category: for example, Boesch's primatological fieldwork could not stand in for European fieldwork, nor did Matsuzawa's synthesis of lab work and fieldwork typify Japanese primatology. Nevertheless, we gain important insights into cultural primatology as a space of possibilities by comparing the four cells of the table, however particular their contents might be.

Boesch's fieldwork on wild chimpanzees and Tomasello's laboratory experiments on captive chimpanzees contradicted each other regarding the capacity of *Pan troglodytes* for culture and cooperation. The ensuing controversy highlighted that controversies follow different logics in the lab and in the field. So far, most controversy studies had focused either on debates between experimental scientists or debates between field scientists, but not on debates across the lab/field divide. Boesch and Tomasello could not agree because the fieldworker doubted the ecological validity of the experimenter's findings, while the experimenter denied that field observations could provide any insights into what caused the observed behaviors, leaving chimpanzee ethnographers unable to rule out alternative explanations of supposedly cultural behaviors. Thus, the controversy between Boesch and Tomasello was also a controversy over how to resolve a controversy.

Matsuzawa's laboratory research at the Kyoto University Primate Research Institute and his field research in Bossou presented an interesting contrast to the disagreements between the two codirectors of the Max Planck Institute for Evolutionary Anthropology because he integrated lab and field in his own work. He studied the social learning of nut cracking through field observation, field experiment, participation observation, and controlled laboratory experiment—and conceived of his synthesis of all these approaches as an expression of Japanese holism. It would not have occurred to either Boesch or Tomasello to frame their scientific methods in terms of national culture. Although Matsuzawa's cosmopolitan assemblage of practices and theorems was very particular, I could not pinpoint a single element that proved uniquely Japanese. Despite his idiosyncratic implementation, the general vision of amalgamating laboratory and field practices had been discussed in Europe and North America for almost a century. Yet the auto-anthropologization of Japanese primate science seems to imply a conception of science as reflecting cultural diversity rarely found among Euro-American natural scientists (and, in contrast to Imanishi, Matsuzawa did understand himself as a natural scientist).

If cultural differences in cosmology indeed informed how primatologists conducted their research, then it might also have influenced their choice of field sites: Boesch preferred Taï Forest, which he had originally imagined as pristine nature, while Sugiyama and Matsuzawa set up shop in the highly anthropogenic environment of Bossou (but remember that the Dutchman Adriaan Kortlandt had conducted field experiments on Mont Gban before Sugiyama came to observe the Bossou chimpanzees naturalistically). If the Japanese understood humans as part of nature and nature as a garden, it would make sense that Matsuzawa built an outdoor laboratory and studied chimpanzee road crossings as "natural" (at least in the Japanese sense) experiments, while Boesch (but not his successor Roman Wittig) refrained from field experiments and generally sought to minimize human interference. The scientific object of Boesch's chimpanzee ethnography was wild cultures.

For four decades, Euro-American science studies scholars have argued against conceptions of nature as distinct from human culture—and have promoted an almost Japanese ontology of naturecultures. It would be easy to dismiss Boesch's naturalism as a Western illusion: Taï is a park, constantly maintained and disturbed by humans; its chimpanzees aren't exactly wild because they have been habituated to human presence, and so on. However, what I took away from my fieldwork is that the imagination of an untouched nature operates as a regulative ideal that led primatologists to pick different field sites, adopt observational rather than experimental approaches, and more. Taï and Bossou were only 350 kilometers apart, but they represented radically different visions of field primatology.

Likewise, Matsuzawa's laboratory in Inuyama and Tomasello's laboratory in Leipzig represented radically different visions of comparative psychology. For Matsuzawa, Sakura, and Morimura, understanding chimpanzee nature meant not only teasing out species differences in strictly controlled experiments, which often involved cutting-edge technology, but also developing a feeling for the organism through participation observation. The time-consuming cultivation of a potentially dangerous intimacy between *Pan* and *Homo* would have been unthinkable at the Max Planck Institute.

Again, the science studies literature encourages a bias privileging the kind of interspecies entanglement showcased by the Japanese approach. Moreover, Tomasello's quest for a *differentia specifica* between humans and other apes has become extremely unpopular in the posthumanities. The confluence of Darwin and Buddha in the doctrine of human-animal continuity could help to combat the belief in human supremacy, which the ecologically woke have come to consider on a par with white supremacy. However, if we want to answer the question of what has enabled modern humans rather than chimpanzees to become a superdominant species across the globe, we have to take seriously Tomasello's emphasis on discontinuity. Philosophers might be well advised to consider Boesch's methodological critique of his colleague's captive studies and to consider why the French Swiss field primatologist pointed to Matsuzawa's approach as a more credible alternative in the laboratory. Although primatologists have not reached agreement about whether to interpret behavioral differences between chimpanzee groups in terms of culture, all claims about human nature based on comparisons with our closest evolutionary relatives now have to consider the variability of both human and chimpanzee ways of life. But even if the controversy is decided in Boesch's favor and chimpanzees are unanimously accredited with culture, I would still hope that primatologists will find a different explanation for why the numbers of *Pan troglodytes* are plummeting while those of *Homo sapiens* continue to skyrocket.

Epistemic Diversity

As I complete *Chimpanzee Culture Wars,* most of the book's protagonists have just retired. McGrew has largely given up chimpanzee research for observing New World monkeys with his wife, a Japanese primatologist studying capuchins in the jungles of South America. Shortly before Matsuzawa passed the laboratory on to Masaki Tomonaga, he told me: "Matsuzawa is already over, he is now becoming history." But he vowed to use the newly gained freedom to meet his obligation of caring for his nonhuman subjects Ai, Akira, and Mari until the end of their lives. He also sought to put his chimpocentrism into perspective by studying snub-nosed monkeys in the mountains of Yunnan in China (which provided a welcome opportunity to exercise his rusty alpine skills). The Max Planck Society replaced Tomasello's Department of Developmental and Comparative Psychology with the Department of Human Behavior, Ecology and Culture. They were also searching for a new director to succeed Boesch, who eventually retired in 2019. Meanwhile, Boesch had begun to follow cultural anthropologists to observe how human foragers cracked nuts in the rainforests of Central Africa. Thus the publication of this book marks the end of an era in chimpanzee culture research. My fieldwork was as much a salvage operation as that of cultural primatologists. In Europe, Japan, and Africa, in the laboratory and in the field, it documented very different knowledge cultures on the wane.

What set the projects at the ethnographic center of *Chimpanzee Culture Wars* apart from the big data methods of the PanAf program or the camera traps that will now record the last days of the Bossou community is that these approaches to chimpanzee culture and cognition were deeply experiential. Comparing and contrasting the very different forms of research experience afforded by Taï Forest and the Leipzig Zoo, Kyoto University Primate Research Institute and the outdoor laboratory on sacred Mont Gban sheds light on what scientific lifeways primatologists could share with their primates. That space of possibilities narrowed with every chimpanzee community that disappeared and every set of safety measures that further regulated how humans and apes could relate to each other. At the same time, however, new technologies also enabled new kinds of epistemic depth, providing different insights than the daily observation of habituated animals or direct interaction with enculturated apes. Knowledge of human and nonhuman primate cultures would continue to accumulate in different conceptual frameworks.

Whether these developments endanger or preserve scientific diversity is for future anthropologists and sociologists of science to find out. Despite all efforts to standardize, internationalize, and unify primatological knowledge production, it seems to me as if the competing thought collectives still cultivated and rearticulated their differences. For example, a new generation of Kyoto School scholars, taking their cues from Junichiro Itani rather than Kinji Imanishi, reconnected Japanese primatology to cultural and ecological anthropology—as a counterbalance to sociobiology.[1] Thus the chimpanzee culture wars might go on as long as there are people who passionately care about how we relate to the apes. Even if methods and findings converge, there is no rational or empirical way of resolving the dispute over competing definitions of the culture concept. I'm not sure whether we should celebrate the corresponding intellectual diversity. It is inseparable from academic identity politics and disciplinary chauvinism. However, while the resulting polemics cause much distress and one-sidedness, they also maximize available alternatives. Japanese primatologist Hiroyuki Takasaki concluded an article on the Kyoto School with an evolutionist plea for preserving cultural differences in science: "Such differences are comparable to the biological diversity and the gene stock that are necessary for the global ecosystem's stability and for evolution of new diversity. There are always some parts of the world that may be viewed better upside down."[2] In this spirit, second-order primatology throws into relief the possibility of thinking differently.

Should we overcome the two-culture divide and make second-order primatology part of primatology proper? Such an integration of the sciences and the humanities is what Edward O. Wilson envisaged.[3] His dream of unifying the two great branches of learning provoked massive opposition on the part of humanities scholars, whom Wilson wanted to assimilate into the knowledge culture of evolutionary biology. Instead of consilience we got the science wars.

Did Japanese primatology provide a more accommodating framework? After all, Imanishi had conceived of it as a form of social research that had more in common with the humanities than with the natural sciences.[4] Moreover, the collapse of the nature/culture dichotomy initiated by the discovery of monkey culture seemed to dovetail with the ontology of important parts of science studies. Historian Julia Adeney Thomas noted in her book on concepts of nature in Japanese political ideology the parallel between Japanese versions of naturecultural states and Bruno Latour's insistence that there were only "natures-cultures."[5] Both diminished modernity's grand vision of complete freedom. Thomas considered this curtailed redefinition of freedom "not all positive."[6] She maintained a critical distance toward attempts at creating consilience across disciplinary divides that enabled scholars from all corners of the university to come together around the embodied mind, a naturalized history, and a cultured nature.[7] In opposition to this zeitgeist, she sought to restore the nineteenth-century division of labor between the natural sciences and the humanities as taking care of facts and values, respectively. While scientists determine what is the case, humanists interpret conflicting conceptions of the human presented by different sciences as revealing multiple ways of life open to us—and they might help us to decide which of these ways to pursue.

At least in the humanities, many have come to value diversity in science for political and epistemological reasons: including the perspectives of marginalized groups increases objectivity and fosters academic excellence.[8] My own preoccupation is aesthetic. I find it exhilarating to discover that other times and places allow for other styles of thinking and doing research—a cognitive passion I share with a whole generation of historians of science who came of age with Thomas Kuhn's *The Structure of Scientific Revolutions*.[9] By contrast, the epistemic monoculture of a "unified science of cultural evolution" is as unappealing to me as a plantation must seem to a field biologist who has experienced the rainforest.[10]

However, visions of unified sciences cannot only smother diversity in science; they are also expressions of the scientific imagination. Knee-jerk reactions against all forms of knowledge that smack of positivism and determinism wall in the contemporary humanities. Precisely because intellectual projects at the disciplinary center of primatology are less compatible with a radically postpositivist knowledge culture that emphasizes all kinds of agency, it is important that they find representation in humanist conversations about what to make of the unprecedented evolutionary success of *Homo sapiens*. In multispecies studies, monkey researchers like Barbara Smuts, Shirley Strum, Thelma Rowell, and Agustín Fuentes have been considered natural interlocutors—and for good reasons. But there are different lessons to learn from William McGrew, Christophe Boesch, Michael Tomasello, and Tetsuro Matsuzawa.

Against the Grain

Throughout this book I have mined the historical and ethnographic material for insights that challenge the doxa of my own fields, cultural anthropology and science studies. I hope readers will have come across a number of observations that surprised and even irritated them. It is such deviations from the habitual that get us to think. For example, when I began this fieldwork, I was so mired in an ontology that had deconstructed the difference between nature and culture that it took me a while to appreciate the efforts of cultural primatologists to disentangle culture from genetics and ecology. Although they recognized gene-culture coevolution as well as the adaptive function of many cultural behaviors, they sought to demonstrate that certain animal species had evolved a second biological mechanism to transmit novel behaviors within groups and across generations: while genetic inheritance of mutations occurred at a very slow pace in long-lived primates, social or cultural transmission allowed the acceleration of adaptions to rapidly changing environments. Much of the chimpanzee culture controversy revolved around the researchers' inability to reach consensus over how to define culture, but they also fought over a very real phenomenon that existed independent of how primatologists would speak about it: How exactly did various primate species pass on newly acquired behaviors? If cultural transmission was biologically distinct from genetic inheritance, including genetically inherited adaptations to particular environments, wouldn't it be time for cultural anthropologists to stop writing against culture and for science studies scholars to reconsider the ontology of naturecultures? And if we reconstructed an analytic that set apart cultural, ecological, and genetic explanations of primate behaviors, including our own, wouldn't we arrive at a richer understanding of the hominoid condition than either genetic determinism or the naturecultural metaphysics of an agency beyond the human could provide?

Studying the epistemic practice of chimpanzee ethnography has also helped me overcome an animus toward positivism, which is almost common sense among interpretive social scientists today. I hope the humanists and posthumanists among the readers of this book will no longer accuse cultural primatologists of an abuse of ethnography but will appreciate their virtue of interpretive restraint. Aspiring to a reduction of the theory-laden nature of observations facilitated scientific collaborations irrespective of theoretical commitments and the creation of an ethnographic archive for future researchers whose conceptual frameworks contemporary chimpanzee ethnographers couldn't possibly anticipate. Aspiring to a reduction of the value-laden nature of observations facilitated collaborations irrespective of moral and political commitments. One lesson science studies scholars should have learned from the science wars is that the moralization and politicization of scientific facts is often possible, but rarely necessary—especially since every fact can be moralized and politicized in multiple ways, and very few if any of them are conducive to working with differently minded people. But maybe that's not the goal?

Even though I do not have the slightest objection to chimpanzee ethnography, its positivism is far removed from my way of doing research. We need to talk across the two-culture divide; we need three, four, and many more cultures (and in fact, we already have them), but we do not need consilience. Thus, the epilogue will return to this book's philosophical point of departure in an unabashedly humanist and interpretative spirit.

Epilogue

THIS BOOK BEGAN with Rousseau and his heirs, who had finally made the philosophical voyage to the monsters of Loango. Of course, other genealogies of the common Enlightenment origin of cultural and biological anthropology would have been possible. My reference to the *Discourse on the Origins and the Foundations of Inequality among Men* served mostly to highlight that we continue to ask many of the questions that the *philosophes* raised in eighteenth-century France, but in a radically different historical context these questions have taken on new significance. The shock of the insight that a primate culture, that of modern humans, could cause global climate change and exterminate other species and their newly discovered cultures accelerated the collapse of a dualist ontology separating nature from culture and society. It fueled a sense of humanity having created a watershed in naturecultural history with one slope leading to unprecedented catastrophe and the other to a new harmony with the world of nature, as Kenyan conservationist Richard Leakey put it.[1] Since the early 2000s, this crisis of our own making has been christened the Anthropocene.[2] As crisis diagnoses go, it serves to judge history as a form of temporality upon which *Homo sapiens* (more than other species) can act.[3] As a participant observer of this unfolding drama, I have often wondered whether the natural historical transition we are witnessing could be framed differently.

Japanese Natural History: Evolution without Crisis

In 1941, Kinji Imanishi proposed his own account of the momentous ruptures in the fossil record, which Darwin and his fellow uniformitarians had denied and which neocatastrophists more recently came to interpret as the sixth mass extinction. Like the latter, Imanishi doubted that the simultaneous disappearance of large numbers of species could be explained in terms of Darwin's theory of natural selection. But he also downplayed the significance of other kinds of external events such as plate tectonics or climate change. Instead he imagined all taxa to follow their own inherent trajectories. Once a community of living things had reached "the summit" of its development, Imanishi contended, "sooner or later it begins to self-destruct and by its collapse, another whole community with different features begins to develop."[4]

In a strictly hierarchical biosociety, in which species coexisted harmoniously (without the fierce competition that only an Englishman at the time of Manchester capitalism could have projected onto nature), every class of animals had its time to rule the world of living things. In this cosmological vision that collapsed taxonomic categories and social strata, Imanishi defined evolution as "the history of the rise and fall of the ruling class."[5] But these natural historical revolutions preserved the structure of a multispecies society, which Imanishi described as fundamentally "conservative."[6] Not only would there always be a class of animals that dominated all others, but this class could not possibly ascend from some zoological proletariat. It had to evolve from the previous ruling class. That's how Imanishi pictured the transition from the Mesozoic to the Cenozoic era: the dinosaurs had been neither outsmarted by mammals nor wiped out by a natural disaster—instead these "reptiles transformed into mammals."[7] For a while, elephants reigned over the animal kingdom. Today, at a time when African elephants seem to be going the way of the mammoth, the scepter has been passed to humans: "Mankind arose within the community of mammals, and has replaced other mammals temporarily as the ruling class of the society of living things. The history after that is the real history of mankind."[8]

What set this Japanese view of nature apart from the Darwinian view prevailing in European and American evolutionary theory was not just its denial of a perpetual struggle for existence. Imanishi also rejected the emphasis on contingency shared by Darwin's natural selection of random mutations and the neocatastrophist assertion that chance, not genes, determined which species would survive and which would succumb to mass extinction events. Instead evolution proceeded in a teleological manner. Although Imanishi referred to the disappearance by transformation of whole groups of animals as "extinction," his account was closer to Lamarck's than Cuvier's.[9] He believed that no class went under defeated: it progressed. New ruling classes always grew out of the old ones. Our own reign would be no exception: "Human development has its own limit," Imanishi remarked. "This should not worry us, however. Those who replace us, though they perhaps should no longer be called mankind, will originate and be created from the human race."[10] Thus, the world of living things would forever maintain its equilibrium and self-completeness.

In contrast to the currently prevailing narrative of the Anthropocene, Imanishi's natural history proceeded without crises—and consequently without critique.[11] While Darwin's gradualism left no room for major crises because extinction and speciation balanced each other out, Imanishi's anti-Darwinian theory of evolution did admit for large-scale transformations of the world of living things, which the neocatastrophists would rehabilitate in Euro-American paleontology a few decades after Imanishi had begun to lay out his Japanese view of nature. But these moments did not represent the turning points that medieval physicians had in mind when they designated as crises the crucial stages of a disease at which a decision had to be made.[12] Nor did they evoke the moral demand to right a deviation from the norm entailed by the modern semantics of crisis. Nothing had gone wrong. No, Imanishi's natural history ran through a deep evolutionary riverbed, in a steady flow without turbulence. Species did not disappear into a vortex of time but evolved from lower to higher forms of life, following an "unconscious design"—and culture was nothing but "the flower which blooms on this destiny," Imanishi maintained.[13] Neither the ascent of humankind to its dominant position nor our anticipated downfall called for self-reproach or activism. Life would simply go its way.

There has been some discussion about whether Imanishi's account actually represents *A Japanese View of Nature*, as the English translation of *Seibutsu no Sekai* (The world of living things) was titled, or whether it should be taken as a highly idiosyncratic perspective, as Imanishi's declaration of the book as a "self-portrait" suggests.[14] Be that as it may, Imanishi's trust in the evolutionary process appears to be in line with social ecologist Stephen Kellert's finding that despite their aesthetic and spiritual appreciation of nature and wildlife, Japan's public and its government invested much less in conservationism than Germany or the United States.[15] In the face of orientalist clichés of Japanese harmony with nature, in the late 1980s, the United Nations Environment Programme rated Japanese environmental concern lowest among fourteen surveyed countries.[16] The paradox can be resolved easily: if humans and nature are essentially the same, both equally transient, and if nature is everything around us, so that garbage is as much a part of it as flowers, as Arne Kalland and Pamela Asquith said about Japanese perceptions of nature, then the preservation of independent nature, untouched by human hands, makes little sense.[17]

Joyous Primatology

Could such a natural history, devoid of ecological critique and crises, inspire us to think differently about the Anthropocene? Moralist condemnations of human exceptionalism haven't changed the recent course of natural history, and there are few indications that they will make much difference to the future of our planet.[18] They did transform how we relate to ourselves, though, infusing human species identity with a sense of self-loathing and *ressentiment*. While McGrew's defiantly professed positivism might be epistemologically no more tenable than Haraway's radicalization of the postpositivist doctrine of the theory- and value-laden nature of observation, it offers a counterweight to the habit of humanities and posthumanities scholars of interpreting human behavior in terms of right and wrong.[19] Exposing the economic exploitation, sexist bias, racist discrimination, and speciesist treatment of nonhuman animals behind primatological research might be a noble cause and has taught us important lessons about how scientific knowledge could become a foundation of inequality among men and women of different origins. But just as there is more to chimpanzee life than dominance rank hierarchies and intergroup conflict, there is much about human life in general and science in particular that escapes this form of critique. The use of academic knowledge to domesticate our own species should not limit the scope of anthropological inquiry.

In cultural anthropology, the moralist condemnation of human supremacy has stymied curiosity about what enabled *Homo sapiens* rather than other primate species to transform the planet so spectacularly.[20] Of course, unsustainable exploitation of limited resources is not a uniquely human trait: in Thailand, stone tools enabled long-tailed macaques to overharvest shellfish, and the Ngogo chimpanzees drove the colobus monkeys in their territory to the edge of extinction.[21] Modern humans, however, have not only diminished local populations but eradicated entire species from the planet. When I asked Shirley Strum how this loss of biodiversity and human-caused climate change affected her research on baboons, she expressed concern about the well-being and survival of the monkeys, but she also saw the Anthropocene as "a fantastic methodological bonus" because the lightning speed of environmental change enabled her to observe the process of adaptation in real time.[22] For anthropologists, the momentous transformations that we are currently witnessing also provide a unique opportunity to rethink the human potential—and that of other primate species. The adaptive nature of cultural evolution contrasts sharply with the teleology of Rousseau's conception of perfectibility.[23] *Chimpanzee Culture Wars* has recounted the controversy between Boesch, Tomasello, Matsuzawa, and their colleagues over whether the capacity for cumulative culture (or language or cooperative breeding) determined our species' exceptional evolutionary trajectory. While the jury is still out, this is a conversation that might get cultural and evolutionary anthropologists talking again.

I share Strum's excitement about living through such anthropologically eventful times, and yet I emerged from the fieldwork for this book feeling glum. I admired the primatologists' intense efforts to preserve the chimpanzee populations they studied, but in the grand scheme of things, they knew they were fighting a losing battle. It is hard not to feel the intense loss, especially after having encountered these hominoid communities in the kaleidoscopic forests of West and Central Africa. The energizing crescendo of pant-hoots during interviews still cheered me up when I listened to the recordings at my desk. Thinking back, I have forgotten all the torment of fieldwork in an environment hostile to unhabituated humans like me. I can hardly remember how stressed and intimidated I was by Jacobo's charges against me. Even if the dream of living in a harmonious hybrid community with our next of kin will always remain just that, I wish the wondrous worlds of Taï, Loango, and Bossou with their casts of colorful characters would always be out there.

It was this thick layer of melancholia that made me long for some comic relief. Historian Hayden White proposed that any series of historical events could be told in different manners: as tragedy, romance, satire, or comedy.[24] Donna Haraway and her anthropological interlocutors followed the dominant emplotment of the Anthropocene as tragedy, although one that offered an opportunity for learning that nature was no longer what "conventional science" had imagined it to be.[25] I wondered whether the savage success of our species could alternatively be cast as a *comédie humaine*—or really a *comédie simiesque*. Or would it be possible to narrate how *Homo sapiens* lapsed into cosmic solitude as we eradicated other cultured apes in the form of satire?[26]

Voltaire's *Candide* recognized the sublime ridiculousness of a determinist universe and made fun of Leibniz's idea that it amounted to the best of all possible worlds. One awful event leads to another, from the Lisbon earthquake to people set to be burned at the stake and freed at the last minute, only to murder someone and subsequently be enslaved. In this endless chain of horrors, every once in a while the greatest metaphysician of his time, Dr. Pangloss, shows up to assure the reader that "everything is necessarily for the best purpose."[27] This famous piece of Enlightenment prose, revealing its author's attitude of profound world rejection, found its way into evolutionary theory.[28] Stephen Jay Gould and Richard Lewontin criticized the adaptationist program of their Anglo-American colleagues as the "Panglossian paradigm": at the height of the sociobiology debate, they rejected "the near omnipotence of natural selection in forging organic design and fashioning the best among possible worlds."[29] Politically, the left-leaning paleontologist and population geneticist worried that naturalizing the human in such a manner would amount to an affirmation of the status quo.[30]

As a primate struck with empathy, I did not gain enough distance to appreciate the folly of the biotic world, in which species and cultures, human and nonhuman, come into being, only to drive each other extinct. Writing *Chimpanzee Culture Wars* as a *Candide*-style satire would have felt quite inappropriate. Nor did rehabilitating the metaphysics of adaptationism appear a viable strategy to overcome the Anthropocene blues. Nietzsche's *Joyous Science* offered a better philosophical blueprint in that it did not derive its gaiety from the pretension that we were living in the best of all possible worlds. In fact, it was born out of despair and hopelessness. But this spiritual exercise aspired to a view of life that overcame self-denial and refrained from crying out in accusation to make others feel ashamed or guilty. Saying yes even to the ugly side of existence was what Nietzsche's formula of *amor fati* encapsulated. A joyous primatology that adopted this brand of life-affirming fatalism would have to tell a story about the disappearance of primate cultures without discerning the historical significance of this event in terms of moral failure.[31]

Therapeutic Fatalism

While the French Revolution inspired faith in politics, in the capacity of a general will to give history entirely new directions, the large-scale environmental transformations we are currently witnessing foster a very different experience of time. Although many in the humanities and posthumanities try hard to rearticulate a belief in human and more than human agency to make history and narrate the Anthropocene as a series of political crises in which humans happen to always take the wrong path, this seems increasingly out of sync with how we experience time. At the beginning of the twenty-first century, literary theorist Hans Ulrich Gumbrecht observed, the nineteenth-century chronotope of history is giving way to a different conception of temporality as the future appears to inevitably come toward us in the form of global warming, mass extinction, and an exploding human population.[32] Whereas we could slow down these processes, we feel incapable of choosing radical alternatives.

Like positivism, fatalism is a swear word by which hardly anyone is swearing. It was coined as a tendentious term in the early eighteenth century to denounce the determinism of Spinoza and Leibniz.[33] The Catholic church declared the idea that we were powerless to do anything other than what we actually did a heresy because the Christian doctrine of sin required the assumption of free will. Friedrich Heinrich Jacobi, who played a significant role in introducing the concept of fatalism to philosophy, could not accept a system of blind physical necessity because it seemed to reduce the soul to the role of a mere onlooker (although Diderot's fictional character Jacques the fatalist appears a good deal more animated than his mechanically acting master).[34] Humanists rejected this worldview because the denial of human freedom undermined their conception of morality and faith in the openness of history. And even contemporary thinkers who claim to have left humanism and its Christian origins behind continue to condemn those who wallow in "sublime despair" instead of "doing many important things better."[35]

In the mid-twentieth century, one of the rare advocates of fatalism, American philosopher Richard Taylor, argued that it was only because we knew more about the past than about the future that we believed in our power to take charge of what will happen tomorrow. It is for this epistemological reason that we are all fatalists with respect to the past, which we can only accept and make the best of, while the fading historicist imagination led us to conceive of future events as possibilities that we select from. By contrast, Taylor noted, "a fatalist . . . thinks of the future in the manner in which we all think of the past."[36]

My own fatalist view of the future of chimpanzee cultures remains confined to a much more narrowly circumscribed metaphysics of apes. While I wholeheartedly support anyone determined to conserve wild primates and their habitats, we also need to step back and acknowledge that no species in natural history has lived forever—and yet life has gone on. Even the behavioral flexibility of culture might not allow us to change this way of the world, but it does provide the possibility of relating to it differently. The realization that *Homo sapiens* has dramatically altered the face of the Earth has created the illusion that this curious ape is master on its own planet and that we could, either on our own or in assemblages with other species, take fate by the throat. Against the overburdening of human freedom and the corresponding exaggeration of guilt a good dose of fatalism might offer a philosophical remedy. Writing this book has served as a self-admonishment to curb my critical impulses and accept the world and its inhabitants with all their apparent faults. Maybe the most important lesson of cultural primatology is that we too are primate, all too primate.[37]

Twilight

If the busy nut-cracking workshop has become the iconic scene of chimpanzee culture, I would like to end this book with a very different image. After following the apes from dawn until dusk, the researchers of the Taï Chimpanzee Project could not set off on the often long way home to camp before the animals had begun to build their nests. As the sun set, the party would stop somewhere. Tired and hungry, I hoped they would quickly climb up their trees and weave some bent branches into a comfortable sleeping platform for the night. Instead they lay down on the forest floor, crossed their legs, stared into space, yawned, farted. As the forest slowly cooled down and the last golden rays of sunlight beamed through the giant trees, they listened intently to the vocalizations of other parties settling down nearby. A few individuals disappeared into the canopy, occasionally uttering soft grunts. But until the primatologists and their assistants heard the rustling and cracking sounds of bed making, protocol did not allow them to take a GPS point to which we would have to return before sunrise, if we didn't want to lose the group. It was always possible that the animals would change their minds and spend the night elsewhere.

I could hardly bear these moments of idling. The forest grew darker by the minute and I would have liked to return to camp before nightfall. Having chased after chimpanzees for twelve hours, even some sourish manioc paste with *sauce graine* and a tin of sardines in oil seemed appealing. But also when well fed, I lacked the patience to observe these long stretches of chimpanzee life that were about doing nothing. If their behavior did not warrant a field note for a short while, I began to rummage about in my head for scraps of past conversations that I could enter into the diary on my iPad. Maybe I could scribble down an idea for an article or an outline for this book. My brand of fieldwork in philosophy is certainly more interpretive than observational.

In the end, however, my inquietude was not just an individual quirk. The primatologists around me were more disciplined and appeared better attuned to the rhythms of the chimpanzees' ways of life, but they did not take it easy either. The best among them never averted their eyes. They constantly watched and collected data to contribute to a growing body of knowledge. Science is a cumulative culture if there ever was one. What the very different kinds of fieldwork conducted by cultural anthropologists and cultural primatologists have in common is that both amount to work, and as such they are transformative. Knowing more, thinking differently, frantically salvaging endangered cultures, making the world a better place, never being content with what they have already achieved—*Homo academicus* partakes in a culture of restlessness promoting interminable advancement and strife.[38] The chimpanzee culture wars raised the question of whether chimpanzee lifeways remained essentially timeless, as an earlier generation of anthropologists had imagined primitive human societies, or whether chimpanzees too had plunged into one of the African jungle creeks that would eventually swell into the torrent of historical time, ever changing, progressing, and accumulating.

Matsuzawa, who entertained the possibility that chimpanzees were capable of cumulative culture, too, concluded his farewell lecture at Kyoto University by giving a very different and highly speculative response to the question of human nature: "What is uniquely human? For me, the shortest answer is *imagination*." Chimpanzees lived mainly in the here and now, he told his students. They didn't seem depressed about the past or anxious about the future. We humans, by contrast, constantly commemorate events from before our birth and work for the time after our death.[39] This grand philosophical claim was based on a single study, which compared the drawings of human children and chimpanzees.[40] How strange, I thought, that it had never gained any prominence in the chimpanzee culture controversy.

As the chimpanzees finally began to build their nests, we hastened back to our field station. Stumbling along the dark forest trails, I wondered whether the way they lived out their lives in one of the last remaining patches of primary forest, oscillating between frenetic activity and languid hours of repose as poachers intruded and planters nibbled away their habitat, offered a lesson on how to find some second-order peace of mind by going with the raging flow that, one day, will wash our kind away as well.

NOTES

Preface

1. Langlitz 2012.

2. Fouts and Mills 1997.

Prologue

1. Battel 2009, 54.

2. Rousseau 1997, 209–10.

3. Battel 2009, 54–56.

4. Rousseau 1997, 205.

5. Lévi-Strauss 1976.

6. Montgomery 2015, 11–26.

7. Hester 1968.

8. Crutzen 2002; Hockings, McLennan, et al. 2015; Chakrabarty 2016.

9. Haraway et al. 2016, 539; see also Malm and Hornborg 2014.

10. Haraway 2015; J. Moore 2017; 2018.

11. Haraway 2016, 100, 116.

12. Martin 1966; 2005.

13. Henrich 2016, 10.

14. Other evolutionary biologists continue to advance the counterargument that climate change rather than humans drove Pleistocene megafauna extinct (e.g., Faith et al. 2018).

15. Wulf 2016, 121.

16. Lévi-Strauss (1955) 1974.

17. Naam 2013.

18. Wynes and Nicholas 2017.

19. Farman 2014.

20. Haraway 2015, 161.

21. Sahlins 1976, 12.

22. de Waal 2001, 236.

23. Gray 2002, 3.

Introduction

1. Imanishi 1957b, 51. A more appropriate transcription would be *karuchua*, but—for the sake of continuity—I will stick to the transcription that both Japanese and European authors have used before me (e.g., Nakamura and Nishida 2006; Pracontal 2010).

2. Sahlins 2008, 105.

3. Lestel 2001, 10, 15.

4. de Pracontal 2010, 19, 21.

5. Shapin 1996.

6. Bonner 1980; Goodall 1973; McGrew and Tutin 1978; Menzel 1973.

7. Washburn and Benedict 1979. Ontology is the branch of metaphysics that determines what kinds of entities exist. Examples that pertain to the subject matter of this book are distinctions between natural and cultural (or artificial or supernatural) phenomena or between humans and animals. Of course, such distinctions become matters of controversy. This can happen in the context of scientific controversies like the debate over primate cultures. The term *political ontology* suggests that metaphysical questions of being are often entangled in political questions. For instance, if human beings are not hierarchically distributed across a great chain of being but all races fit into the same ontological category, this might change the way we think about social justice. Or, if there is no fundamental ontological difference between humans and other animals, how do we justify a fundamentally different treatment of animals?

8. Galef 1990; Tomasello 1990.

9. McGrew 2003.

10. Hunter 1991.

11. Clifford 1988; Abu-Lughod 1991.

12. Ingold 2000, 364.

13. Traweek 1988, 162.

14. Boesch 2007; Tomasello and Call 2008; Boesch 2008.

15. Whiten et al. 1999.

16. Boesch 2012b, 237.

17. Montgomery 2015; Radick 2007; A. Rees 2009a; M. Thomas 2006.

18. Langlitz 2017b; Matsuzawa 2006.

19. Haraway 2003; see also Fuentes 2010.

20. Eggan 1954.

21. E.g., Mead 1928; Strathern 1988; Viveiros de Castro 1998.

22. Asquith 1981.

23. Boyd 2018; Boyd and Richerson 2005; Boesch 2012b; 2012a; Henrich 2016; Hoppitt and Laland 2012; Laland and Galef 2009; Laland 2017; Matsuzawa et al. 2001; McGrew 1992; 2004; S. Perry 2006; Tomasello 1999; 2014; de Waal 2001; Whiten et al. 2012; Whiten 2017; Wrangham et al. 1996.

24. Musil 1995, 50.

25. Rheinberger 2015, 11–21.

26. Clifford 1988; Haraway 1989; Strathern 1992.

27. Latour 1993; Pickering 1995.

28. Churchland 1986; see also Langlitz 2015b; 2015a.

29. E. Wilson 1975; 1998.
30. Mesoudi, Whiten, and Laland 2006.
31. Descola 2013, 181.
32. Snow 1964.
33. Ingold 2001; Tehrani and Carrithers 2015, 471; Ingold and Palsson 2013; 2016.
34. Despret 2006; Haraway 2008; Kirksey 2015.
35. Ingold 2001, 337.
36. Candea 2016.
37. T. Gruber et al. 2015, 9.
38. de Waal 2003, 293.
39. de Waal 2001, 361.
40. Haraway 1989; Cartmill 1991; Reynolds 1991; Stanford 1991.
41. Luhmann 1995, 54.
42. Luhmann 1998, 47–48; see also Langlitz 2007; Rabinow 2008, 51–72.
43. Rabinow 2003, 6.
44. A. Rees 2009a, 213–18; 2009b, 455.
45. A. Rees 2009a, 2.
46. Radick 2007; M. Thomas 2005; 2006; 2016.
47. Latour 1987.
48. Langlitz 2019; forthcoming.
49. Ingold 2008, 82.
50. Whiten 2010, 98.

Chapter 1

1. Traweek 1988, 162.
2. T. Gruber et al. 2015.
3. Asquith 1981; 1986a.
4. Houdart 2007; Lock 1993; Traweek 1988.
5. Asquith 1981.
6. Traweek 1988, 146.
7. Asquith 2007, 636.
8. de Pracontal 2010, 21.
9. de Waal 2003, 295; Lestel 2001, 10; de Pracontal 2010, 19.
10. Lévi-Strauss 2013, 7–8.
11. Jensen and Blok 2013.
12. Asquith 1986a, 64.
13. Daston 2005, 39.
14. Asquith 1983.
15. Cited in Asquith 1981, 347.
16. M. Kawai 1965, 22.
17. Koyama (1980) 2012, 43.
18. Malinowski (1922) 1961, 19 (citation refers to the 1922 edition).
19. Cited in Hirata, Watanabe, and Kawai 2001, 488.

20. Asquith 1981, 354; 1986a; de Waal 2003.

21. Kawade 1998, 284.

22. Morris-Suzuki 1995, 761–64; 1998, 60–78.

23. Imanishi (1941) 2002, liii.

24. Imanishi (1941) 2002, 81.

25. Imanishi (1941) 2002, 83.

26. Imanishi (1941) 2002, 84.

27. Kawade 1998.

28. Izumi 2006, 16–17.

29. Cited in Nakamura and Nishida 2006, 35.

30. Cited in Nakamura and Nishida 2006, 36.

31. Ohnuki-Tierney 1987, 6.

32. Ohnuki-Tierney 1987, 62.

33. Ohnuki-Tierney 1987, 121–23.

34. de Pracontal 2010, 180–82.

35. See also Nakamura 2010. The dissociation of culture from intelligence also favored by Nakamura contrasts sharply with Carel van Schaik's cultural intelligence hypothesis (Nakamura 2010; Whiten and van Schaik 2007; van Schaik and Burkart 2011).

36. Daston 2005, 51.

37. Imanishi (1941) 2002, 7.

38. Imanishi (1941) 2002, 7–8.

39. Imanishi (1941) 2002, 6–7.

40. Jensen and Blok 2013; Lévi-Strauss and Eribon 1991, 138.

41. Imanishi 1957b, 51; Kawamura 1959; M. Kawai 1965.

42. M. Kawai 1965, 6. Whether Imo, the youthful inventor, constituted a good paradigm case of primate culture later became a matter of debate. Based on a review of the literature, Hopplitt and Laland (2012, 52) found "a greater reported incidence of innovation in adults than in nonadults, which may reflect the greater experience and competence of older individuals."

43. Nakamura and Nishida 2006, 36.

44. Haraway 1989, 251.

45. Imanishi 1957a, 2; 1957b, 48.

46. M. Kawai 1965, 25.

47. Imanishi 1957a, 3.

48. In contrast to imitation, Imanishi (1957a, 2) proposed, such identification introjected the observed behavior and enabled the monkeys to behave accordingly, even in the absence of the model. This seemed especially important with respect to behaviors that would become relevant only at a later point in life, such as when they became leaders themselves.

49. Imanishi 1957b, 53.

50. Cited in Nakamura and Nishida 2006, 36. Half a century later, Frans de Waal's socioemotional theory of Bonding- and Identification-based Observational Learning (BIOL) developed Imanishi's conception of culture as group identity into the claim that social learning was reinforced not by tangible benefits of the newly acquired behavior but by the desire to fit in with individuals with whom the learner had bonded and identified. Conformism, not survival, drove

cultural transmission, even if conformist tendencies could ultimately contribute to survival, de Waal (2001, 216) maintained.

51. Hirata, Watanabe, and Kawai 2001, 489.

52. Imanishi 1957b, 48.

53. Frisch 1959, 590.

54. Frisch 1959, 593.

55. Nakamura 2009, 142. In an interview, Michio Nakamura told me that, like the ethnographies of cultural anthropologists, these descriptions of monkey troops sought to provide "a complete picture of the target people."

56. M. Kawai 1965, 27.

57. M. Kawai 1965, 24–25.

58. Itani 1958.

59. Imanishi 1960, 393.

60. DeVore and Hall 1965, 27; de Waal 2001, 192; Haraway 1989, 248.

61. Goodwin and Huffman 2017, 548; Knight 2011, 145.

62. Knight 2011, 147.

63. Kawamura 1959.

64. Matsuzawa and Yamagiwa 2018, 320; Asquith 1989, 132; de Waal 2001, 191–92.

65. Cited in Knight 2011, 150.

66. M. Kawai 1965, 2.

67. Asquith 1989, 149.

68. Asquith 1989, 149–50.

69. Frisch 1959, 594.

70. Imanishi 1960, 394; 1957b, 48.

71. Imanishi (1941) 2002, 4.

72. J. Thomas 2001, 170.

73. Marcon 2015, 17.

74. J. Thomas 2001, 170. Casper Bruun Jensen and Atsuro Morita (2017, 5) pointed out that "in stark contrast with the Western idea of a passive nature, the meaning of *shizen* can thus be roughly translated as 'spontaneous becoming'. Here is a key difference, for whereas nature, seen as a resource for human ingenuity, 'matches' with culture, *shizen* and its opposite *sakui* are mutually incompatible: wherever there is human effort, there is by definition no *shizen*."

75. Cross 1996; Shimao 1981; Unoura 1999.

76. J. Thomas 2001, 162.

77. J. Thomas 2001, 171.

78. J. Thomas 2001, 179–208.

79. Benedict 1946.

80. Befu 2001; Yoshino 1992.

81. Iida 2015, 550–51.

82. Lock 2002, 149–50; J. Thomas 2001, 180, 187; Asquith 2002, xxxiv–xxxv.

83. Dale 1986, 191.

84. Cited in Dale 1986, 193.

85. Darwin (1871) 1981.

86. Imanishi (1941) 2002; 1984.

87. Imanishi 1960, 401.

88. Kappeler and Watts 2012, 5.

89. Asquith 1981, 434.

90. Tsunoda 1985.

91. Sleeboom 2004, 48.

92. Dale 1986, 199. Itani told de Waal (2001, 381) that even in the 1950s, Imanishi spoke of "cultural biology" to stimulate exchange between cultural anthropologists and zoologists.

93. Cited in Dale 1986, 192.

94. Asquith 1981, 322.

95. Frisch 1959, 595.

96. Frisch 1959, 595.

97. Hallowell 1961, 247.

98. Hallowell 1961, 248.

99. Tomasello 1999.

100. Hallowell 1961, 247.

101. Carpenter 1960, 402.

102. Chance 1960, 404; Emlen 1960, 405.

103. Emlen 1960, 404; Schultz 1960, 405.

104. Imanishi 1960, 406.

105. A. Rees 2009a, 79–81.

106. Jolly 2000, 77–78.

107. Jolly 2000, 82.

108. Asquith 1981.

109. Asquith 1999, 37.

110. Asquith 1999, 36.

111. T. Nishida 1976.

112. Nakamura 2009, 143.

113. Ohnuki-Tierney 1987, 66.

114. Kuwayama 2004.

115. Asquith 2000; see also Kutsukake 2010.

116. Johnstone 1987, 74.

117. In the 1950s and 1960s, de Saussure's reception in France had profoundly transformed the French *sciences humaines,* from Claude Lévi-Strauss's anthropology to Jacques Lacan's psychoanalysis (Langlitz 2005). Sibatani was not the first to open Imanshi's evolutionary biology to structuralism. In the 1970s, Itani had already clashed with his mentor Imanishi when he adopted Lévi-Strauss's work on the incest taboo to interpret his observations of Japanese macaques (Izumi 2006, 21).

118. Ikeda and Sibatani 1995, 84.

119. Sibatani 1983, 337; see also Sibatani 1972.

120. Imanishi (1941) 2002, 48.

121. Sibatani 1983, 339; Darwin (1859) 2009, 87.

122. Sibatani 1983, 338.

123. Sibatani 1983, 337.

124. Haraway 1984; Morris-Suzuki 1995, 774.

125. Johnstone 1987, 75.

126. Cited in Burgess 1986.

127. Johnstone 1987, 74.

128. Halstead 1985.

129. Halstead 1987, 21.

130. Bloor 1976; Latour 1987.

131. Halstead 1985, 588; 1987, 21.

132. Halstead 1985, 588.

133. Halstead cited contested ecological field experiments that had demonstrated interspecific competition in 90 percent of all cases (Schoener 1983).

134. Halstead 1985, 588.

135. Halstead 1985, 589.

136. Asquith 2006, 204; see also Baxter 2006, x–xi.

137. de Waal 2001; 2003; Ikeda and Sibatani 1995; Asquith 2006, 203.

138. de Waal 2010; Langlitz 2016a.

139. Ikeda and Sibatani 1995.

140. Cf. Houdart 2007, 54–57.

141. Sibatani 1983; Imanishi 1984, 365.

142. Imanishi (1941) 2002, 83.

143. Carpenter 1960, 403.

144. Windelband 1904; Imanishi 1960, 406.

145. Imanishi 1984, 363.

146. Popper 1957.

147. Imanishi 1984, 358.

148. Halstead 1985, 588; Hokkyo 1987, 378.

149. Nakamura 2009, 145.

150. Sugiyama 1965, 460. Nakamura's and Sugiyama's characterizations of Japanese primatology appear to be at odds with Imanishi's nomothetic commitment to particulars (Imanishi 1960, 406; see also Asquith 1981, 228).

151. Ohnuki-Tierney 1987, 17.

152. Morris-Suzuki 1995, 775–76.

153. Latour and Strum 1986; see also Landau 1991.

154. Imanishi 1984, 366.

155. Imanishi 1984, 366.

156. Itō 1991, 145.

157. Itō 1991, 149.

158. Asquith 1986b, 676.

159. Itō 1991, 145.

160. Sakura et al. 1986.

161. Sakura 1995.

162. E. Wilson 1975.

163. Izumi 2006, 22–25.

164. E. Allen et al. 1975.

165. Sakura 1998, 352.

166. Sakura 1998, 351; Itō 1991, 150.

167. Sakura 1998.

168. Itō 1991, 151.

169. Sakura 1998, 343.

170. Sakura 1998, 355.

171. E. Wilson 1980, 4.

172. Haraway 1989, 244.

173. Yagi 2002, xiii.

174. Yagi 2002, xv.

175. Sakura 2005, 288; see also Jensen and Blok 2013, 100.

176. de Waal 2001, 190–94; Kappeler and Watts 2012, vii, 5; de Pracontal 2010, 524.

177. Dale 1986, 190–91. Imanishi (1960, 393) had said as much about identification. In a colony of free-ranging but hand-fed rhesus macaques on Cayo Santiago island, which has been the object of still ongoing long-term observation since 1938, Carpenter (1942) had studied social interactions by first tattooing individual monkeys. But he soon learned to tell them apart without the markings (see also Asquith 1981, 218; 1989, 141–42; Montgomery 2005, 524).

There had also been many instances of provisioning in the history of Western primatology, John Knight (2011, 142) noted in his history of Japanese monkey parks, but nowhere had provisioning been used as systematically as in Japan. At the same time, Yakushima was hardly the only Japanese study site that refrained from the supposedly Japanese method of provisionization (Asquith 1989). Maybe the cultural difference was one of degree, not kind. Or, as Haraway (1989, 248–49) argued, it was less about *what* Japanese and Western primatologists did than *how* they did it: while Europeans and Americans sought to minimize interference and apologized for any violation of their ideally neutral relation to the animals, the Japanese affirmed the participatory dimension of their field observations.

178. Sakura 2005, 288.

179. Asquith 2007.

180. Kropotkin 1902; Mitman 1992, 1–9, 64–71.

181. Elton 1930, 84–88.

182. Hart and Pantzer 1925; Kroeber 1928.

183. Yerkes 1943, 51–52.

184. J. Fisher and Hinde 1949.

185. de Waal 2003, 295.

186. Itani 1985.

187. McGrew 2003.

Chapter 2

1. Perpeet 1976, 1309; Williams 1983, 87.

2. Williams 1960.

3. Tylor 1871.

4. Kroeber and Kluckhohn 1952; Kuper 1999, 59–68; Stocking 1968, 199–203.

5. Radick 2007, 189–98.

6. Kroeber and Kluckhohn 1952, 36.

7. Kroeber and Kluckhohn 1952, 3.

8. Eggan 1954, 760.

9. Hunter 1991.

10. E.g., Bonner 1980; Menzel 1973.

11. McGrew 2003, 439.

12. McGrew 2003, 419.

13. Hart and Pantzer 1925; Kummer 1971; M. Kawai 1957; Goodall 1973. Jolly (2000, 75) traces Goodall's interest in "innovative, learned, cultural behavior" back to a talk she gave in 1962. However, the publication Jolly cites suggests only that chimpanzee infants learn nest building from their mothers, but Goodall (1962, 66–67) did not write about culture. Was she less guarded at the New York conference that Jolly attended?

14. Goodwin and Huffman 2017, 551.

15. McGrew 2004, 137.

16. Nakamura and Nishida 2006, 35–36.

17. Kroeber 1928, 331; Köhler 1921, 222–23.

18. McGrew and Tutin 1978, 245, 247–48.

19. Cf. Durkheim 1895.

20. McGrew 2004, 48.

21. Goodwin and Huffman 2017, 548–49.

22. Frisch 1959, 594.

23. McGrew, Tutin, and Baldwin 1979, 212.

24. E.g., Callon and Latour 1981; Fuentes 2010; Kirksey and Helmreich 2010; E. Wilson 1975.

25. Fuentes 2010; Haraway 2003, 1–5.

26. Boesch 2012b.

27. McGrew and Tutin 1978, 247.

28. Daston and Vidal 2004.

29. Fabian 1983; Clifford and Marcus 1986.

30. McGrew 2004, 89.

31. Stocking 1992, 54; Malinowski (1922) 1961 (citation refers to the 1961 edition).

32. Sanjek 1991.

33. Trigger 1981.

34. Fabian 1983.

35. Pina-Cabral 2000.

36. MacClancy and Fuentes 2010.

37. McGrew and Tutin 1978.

38. Corbey 2005, 62–75.

39. Sebastiani 2015, 105.

40. Washburn 1951.

41. Washburn 1973.

42. Washburn and Moore 1973, 77.

43. Washburn and Benedict 1979, 163.

44. Washburn and Moore 1973, 163–64.

45. Haraway 1989, 187.

46. Washburn 1944.

47. Washburn 1944, 65, 72.

48. Washburn 1944, 71.

49. UNESCO 1952, 5.

50. Washburn 1983, 19.

51. Washburn 1963, 521, 524.

52. Washburn 1963, 530.

53. DeVore 1992, 422.

54. Jolly 2000, 75.

55. Leakey and Lewin 1995, 81.

56. DeVore 1992, 422.

57. Washburn and Benedict 1979, 164.

58. Kuper 1999, 60–62; Stocking 1968, 195–233.

59. McGrew and Tutin 1978.

60. A few years later, Gordon Hewes (1994) compared chimpanzee cultures with two human cultures that he considered even more primitive than the Parlevar and whose forest habitats were more like those of most chimpanzee groups: the Mrabri of Thailand and the Tasaday of the Philippines.

61. McGrew 1987, 251.

62. McGrew 1987, 256.

63. Cove 1995, 28.

64. Cove 1995, 185–86.

65. Cove 1995, 20.

66. McGrew 2004, 176; Reichenbach 1947, 2.

67. Williams 1983, 239.

68. Ingold 2001, 337.

69. Ingold 2000, 364.

70. Sperling 1991, 209.

71. Ingold 2000, 364.

72. Lestel 2001, 148.

73. McGrew 2003, 419.

74. McGrew 2003, 439.

75. Hunter 1991, xii.

76. Zimmerman 2001, 38–61.

77. Hunter 1991, 42–48, 86–88.

78. E. Wilson 1975.

79. E. Allen et al. 1975, 185. Among field primatologists, the sociobiology debate gave rise to the infanticide controversy: Was the killing of infants a social pathology that reflected abnormal ecological conditions at a particular field site, or was it normal behavior that increased the killers' reproductive fitness, as sociobiologically oriented primatologists claimed? Supporters of the sociobiological interpretation complained that especially colleagues trained in American anthropology departments had singled out infanticide for attack, "not because the evidence for this behavior was significantly weaker than that for, say, predation on primates, but because infanticide was morally wrong, and to say that it was both natural and to be expected for nonhuman primates implied that it was an appropriate behavior for humans as well," noted sociologist

and historian of science Amanda Rees (2009a, 19). But Rees's study of this controversy also shows that it reflected disciplinary disagreements over the standards of primatological fieldwork.

80. Sahlins 1976.

81. Sahlins 1976, 6.

82. E. Wilson 1994, 347.

83. E. Wilson 1978, 48.

84. E. Wilson 1978, 111.

85. Segerstråle 2000, 3; Jumonville 2002.

86. Rorty 1999.

87. Clifford 2000.

88. Segerstråle 2000, 32.

89. Hollinger 1995, 51–77.

90. McGrew 2003, 438–39.

91. McGrew 2010, 171.

92. Lestel 2001, 329.

93. McGrew 2003.

94. Itani 1985, 595; Menzel 1973.

95. Menzel 1973, xii–xiii.

96. Menzel 1973, xii.

97. Bonner 1980; McGrew 1992, 86–87; Menzel 1973, xiii; Sahlins 1976, 6–16; Washburn and Benedict 1979, 164.

98. T. Nishida 1987, 473.

99. Lestel 2001, 149.

100. Galef 1990, 87.

101. Visalberghi and Fragaszy 1990.

102. M. Kawai 1965, 14; see also Itani and Nishimura 1973, 34.

103. M. Kawai 1965; Galef 1990, 89.

104. Green 1975, 309; Galef 1990, 87.

105. Galef 1990, 87.

106. Galef 1990, 81, 90–91.

107. Darwin (1871) 1981, 161. In his *Poetics*, written around 355 BC, Aristotle (1997, 57) had already declared: "To imitate is, even from childhood, part of man's nature (and man is different from the other animals in that he is extremely imitative and makes his first steps in learning through imitation)." But the Aristotelian concept of mimesis, which has been translated as imitation, carries too many other connotations to draw a direct line between his literary theory and contemporary evolutionary theory, as Hoppitt and Laland (2012, 16) have done. Maybe some future intellectual historian will connect the dots.

108. Hoppitt and Laland 2012, 18–20.

109. Radick 2007, 211.

110. Tomasello et al. 1987, 181.

111. Tomasello 1990, 284.

112. Tomasello 1990, 283.

113. Whiten et al. 1996, 4, 13.

114. Tomasello 1999, 30.
115. de Waal 2001, 237.
116. Matsuzawa 1999.
117. Matsuzawa et al. 2001, 572.
118. Myowa-Yamakoshi and Matsuzawa 2000, 381.
119. Matsuzawa 1999, 645.
120. Hoppitt and Laland 2012, 24.
121. Tomasello 1990, 305–6.
122. Dobzhansky 1955, 340; Hallowell 1961, 253.
123. Tomasello 1999.
124. E.g., Boyd 2018; Henrich 2016; see also Laland 2017.
125. Rousseau 1997, 208.
126. Boesch 2012b, 67–72.
127. Sasaki and Biro 2017; McGrew 2017, 142.
128. McGrew 2003, 431.
129. McGrew 2015, 45–46.
130. McGrew 2004, 168–69.
131. Tomasello 2001, 9.
132. Galef 1992, 158; Gyger and Marler 1988; de Waal 1982.
133. E. Wilson 1975, 168.
134. Galef 1990, 85; 1992, 159.
135. Galef 1992, 158.
136. C. Morgan 1894, 59.
137. Galef 1990, 91.
138. de Waal 2001, 207.
139. de Waal 2001, 211.
140. Tomasello 1990, 274.
141. E.g., Dolhinow 1968; Rowell 1967.
142. A. Rees 2009a.
143. Tomasello 1990, 283.
144. Tomasello 1990, 275.
145. Tomasello 1990, 282; Galef 1990, 91.
146. McGrew 2003, 437.
147. McGrew 2003, 424.
148. Povinelli 2000, xi–xii.
149. Povinelli and Povinelli 2001, 463.
150. Povinelli and Povinelli 2001, 464.
151. Boesch and Boesch-Achermann 2000, 228.
152. Povinelli and Povinelli 2001, 463.
153. Segerstråle 2000, 255–63.
154. McGrew 2004, 172; E. Wilson 1975, 168; R. Dawkins 2004, 91.
155. McGrew, Tutin, and Baldwin 1979, 212.
156. Whiten et al. 1999.
157. Neumann 2012.

158. van Schaik et al. 2003; S. Perry et al. 2003; Rendell and Whitehead 2001; Krützen et al. 2005.

159. Angier 1999; de Waal 2001, 269.

160. Beck 1982.

161. Whiten et al. 1999.

162. Boesch et al. 1994.

163. Whiten et al. 1999, 685.

164. E.g., Byrne 2007; Laland and Janik 2006; Lycett, Collard, and McGrew 2007.

165. Tomasello 1990, 282.

166. Whiten et al. 1999; Laland, Kendal, and Kendal 2009, 188.

167. Whiten et al. 1999.

168. Boesch 1994.

169. B. Morgan and Abwe 2006.

170. Wrangham 2006, R635.

171. It should be noted that the chimpanzees of Ebo Forest belong to a fourth, by now very rare and critically endangered subspecies of chimpanzee called *Pan troglodytes vellerosus*, which makes them genetically distinct from *Pan troglodytes verus*.

172. Laland and Janik 2006, 544.

173. Langergraber et al. 2011, 414.

174. Laland, Kendal, and Kendal 2009, 185.

175. Laland, Kendal, and Kendal 2009, 174.

176. McGrew 2010, 169.

177. Laland, Kendal, and Kendal 2009, 186.

178. Laland and Janik 2006; Krützen, van Schaik, and Whiten 2007.

179. Kummer 1971, 11.

180. Krützen, van Schaik, and Whiten 2007, 6.

181. Sahlins 1976, x.

182. Sahlins 1999, 400; L. White 1949, 24.

183. L. White 1949, 24.

184. L. White 1949, 26.

185. Boesch 1991b.

186. Boesch 2012b, 110–12.

187. Boesch 2012b, 125.

188. Boesch 1991b, 86.

189. Sahlins 1976, 12.

190. McGrew and Tutin 1978.

191. Rendell and Whitehead 2001; van Schaik et al. 2003.

192. Clifford 2005.

193. Whiten et al. 1999.

194. Watson, Buchanan-Smith, and Caldwell 2014, 163.

195. Boesch 2012b, 57–58.

196. McGrew 2007; see also Whiten 2010, 96–98.

197. McGrew 2007, 175.

198. McGrew 2007, 175.

199. Hermeneutics refers to the branch of knowledge that deals with interpretation. Originally, it was a religious practice focusing on critical exegesis of the Bible. In the nineteenth century, however, the humanities applied it to literature and historical sources before mid-twentieth-century anthropologists adapted the approach to understanding other cultures as "webs of significance" (Geertz 1973, 5).

200. von Herder 2002.

201. Winkler 2007; Cabral 1973.

202. Hamill 2004, 168.

203. Strathern 1980.

204. Viveiros de Castro 1998; 2003.

205. T. Gruber et al. 2015.

206. Wagner 1981, 31.

207. Turner 1991, 304–5.

208. Huntington 1993, 22.

209. Rousseau 1997, 205.

210. Clifford 1988, 95.

211. Clifford 1988, 93, 95.

212. Clifford 1988, 10.

213. Holmes 2000.

214. Abu-Lughod 1991, 137–38.

215. Abu-Lughod 1991, 141.

216. Abu-Lughod 1991, 144.

217. Abu-Lughod 1991, 147.

218. Said 1978; Abu-Lughod 1991, 146, 153.

219. Kuper 1999, xi.

220. Kuper 1999, xiii.

221. Kuper 1999, 247.

222. McGrew 2003, 437.

223. King 2001, 442.

224. Whiten, Horner, and Marshall-Pescini 2003. But see Brumann 1999; Sahlins 2000. When I interviewed orangutan researcher Carel van Schaik, he also regretted that no exchange with cultural anthropologists ensued: "They were so opposed to biology—and there were historical reasons for that, so it wasn't as crazy as it sounds now—that they didn't recognize the debate as an opportunity to find out why culture had even evolved and how it had changed during human evolution. In another place and time, the culture controversy would have developed in a completely different way."

225. de Waal 2001, 214.

226. Haraway 1989, 10.

227. Haraway 1989, 10.

228. Haraway 1989, 13.

229. Haraway 1989, 244–75, 277.

230. A. Rees 2009a, 182–85; Segerstråle 2000, 333–47.

231. Haraway 1984.

232. Foucault 1998, 242.

233. Reynolds 1991, 167.

234. Stanford 1991, 1031.

235. Cartmill 1991, 67–68.

236. Haraway 1989, 3.

237. Kitcher 1998; Zammito 2004.

238. Quine 1960; Feyerabend 1962; Kuhn 1962.

239. Haraway 1989, 288.

240. Cartmill 1991, 74.

241. Cartmill 1991, 74.

242. Stanford 1991, 1031–32.

243. Reynolds 1991, 167; see Langlitz 2019; Strum 2017.

244. Reynolds 1991, 168.

245. Reynolds 1991, 168.

246. Cartmill 1991, 74–75.

247. Haraway 1989, 13.

248. Daston and Galison 2007.

249. Daston 1995.

250. Haraway 1989, 255.

251. Cartmill 1991, 70–71.

252. Stanford 1991, 1031.

253. Haraway 1989, 244–58.

254. Haraway 1989, 244.

255. Haraway 1984, 510; 1991, 67.

256. Rodman 1990, 484.

257. Luhmann 1991, 86.

258. Cartmill 1991, 69, 72.

259. Stanford 1991, 1031; Cartmill 1991, 73.

260. Strum and Latour 1987; Latour and Strum 1986; Strum 2017; Langlitz 2019; for an extensive discussion of the tensions between science and primate studies, see Strum and Fedigan 2000.

261. E.g., Hacking 1999; Zammito 2004.

262. Langlitz 2007.

263. Luhmann 1991, 90.

Chapter 3

1. Boesch 2012b.

2. Genette 1997, 144–60.

3. Boesch 2012b, 9; Laërtius 1901, 231.

4. Boesch 2012b, 155; Voltaire 1901, 174.

5. Boesch 2012b, 47; Rousseau 1997, 208.

6. Rousseau 1997, 209.

7. Rousseau 1997, 211.

8. Boesch 2012b, 6.

9. Lorenz 1952.

10. Schaller 1964.

11. Corbey 2005, 161.

12. Darwin (1871) 1981, 51; Savage and Wyman 1843, 383.

13. Beatty 1950.

14. Corbey 2005, 162; Peterson 2014, 212.

15. Struhsaker and Hunkeler 1971.

16. Boesch 1978.

17. Boesch and Boesch-Achermann 2000, v; Boesch 2012b, 18.

18. Boesch 2009, 71–72.

19. E.g., Lorenz 1950.

20. Struhsaker and Hunkeler 1971.

21. B. Allen 1997, 34; see also B. Allen 2008, 61–71.

22. It might be no coincidence that Barry Allen (2008) enthusiastically embraced Bruno Latour's philosophy of technology. In his work with baboon researcher Shirley Strum, Latour also presented the human use of material objects as enabling the extension of group size. In other publications, Latour (1987; 1993) spoke of networks that could grow longer as more and more nonhuman entities came to mediate social relations between humans. Strum and Latour (1987) contrasted this human capacity with the inability of baboons to build larger troops with the help of material resources. Strum and Latour did not consider that chimpanzees also used objects, but for different purposes than mediating social relations (Langlitz 2019; Lestel 2001, 67, 98). Their understanding of what made human tool use unique anticipated Allen's without calling for such an unusually demanding redefinition of a tool.

23. Engels 1987, 460.

24. Boesch 2012b, 23.

25. Boesch 2009, 113.

26. A. Rees 2009a, 214.

27. Boesch 2009, 114.

28. Dart 1925, 199.

29. Boesch 2009, 115.

30. Boesch 2009, 115.

31. Boesch 2009, 112.

32. Dolhinow 1968; Rowell 1967.

33. Kummer 1971, 125.

34. Kummer 1971, 9. Especially in long-lived animals, tradition learning provided a much faster mechanism of adaptation than mutation and selection, Kummer (1971, 126) argued. It also proved superior to individual learning if a behavior was difficult to come up with on one's own (not every macaque was as inventive as Imo), if experimenting with the environment proved dangerous (not every animal needed to try a potentially poisonous food plant), or if environmental situations occurred infrequently (not every troop member knew where to find a lasting water hole in case of a drought).

35. Boesch and Boesch 1982, 284.

36. Sugiyama and Koman 1979.

37. Boesch and Boesch 1989, 568–69.

38. Boesch et al. 1994; Boesch and Boesch 1990, 97.

39. Boesch 2012b, 45.

40. Grützmacher et al. 2018.

41. Grützmacher et al. 2018, 2.

42. Boesch 2012b, 11–12.

43. Formenty et al. 1999; Le Guenno et al. 1995; Quammen 2012, 79–80.

44. Ayouba et al. 2013.

45. Sponsel 1997.

46. Haraway 2008, 3–4.

47. Malinowski (1922) 1961 (citation refers to the 1961 edition); Mead 1928.

48. Huffman 2014, 63.

49. Goodall 1971, 201.

50. Strier 2003, 18.

51. Jampel 1993.

52. Boesch 2012b, 21.

53. Montgomery 2005, 508–9.

54. Williamson and Feistner 2003, 35.

55. Williamson and Fesitner 2003, 25; see also Estep and Hetts 1992, 11.

56. Candea 2010, 245; Quiatt 1997, 235.

57. See also A. Rees 2007, 888.

58. Alcayna-Stevens 2016, 848.

59. See also Boesch 2009, 45.

60. Alcayna-Stevens 2016.

61. Samuni et al. 2014.

62. Fothergill and Linfield 2012; see also Boesch and O'Connell 2012.

63. McGrew 2003, 425.

64. Similar concerns have been raised at other primate field sites, though; see A. Rees 2006, 324.

65. Williamson and Feistner 2003, 35.

66. For further breaches of this secrecy, see Strum 1987, 37; Williamson and Feistner 2003, 35; A. Rees 2006, 324.

67. Haraway 2008, 23–25; Despret 2013, 68.

68. Smuts 2001, 297–98.

69. Koyama (1980) 2012, 43.

70. Radcliffe-Brown 1940, 2.

71. Hoppitt and Laland 2012, 33.

72. McGrew 2004, 176; Hacking 1983, 41.

73. Nakamura 2010, 165.

74. Cf. Strum 1987, 36; Despret 2013, 65–66.

75. Williamson and Feistner 2003, 36.

76. A. Rees 2006, 316–17.

77. Daston 2011.

78. A. Rees 2006, 325; see also Montgomery 2015, 106–22.

79. Montgomery 2015, 7.

80. Daston 2014, 38.
81. Altmann 1974, 235.
82. Haraway 1989, 304–15.
83. Altmann 1974; Quiatt 1997, 221; Fisler 1967; Haraway 1989, 309.
84. Altmann 1974, 236.
85. A. Rees 2009a, 132.
86. Boesch 2009, 8.
87. Hacking 1983, 41–57.
88. Daston and Galison 2007.
89. McGrew 2004, 10.
90. McGrew 2004, 9.
91. Boesch 2012b, 4; Bourdieu 1977, 16–22; 2000, 4.
92. Bourdieu and Wacquant 1992, 7–11.
93. Bourdieu 1999, 607–26.
94. Hinde 1973, 396–97.
95. Boellstorff 2008, 71; Cohen 1998, xvi; Strathern 2004a, 5–6.
96. T. Nishida et al. 2010, 8.
97. Wittig 2011.
98. Hinde 1973.
99. Wittig 2011, 498.
100. Whiten and Barton 1988.
101. Liebenberg 2013.
102. Candea 2014.
103. Crist 1999.
104. M. Dawkins 2007, 9.
105. M. Dawkins 2007, 74.
106. E.g., Fortun 2015, 155; T. Rees 2016, 10, 98, 237n18.
107. Boesch 2009, 8.
108. Ingold 2001, 337.
109. Sahlins 1976, 6.
110. Ingold 2011, 215.
111. Nakamura 2010, 156, 163; see also Nakamura 2009, 145–46. The convergence of Ingold's and Nakamura's accounts puts a question mark behind the latter's claim to the "uniqueness" of "Japanese" interaction studies.
112. Ingold 2001, 337.
113. Ingold 2001, 337.
114. Ingold 2007, 17.
115. King 2004.
116. Strathern 2004b, 11.
117. King 2004, 196.
118. King 2004, 197; Whiten et al. 1999.
119. Boesch and Boesch-Achermann 2000, 249.
120. McGrew 2004, 87; Goodall 1971.
121. E.g., Galef 1990, 91; Povinelli and Povinelli 2001; Tomasello 1990, 275.

122. McGrew 2003, 438.

123. Boesch 2012b, 125.

124. Porter 2012.

125. Whiten et al. 1999.

126. Latour 1987, 245.

127. Whiten et al. 1999; Boesch and Boesch-Achermann 2000; Goodall 1971. On 14 March 2018, Google Scholar found 2,140 citations of Whiten's "Cultures in Chimpanzees" compared with 1,639 citations of Goodall's *In the Shadow of Man* and 1,476 citations of the Boesches' *The Chimpanzees of the Taï Forest*.

128. Lévi-Strauss 1969; Descola 2013.

129. McGrew 1992.

130. Lestel 2001, 4, 331.

131. Lestel 2001, 13.

132. Herzfeld and Lestel 2005, 644.

133. Lestel 2001, 14.

134. Lestel, Brunois, and Gaunet 2006, 165–70.

135. von Uexküll (1934) 2010.

136. Lestel, Brunois, and Gaunet 2006, 170.

137. Herzfeld and Lestel 2005, 647; Herzfeld 2016.

138. Lestel 2002, 57.

139. Lestel 2002, 55.

140. Savage-Rumbaugh et al. 2005.

141. Kortlandt 1986, 77.

142. Shapin 1994.

143. Tomasello, Savage-Rumbaugh, and Kruger 1993.

144. Laland and Janik 2006; Lycett, Collard, and McGrew 2007.

145. Nakamura 2009, 142–43.

146. Nakamura 2009, 143.

147. McGrew, Tutin, and Baldwin 1979, 212.

148. Boesch 2012b, 21.

149. Boesch 2012b, 193, 21.

150. Boesch 2012b, 18.

151. Corbey 2005, 145.

152. Rousseau 1997, 134–35, 208; Wokler 1978, 118.

153. Schiavenato, Rottenburg, and Langlitz 2016, 138–39.

154. Blanckaert 1993; Wokler 1978, 114.

155. Corbey 2005, 43–48, 146–47.

156. Corbey 2005, 37–43; Sebastiani 2019.

157. Hughes 1994, xi, 46–48.

158. Boesch 2012b, 16–17.

159. Rabinow 2011, 177–87.

160. Markus 1987, 29–40.

161. Price 1970, 9.

162. A. Rees 2009b.

163. Kuper 1994, 1–18.
164. Markus 1987, 34–35.
165. Markus 1987, 9.
166. E.g., Collins 1983; Shapin and Schaffer 1985; Latour 1987; see also Langlitz 2019, 6–7.
167. Hinde 2000, 105, 115.
168. Markus 1987, 34.
169. Boesch 2012b, 193.
170. Cf. Lévy 2010, 33–56.

Chapter 4

1. Schmuhl 2008.
2. Habermas 1997, 39; Tanner 2004, 57–58.
3. Pääbo 2014, 83.
4. Balter 2001.
5. Tomasello, Kruger, and Ratner 1993; Boesch 1993.
6. Boesch and Tomasello 1998, 592.
7. Tomasello 1999, 28, 53.
8. Boesch 2007; Tomasello and Call 2008, 451.
9. Pääbo 2014, 84.
10. Eggan 1954, 747.
11. E.g., Mead 1928; Strathern 1980; Viveiros de Castro 2004.
12. Luncz, Mundry, and Boesch 2012.
13. Tomasello 2006, 7–9.
14. Wundt 1916.
15. Tomasello 2006, 7.
16. Tomasello 2006, 9.
17. For a related argument regarding the natural history of the humanities, see E. Wilson 2015.
18. Kusch 1995.
19. Woods 2007.
20. Langlitz 2015b; 2015a; 2016b.
21. Bennett and Hacker 2003.
22. R. Moore et al. 2015; R. Moore, Call, and Tomasello 2015.
23. R. Moore 2013; Fridland and Moore 2014.
24. Kuhn 1962.
25. Sterelny 2012, xi; 2003, 3–5.
26. Weber 1958, 134–35.
27. Segerstråle 2000, 255–63.
28. Kusch 1995.
29. But see the collaboration of Russon and Andrews 2010, and Langlitz 2016b. In recent years, the label *experimental philosophy* has been applied mostly to an approach that is actually not based on experiments but continues the tradition of intuitive conceptual analysis by collecting survey data on other people's intuitions instead of relying solely on the philosopher's own intuitions. Here, I use *experimental philosophy* more liberally—or, rather, more strictly—to

designate philosophical projects that involve laboratory experiments of the kind that Wundt and Moore conducted (Prinz 2008).

30. Habermas 2013, 166.

31. Habermas 2013, 166–67.

32. Hagner 2015, 245.

33. Tomasello 1999.

34. Habermas 2013, 173.

35. Tomasello 2006, 9.

36. Habermas 2013, 167; 2015, 132; Honneth 2008, 42–44.

37. Habermas 2008, 170–71.

38. Tomasello 1999, 62–66.

39. Habermas 1985, 3–42.

40. Honneth 2008, 43; Tomasello 1999.

41. Petherbridge 2013, 123–64.

42. Honneth 2008, 40, 44.

43. Adorno 2005, 154.

44. van Schaik and Michel 2016, 17–25.

45. Boyd and Richerson 1985.

46. Tomasello 1999, 1–12, 37–40.

47. Welsch 2007, 752; Kuhn 1962. In a related vein, Hoppitt and Laland (2012, 46) reminded their readers that "information can be gained as well as lost in transmission"—a process examined experimentally in transmission chain studies.

48. Fischer 2015, 322.

49. In an interview with me, Tomasello agreed that he tended toward a gradualist perspective. But he didn't want to rule out discontinuities in favor of continuities: "My view is the boring one: it's both. If there is progress or growing complexity there have to be changes. Whether you see them as big or little I don't know."

50. Habermas 2013, 168.

51. Boesch 2012b, 66–72; McElreath et al. 2017; McGrew 2004, 23–24.

52. Boesch, Head, and Robbins 2009.

53. But see Joulian 1996; Haslam et al. 2009.

54. Tomasello 2009b; 2014; Schiavenato, Rottenburg, and Langlitz 2016.

55. Nungesser 2016, 137–42; Wunsch 2015, 281.

56. Tomasello 1999, 5–6.

57. Call et al. 2004; Tomasello 2008, 45–47.

58. Tomasello 2014.

59. Habermas 2013, 69.

60. Habermas 2009, 45.

61. Habermas 2008, 171–72.

62. Habermas 2013, 167–68.

63. But see Langlitz 2015b, 744–45.

64. Jay 1973, 56.

65. Adorno 1967, 117; 2005, 167; Johannssen 2013.

66. Habermas 2013, 167.

67. Habermas 2008; Honneth 2008, 94–95; Assheuer 2005; Geyer 2004; Hartmann 2011.
68. Habermas 2013, 168.
69. Kropotkin 1902; Geulen 2010, 158–59.
70. Tomasello 2009a.
71. Tomasello 2009a.
72. Huxley 1948.
73. Sommer 2009.
74. Boesch 1994; 2012b, 81–107.
75. Thies 2017, 107–8.
76. Schnädelbach 1984, 66–108.
77. Kant (1800) 1992, 538.
78. Schnädelbach 1984, 220–21.
79. Schnädelbach 1984, 224.
80. E.g., Moderlak 2016, 5.
81. Fischer 2015; Schnädelbach 1984, 95.
82. Fischer 2015, 336.
83. Rehberg 2016.
84. Fischer 2015, 327.
85. Fischer 2015, 337–41.
86. Thies 2017, 118.
87. Bradford and Blume 1992.
88. See also Boesch 2012b, 203–4.
89. Boesch 2007, 233.
90. M. Thomas 2005, 443.
91. Boesch 2008, 453.
92. Tomasello and Call 2008, 449–50.
93. Boesch 2008, 453; 2012b, 152.
94. A. Rees 2009a.
95. Boesch 2012b, 41.
96. Boesch 2007, 229; 2008, 454.
97. Boesch 2012b, 41.
98. However, see Hacking 2005; 2007.
99. Boesch 2007, 228; Descartes (1637) 2003, 39.
100. Locke (1689) 2000, 67.
101. Pinker 2002.
102. Boesch 2007, 229.
103. Boesch 2007, 228–29.
104. Boesch 2012b, 213.
105. Boesch 2012b, 105.
106. Schickore 2017, 151–55.
107. Boesch 2008, 454.
108. Boesch et al. 2017.
109. Ingold 2000, 364; McGrew 1987.

110. Tomasello and Call 2008, 449.

111. Tomasello and Call 2008, 449–50.

112. Boesch 2008, 454; 2012b, 198–200.

113. Inoue and Matsuzawa 2007.

114. Boesch 2008, 454.

115. Tomasello and Call 2008, 451.

116. Tomasello, Savage-Rumbaugh, and Kruger 1993, 1690.

117. Tomasello 1999, 34–36.

118. Tomasello and Call 2008, 451.

119. Tomasello 2009a.

120. Boesch 2012b, 238.

121. Tomasello 1999, 29–30.

122. Tomasello, Savage-Rumbaugh, and Kruger 1993; Tomasello 1999, 34–36.

123. Radick 2007, 15–49.

124. Rakoczy and Tomasello 2008, 2.

125. Tomasello 1999, 4, 7.

126. Boesch 2007, 229.

127. Tomasello and Call 2008, 451.

128. Tomasello and Call 2008, 451.

129. Boesch 2005, 693; Boesch 2012b, 176–214.

130. E.g., Boesch 2012b, 209; Horner and Whiten 2005; Matsuzawa 2006.

131. Collins 1985, 19.

132. Collins 1985, 2.

133. Callon and Latour 1992, 355; Latour 2000.

134. A. Rees 2009a, 191–92.

135. Montgomery 2005; 2015.

136. A. Rees 2009a, 10, 191–92.

137. Boesch 2012b, 202–14.

138. See also C. Allen 2002.

139. Boesch 2012b, 206.

140. Boesch 2012b, 37–38.

141. Boesch 2012a, 689–90; Boesch 1996, 258. Boesch pursued a similar argumentative strategy with respect to the chimpanzee capacity for teaching. He reported having observed active teaching in wild chimpanzees (Boesch 1991c), but few of his colleagues trusted this report because neither Boesch nor anyone else had been able to corroborate it subsequently. Tomasello (1999, 34) had publicly called into question Boesch's claim and maintained that "along with imitative learning, the process of active instruction is very likely crucial to the uniquely human pattern of cultural evolution as well." In response, Boesch (2012b, 144) rejected the focus on active teaching as too narrow and ethnocentric. Citing anthropologist Margaret Mead, he noted that children who did not go through the modern Western school system effortlessly acquired culture without active teaching.

142. Boesch 2012b, 203.

143. A. Rees 2009a, 205.

144. Boesch 2007; 2008.

145. Boesch 2012b, 213.

146. But see Bonnie and de Waal 2006; J. Thompson 1994. Kristin Bonnie and Frans de Waal examined the difference in handclasp grooming between two groups of captive chimpanzees at the Yerkes National Primate Research Center. They noted that a significantly higher proportion of females engaged in this peculiar form of grooming than in the wild and explained this difference as the "product of different social or ecological demands of captive versus wild environments" (Bonnie and de Waal 2006, 33). A number of studies also documented cultural differences in behavior among four semicaptive groups at Chimfunshi Wildlife Orphanage in Zambia, but they did not explain them with reference to the anthropogenic conditions of captivity (van Leeuwen et al. 2012; van Leeuwen, Cronin, and Haun 2014; Rawlings, Davila-Ross, and Boysen 2014).

147. Savage-Rumbaugh et al. 2005.

148. Despret 2004; Boesch 2007, 235.

149. Boesch 2008, 453.

150. Montgomery 2015, 85.

151. Strier 2003, 23.

152. Altmann 1974.

153. Strum 2017, 163.

154. Porter 2012.

155. Taking the context of statistical data into consideration presupposed that fieldworkers could identify each observed animal. Mundry emphasized that processing large numbers did not preclude attention to the individual. In fact, the statistical analysis of chimpanzee cultures required the recognition of individual differences. "If we have data sets based on repeated observations of the same individuals but we don't know which observations stem from which individuals, we have a real problem," Mundry explained. "If I compare chimpanzees from West and East Africa without knowing whether some of the data points are derived from repeated observations of the same individuals, it might well be that the 40 data points from West Africa and the 40 data points from East Africa actually come from only two individuals. In the case of such pseudo-replication, differences between individuals artificially blow up differences between populations. That's why, from my perspective as a statistician, an understanding of individual differences based on behavioral observations is a prerequisite for analyzing such data sets."

Even stories about singular events could advance the crunching of big numbers. At one point, Mundry had considered starting a journal for anecdotes in the behavioral sciences. In ornithology, such journals already existed. "They have titles like *The Birds of West Berlin*, come out twice a year, are read by about 80 people, a quarter of whom have been involved in their production," Mundry told me. "They publish articles like: 'On this or that day, I went for a walk with my wife and our dachshund and heard something I had never heard before.' That's totally uninteresting. But then somebody surveys all these anecdotes, for example, about first observations of migratory birds, which had previously only been observed in the Mediterranean, but are beginning to show up in Central Europe—and suddenly you can see climate change!" Thus, cultural primatology had developed its own set of qualitative-quantitative methods combining thick and thin descriptions to reconnect microinteractions and macrostructures (cf. Venturini and Latour 2010).

156. Luncz, Mundry, and Boesch 2012; Luncz and Boesch 2015; for a discussion of this method of exclusion, see chapter 2.

157. Laland et al. 2011.

158. Luncz, Mundry, and Boesch 2012.

159. Kohler 2002, 1–2, 195; Montgomery 2005, 520; A. Rees 2009a, 31, 34.

160. Laland et al. 2011, 1516.

161. Mayr 1961.

162. Mayr 1961, 1502.

163. Laland et al. 2011, 1516.

164. Laland et al. 2011. Laland's philosophy of biology entailed an array of disciplinary, epistemological, and methodological consequences. The insight that cultural learning in humans and other animals was part of both their proximate *and* their evolutionary biology would change how life scientists related to other disciplines, including the humanities (Laland et al. 2011, 1516). Laland envisaged a "unified science of culture" that applied the methods of evolutionary biology to human and animal cultures alike (Mesoudi, Whiten, and Laland 2006; for a spirited critique of this endeavor, see Ingold 2007).

165. Laland et al. 2011, 1512.

166. Shipley 2000.

167. Laland et al. 2011, 1514.

168. Hoppitt and Laland 2012, 158.

169. Hoppitt and Laland 2012, 161.

170. Fisher 1935, 49.

171. Fisher 1935, 10.

172. Fisher 1925.

173. Shipley 2000; Laland et al. 2011.

174. But see Hoppitt and Laland 2012, 167–68.

175. Boesch 2012b, 21.

176. Hacking 1990, xi, 1.

177. Daston and Galison 2007, 41–42; Langlitz 2012, 243–52; 2017c.

178. Tomasello and Call 2008, 449.

179. McGrew 1987.

180. Ingold 2000, 364.

181. Tomasello, Savage-Rumbaugh, and Kruger 1993.

182. Tomasello 1999, 36.

183. Boesch 2007; 2008.

184. Boesch et al. 2017.

185. Winch 1990, 71–75.

186. Strum 2012, 19; Langlitz 2019.

Chapter 5

1. Lestel 2002, 51, 55.

2. Lestel 2014.

3. Savage-Rumbaugh et al. 2005.

4. Matsuzawa 2011b, 157.

5. See also Knight 2011, 39–40.

6. J. Thomas 2001, 171.

7. Imanishi 1984, 357, 360.

8. See also Asquith 1991, 88–89. Scientists at KUPRI also had the advantage of being able to focus on their research and were not burdened by teaching and administration like their colleagues on the main campus.

9. Matsuzawa 2003, 203.

10. Matsuzawa 2016, 443.

11. Nakamura 2009.

12. K. Nishida 1992.

13. Hayes and Hayes 1951; Gardner and Gardner 1989.

14. Terrace 1980.

15. Asano et al. 1982; Matsuzawa 1985b; 1990.

16. Oden, Thompson, and Premack 1988, 144.

17. Premack and Premack 1983, 56.

18. Matsuzawa 2003, 203.

19. Fechner 1966, 1.

20. Rumbaugh 1977.

21. Savage-Rumbaugh 1999; Hanzel 2012.

22. Cited in Asquith 1981, 318.

23. Asano et al. 1982; Matsuzawa 1985a.

24. Skinner 1957.

25. Berlin and Kay 1969.

26. Sapir 1921; Whorf 1956.

27. Berlin and Kay 1969, 109–10; Chomsky 1965.

28. Skinner 1957; Chomsky 1959; Radick 2016.

29. Grether 1940; Riesen 1970.

30. Matsuzawa 1985a, 290.

31. Lestel 2002, 55.

32. de Melo Daly 2018.

33. Matsuzawa 1996, 199.

34. Lestel 2001, 15.

35. Biro 2013.

36. In the future, Matsuzawa also planned to establish so-called interaction booths, which could be used by several chimpanzees simultaneously. In Kumamoto, Matsuzawa's former student Satoshi Hirata and Naruki Morimura had already piloted such collective testing with the entire community, although without face recognition. This way they could ensure the participation of individuals who did not tolerate separation from their group, which the laboratory experiments in the basement of KUPRI required.

37. Inoue and Matsuzawa 2007.

38. Silberberg and Kearns 2009.

39. Kuklick and Kohler 1996, 3.

40. Langlitz 2012, 83–131, 166–203; Boesch 2007; 2008.

41. Lestel 2001, 56; 2011, 95.

42. McGrew 2004, 17.

43. Agar 1980; Spradley 1980.

44. Malinowski (1922) 1961 (citation refers to the 1961 edition); Mead 1928.

45. Lindeman 1924, 191.

46. DeWalt and DeWalt 2011, 5–10.

47. A. Rees 2007, 891–93.

48. E.g., Hayes and Hayes 1951; Terrace 1980; Fouts 1997; for an opinionated but thorough overview, see Wallman 1992; Dupré 2002.

49. Idani and Hirata 2007, 32.

50. Matsuzawa 2003, 208; 2006, 13–15.

51. Matsuzawa 2006, 14.

52. Matsuzawa 2003, 207–8.

53. Matsuzawa 2006, 14.

54. For a detailed etho-ethnographic account of this interspecies socialization, see de Melo Daly 2018; cf. Herzfeld 2016, 34–35.

55. Holmberg 1955.

56. Tomasello 1999, 39.

57. Matsuzawa 1999, 645.

58. Savage-Rumbaugh et al. 2005.

59. Matsuzawa 2003, 208. Either the mother would be with her child during testing, or, as the infants grew bigger, the bipartite structure of the experimental booth allowed the guillotine door between the rooms to be lowered just enough for the infant to go back and forth between the mother and the researchers while keeping the mother in the neighboring compartment.

60. Hayashi, Mizuno, and Matsuzawa 2005, 93. At the Great Ape Research Institute, Matsuzawa's former student Satoshi Hirata, along with Naruki Morimura and Chiharu Houki (2009), conducted a similar experiment, but they taught nut cracking to Loi, the most dominant male, so he could subsequently serve as a conspecific model for the other four members of his community of chimpanzee infants (at the time of the study, they were between four and seven years old). Unlike chimpanzee mothers, Loi turned out to be a rather intolerant master when the two lowest-ranking apprentices wanted to observe his new craft—they had to wait until the second in the dominance rank hierarchy had picked it up, so they could learn it from him. However, Hirata found no evidence of immediate imitation. Instead he argued that observing a conspecific model influenced the learning process over the following days as the chimpanzees engaged in their own trials and errors, suggesting a combination of individual and social learning (Hirata and Hayashi 2011).

61. Hayashi, Mizuno, and Matsuzawa 2005, 100–101.

62. Boesch 2007; 2008.

63. Idani and Hirata 2007; Morimura, Idani, and Matsuzawa 2010.

64. Idani and Hirata 2007, 29.

65. Idani and Hirata 2007, 31.

66. Boesch 1991c.

67. Matsuzawa 2011a, 201, 206.

68. Tomasello 1999, 34; 2014, 82.

69. de Waal 2001, 23–24.

70. Hirata estimated that it would take three to four years to develop a relationship with an already grown-up chimpanzee that would allow a researcher to conduct participant observation, but very few people had the patience and a strong enough desire to get to this point—and some adult chimpanzees would never tolerate a human in the room.

71. Around 2000, Masayuki Tanaka had also entered the booth with mother and infant. At the time of my fieldwork, he had already left KUPRI.

72. Herzfeld 2016, 23–24.

73. Before becoming director of the Primate Research Institute in 2006, Matsuzawa had also conducted participant observation with Chloé, Pan, and other chimpanzees. But, he explained to me, "because of my time-energy budget I have to cut things, cut, cut, cut, cut." His administrative responsibilities no longer allowed him to invest the time necessary for maintaining relations of mutual trust with so many individuals.

74. Geertz 1973.

75. Savage-Rumbaugh et al. 2005, 313.

76. Keller 1983, 197–207. Reflecting on his face-to-face interactions with orangutans in sanctuaries, Carel van Schaik told me: "You are changing your animal, and I think that's fantastic—it gives you a feel for the phenotypic plasticity and that's data as long as you realize that that's what you're doing."

77. Savage-Rumbaugh et al. 2005.

78. Segerdahl, Fields, and Savage-Rumbaugh 2005, 3.

79. Boesch 2012b, 202–12.

80. There was a third ape group at Kumamoto. In 2013, the sanctuary received six bonobos from San Diego Zoo. But there were no plans to conduct participant observation with them. Such direct interactions might have violated the Bonobo Species Survival Plan, Hirata surmised, and Matsuzawa worried that they lacked experience with this species so closely related and yet so different in its behavior from common chimpanzees. More importantly, however, very few bonobos lived in captivity. The ones at Kumamoto were the only specimens of *Pan paniscus* in Japan. To preserve the captive population, individuals had to be exchanged between zoos for breeding purposes. This required international standards for how to deal with them. Although Sue Savage-Rumbaugh had proved that participant observation with bonobos was possible, the practice would hardly be adopted by zookeepers around the globe, and the Japanese had to interact with them like everybody else.

81. Sakai et al. 2012; Takeshita, Myowa-Yamakoshi, and Hirata 2006.

82. Ueno et al. 2010; 2008.

83. Kano and Tomonaga 2013.

84. Krupenye et al. 2016; Kano et al. 2017.

85. Lindeman 1924, 191.

86. Wieder 1980.

87. Takada 2013, 16.

88. Asquith 1981; 1986a; Biro et al. 2003; Hayashi, Mizuno, and Matsuzawa 2005; Hirata and Celli 2003; Matsuzawa 2012; Myowa-Yamakoshi and Matsuzawa 1999. In the case of cumulative culture, however, Matsuzawa and his students supported Boesch: Carvalho et al. 2009, S104; Hockings 2011b; Yamamoto, Humle, and Tanaka 2013.

To my knowledge, Matsuzawa refrained from ever using the loaded term *cumulative culture* in his publications. This is in line with his nonconfrontational navigation of the chimpanzee culture controversy. However, comparing the Bossou chimpanzees' use of *movable* hammer and anvil stones with the exclusive use of fixed anvils by other communities, he interpreted "this advanced percussive technology as a case of progressive problem-solving." He further corroborated this claim with reference to his own discovery that the Bossou chimpanzees occasionally used wedge stones to stabilize their mobile anvils (Matsuzawa 1994; Carvalho et al. 2009, S104). Although Matsuzawa avoided the polemic between Boesch, Tomasello, Whiten, and McGrew, he entertained the possibility of a modest ratchet effect in chimpanzee cultural history.

89. Segerdahl, Fields, and Savage-Rumbaugh 2005, 8.

90. Hirata et al. 2001, 506.

91. Idani and Hirata 2007, 31; Hirata, Watanabe, and Kawai 2001, 490.

92. Idani and Hirata 2007, 31; Hirata, Watanabe, and Kawai 2001, 490.

93. Matsuzawa 2012.

94. Savage-Rumbaugh et al. 2005.

95. Imanishi (1941) 2002, 63; Durkheim 1964.

96. Savage-Rumbaugh 1999, 187.

Chapter 6

1. M. Wilson et al. 2014.

2. Kortlandt and Kooji 1963, 62.

3. Kortlandt 1986, 77.

4. Verroux 2003, 30–31.

5. Humle 2011, 21.

6. Kirksey, Schuetze, and Helmreich 2014, 18.

7. Hockings et al. 2010, 889.

8. Verroux 2003, 100.

9. Germain 1984, 199. Over time, different informants related this mythical origin story to different researchers in slightly altered forms. For some variants, see Kortlandt (1986, 91) and Yamakoshi and Leblan (2013, 6).

10. The children resulting from marriages between two totemic groups adopted the father's totem.

11. Koops 2011, 280.

12. Matsuzawa 1994; Biro, Carvalho, and Matsuzawa 2010, 145.

13. Carvalho et al. 2009, S104.

14. Kortlandt and Kooij 1963, 62.

15. Goodall 1963. Like the armchair anthropologists of the eighteenth and nineteenth centuries, Kortlandt and Kooij (1963, 80) had sent out a survey to monkey and ape researchers, in response to which only nine fieldworkers could provide examples of "non-agonistic use of tools in the wild" like honey dipping and termite fishing. At the same time, fifty-eight primatologists reported agonistic uses of objects in conflicts. Kortlandt and Kooij (1963, 80) concluded: "Nothing more convincingly demonstrates that the technological age on earth started with the emergence of weapons rather than gadgets."

16. Beatty 1950.

17. Kortlandt and Kooij 1963, 62.

18. Sugiyama and Koman 1979.

19. Kortlandt 1986, 77.

20. If "wild" was the right designation for the Bossou community; cf. Reynolds 1975.

21. For some background to David Premack's decided lack of interest in wild primates, see Radick (2007, 326–27).

22. McGrew 2004, 45.

23. Matsuzawa 2011b, 160.

24. Matsuzawa 2011b, 159. For a very similar problematization of the opposition of naturalism and experimentalism by American primatologists in the 1960s, see Montgomery (2015, 87–90).

25. Yamakoshi 2011, 37–40.

26. McGovern 2012.

27. Kortlandt 1986, 90–94.

28. E.g., Asquith 1981; de Waal 2001, 191–92; Frisch 1959; Imanishi 1960; Matsuzawa and McGrew 2008.

29. French anthropologist Vincent Verroux (2003, 103), however, documented field assistants feeding chimpanzees with bananas.

30. Zuberbühler 2014.

31. Hayashi, Mizuno, and Matsuzawa 2005, 92.

32. Sugiyama 1999; Boesch 1991c.

33. Matsuzawa 2011b, 159.

34. Inoue-Nakamura and Matsuzawa 1997.

35. Boesch 1991c.

36. Inoue-Nakamura and Matsuzawa 1997, 172.

37. Inoue-Nakamura and Matsuzawa 1997, 172; see also Matsuzawa et al. 2001, 571–72.

38. Boesch 1991a, 556; Boesch-Achermann and Boesch 1993; Tomasello et al. 1987.

39. Myowa-Yamakoshi and Matsuzawa 1999; 2000.

40. Matsuzawa et al. 2001, 572–73.

41. de Waal 2001, 23f.

42. Verroux 2003, 32, 47.

43. Clifford and Marcus 1986; Clifford 1988.

44. Matsuzawa 1994, 359–60; 2011c, 78.

45. Viveiros de Castro 2003.

46. Biro et al. 2003, 222.

47. Matsuzawa et al. 2001, 570.

48. Weber (1909) 1978, 300; Huntington 1996, 134.

49. Kroeber 1939.

50. Gellner 1998, 172.

51. Wrangham 2006, R634.

52. Strum 1987.

53. E.g., Boesch and Boesch-Achermann 2000; T. Nishida 2012; Reynolds 2005.

54. Vetter 2011, 1–16.

55. Radick 2007, 364.

56. Radick 2007, 368.

57. Matsuzawa 2011b, 157.

58. Kohler 2002.

59. Montgomery 2015, 33–46.

60. McGrew 2004, 182.

61. Matsuzawa 2011b, 163.

62. Hayashi, Mizuno, and Matsuzawa 2005.

63. Hayashi, Mizuno, and Matsuzawa 2005, 92.

64. Hayashi, Mizuno, and Matsuzawa 2005, 100.

65. Koops, McGrew, and Matsuzawa 2013, 181.

66. Whiten 2002, 385.

67. Whiten 2002, 385.

68. Mesoudi and Whiten 2008, 3486.

69. Mesoudi and Whiten 2008; Hoppitt and Laland 2012, 38–50. In collaboration with Frans de Waal, Whiten had used transmission chains to simulate the intergenerational transmission of cultural traits in the laboratory (Horner et al. 2006).

70. Hegel (1807) 1977; Kohler 2002.

71. Considering that the German researchers who established the bonobo field site LuiKotale in the Democratic Republic of Congo had also introduced secateurs to clip vines and saplings, Matsuzawa's culturalization of this practice again raises the question of how much national cultures determined primatological practices (Alcayna-Stevens 2016, 840).

72. Quammen 2012.

73. Yamakoshi 2002, 2.

74. Yamakoshi 2002, 1.

75. Fuentes 2010. In the 1960s, Sugiyama had already used road crossings for demographic surveys of langurs in India (A. Rees 2009a, 80).

76. Matsuzawa 2006.

77. Hockings 2011a, 225–27; Boesch 1994.

78. Hockings 2011a, 227.

79. Cibot et al. 2015.

80. Quoted in "Wild Chimps Look Both Ways," 2015.

81. Stallard 2015.

82. de Waal 2005, 32.

83. Sponsel 1997, 143.

84. B. Taylor 2009; Sponsel 2012.

85. Sponsel 1997, 164.

86. Sponsel 1997, 160.

87. Sponsel 1997, 145.

88. Fuentes 2010; Fuentes and Hockings 2010.

89. Fabian 1983; Fuentes and Hockings 2010.

90. Fuentes 2010, 600.

91. Fuentes 2010, 610.

92. Fuentes 2010, 603.

93. Imanishi 1984, 198.

94. Anderson and Valente 2002.

95. Imanishi (1941) 2002, 73.

96. Imanishi (1941) 2002, 30–31.

97. Imanishi (1941) 2002, 74.

98. Kummer 1971, 36.

99. Wrangham 2009.

100. Hockings 2007, 92.

101. Hockings, Bryson-Morrison, et al. 2015.

102. Hockings 2007, 209–10.

103. Hockings 2007, 210.

104. Leblan 2013, 3; see also 2017, 31–33.

105. Hockings et al. 2010, 892.

106. Richards 1993, 145–46.

107. Hockings et al. 2010.

108. Yamakoshi 2002, 3.

109. Hockings et al. 2010, 893; Verroux 2003, 38.

110. Richards 1993, 146–48.

111. Evans-Pritchard 1937.

112. Verroux 2003, 104–7.

113. McGovern 2012.

114. Hockings 2007, 211.

115. Yamakoshi 2002, 1.

116. Leblan and Bricka 2013.

117. Yamakoshi and Leblan 2013.

118. Yamakoshi 2002, 2.

119. Yamakoshi and Leblan 2013; Leblan 2016.

120. Kortlandt 1986, 83.

121. Yamakoshi and Leblan 2013.

122. Verroux 2003, 107.

123. Yamakoshi 2002, 3.

124. Yamakoshi and Leblan 2013, 15.

125. M. Thompson et al. 2007.

126. Nevertheless, a subsequent study of the Bossou community's habitat use suggested that they preferred mature forest to cultivated fields and other anthropogenic landscapes (Bryson-Morrison et al. 2017).

127. But, for an account of unhabituated chimpanzees mobbing researchers, see McLennan and Hill (2010).

Chapter 7

1. Sepkoski 2015.

2. Rudwick 1997.

3. Darwin (1859) 2009, 234–37.

4. Darwin (1871) 1981, 238.

5. Sepkoski 2015, 67, 71.

6. Sepkoski 2015, 71.

7. Leakey and Lewin 1995, 235.

8. Leakey and Lewin 1995, 245.

9. Estrada et al. 2017.

10. Malinowski (1922) 1961, xi (citation refers to the 1922 edition).

11. Sahlins 2000.

12. Rosaldo 1989.

13. Lemov 2015.

14. Sepkoski 2015, 77–82.

15. McGrew 2003, 438.

16. Fairet et al. 2014.

17. Fossey 1983.

18. Kolbert 2014; Leakey and Lewin 1995; Martin 1966; 2005.

19. Shipman 2015.

20. Boesch 2012b, 241; Campbell et al. 2011; Köndgen et al. 2008.

21. McGovern 2011.

22. Guengant and May 2013.

23. Malthus 1798.

24. Langer 1963.

25. Bodley 2008, 191–223.

26. Engelman 2016.

27. Darwin (1859) 2009, 84. Kühl's account left me wondering whether in the end, these market forces wouldn't have to compete with other market forces that pushed for the exploitation of natural resources, which would also get scarcer as more humans demanded their share.

28. Gray 2002, 6.

29. Martin 1966.

30. Martin 2005, 167.

31. Kolbert 2014; Leakey and Lewin 1995; Martin 2005; Shipman 2015.

32. Crutzen 2002.

33. Martin 1966, 342.

34. Hawkes et al. 1998.

35. This quite exceptional but not unique trait made humans look less like apes than they looked like orcas (Croft et al. 2017).

36. Hrdy 2008, 101–2.

37. Matsuzawa 2012, 290.

38. M. Thompson et al. 2007; Boesch 2009, 70–71.

39. Matsuzawa 2018, 110.

40. Matsuzawa 2018, 111.

41. Takasaki 2000, 157–59.

42. Giono 1984.

43. Matsuzawa et al. 2011, 366–67.

44. Haraway 1989, 244–75.

45. Haraway 1989, 273.

46. Haraway 1989, 265.

47. Haraway 1989, 262.

48. Montgomery 2015, 106.

49. For a similar account of the descent of animals from humans in Amerindian cosmologies, see Viveiros de Castro (1998).

50. Quoted in Morris 1979, 68.

51. Montgomery 2015, 11–26.

52. In the nineteenth century, Thomas Savage and Jeffries Wyman (1843, 385) found that the conception of chimpanzees as ex-humans did not necessarily protect them against predation: "It is a tradition with the natives generally here, that [chimpanzees—the authors still called them *Troglodytes niger* or Black Orang of Africa] were once members of their own tribe; that for their depraved habits they were expelled from all human society, and, that through an obstinate indulgence of their vile propensities they have degenerated into the present state and organization. They are, however, eaten by them, and, when cooked with the oil and pulp of the palm nut, considered a highly palatable morsel."

53. Ouattara, Lemasson, and Zuberbühler 2009.

54. Krou 2013.

55. E.g., Marcus and Fischer 1986.

56. Viveiros de Castro 2003.

57. E.g., Hockings, McLennan, et al. 2015.

58. Campbell et al. 2008

59. Kühl et al. 2017, 20.

60. Daston 2011, 84; Boesch 2012b, 21.

61. Malinowski (1922) 1961, 16 (citation refers to the 1961 edition).

62. Kühl et al. 2007.

63. Arandjelovic et al. 2016.

64. Kühl and Burghardt 2013; Crunchant et al. 2017.

65. Oelze et al. 2016.

66. Vetter 2011.

67. Kühl et al. 2016.

68. Quoted in J. Gruber 1970, 1294.

69. J. Gruber 1970.

70. Lévi-Strauss (1955) 1974.

71. Lévi-Strauss (1955) 1974, 310.

72. Lévi-Strauss (1955) 1974, 319.

73. Lévi-Strauss (1955) 1974, 397.

74. Lemov 2015.

75. Vidal and Dias 2015, 1–38.

76. Daston 2017a, 332.

77. Weber 1958; Daston 2017b, 175–76.

78. Quine 1960; Feyerabend 1962; Kuhn 1962.

79. Fisher 2002, 213.

80. Lévi-Strauss (1955) 1974, 397.

81. Weisman 2007.
82. Boesch 2012b, 242.

Conclusion

1. K. Kawai 2013; 2017; Nakamura 2009; 2010.
2. Takasaki 2000, 164.
3. Wilson 1998.
4. Imanishi 1984.
5. Latour 1993.
6. J. Thomas 2001, 219.
7. J. Thomas 2015, 254.
8. Harding 2015; Lamont 2009.
9. Kuhn 1962.
10. Mesoudi, Whiten, and Laland 2006.

Epilogue

1. Leakey and Lewin 1995, 224.
2. Crutzen 2002.
3. Koselleck 1988; Roitman 2013.
4. Imanishi (1941) 2002, 65–66.
5. Imanishi (1941) 2002, 71.
6. Imanishi (1941) 2002, 66.
7. Imanishi (1941) 2002, 77.
8. Imanishi (1941) 2002, 71.
9. Imanishi (1941) 2002, 67; Kuper 1994, 7–8.
10. Imanishi (1941) 2002, 72.
11. Cf. Roitman 2013, 8–9.
12. Koselleck 1988, 104.
13. Imanishi (1941) 2002, 84.
14. Imanishi (1941) 2002, liii; Sprague 2004.
15. Kellert 1993.
16. Kellert 1991, 305.
17. Kalland and Asquith 1997, 3–4, 15.
18. E.g., Haraway 2008, 11.
19. McGrew 2004, 176; Haraway 1989, 288.
20. Langlitz 2019.
21. G. Perry and Codding 2017; Watts and Amsler 2013.
22. Strum 2017, 165.
23. Rousseau 1997, 148, 207.
24. H. White 1973.
25. Haraway 2016, 535.
26. Marcus and Langlitz 2014.

27. Voltaire 2006, 4.

28. Magee 2010, 462.

29. Gould and Lewontin 1979, 150.

30. E. Allen et al. 1975.

31. Nietzsche (1882) 2001, 157.

32. Gumbrecht 2014, 31.

33. Ruhnau 2018.

34. Jacobi 2017, 14; Diderot (1796) 2009.

35. Haraway 2016, 4, 51.

36. R. Taylor 1962, 56.

37. A sociologist of science who reviewed the manuscript of *Chimpanzee Culture Wars* for the press worried that my plea for a joyous primatology risked reigniting the science wars: "After the impact that Haraway's *Primate Visions* had on relationships between historians of science and primatologists, this is an unnecessary hostage to fortune." The concern is ironic because *Primate Visions* provoked the ire of primatologists by moralizing their work whereas *Chimpanzee Culture Wars* aims at its demoralization. At the same time, I understand the reviewer's concern: my plea for fatalism calls into question the one point about which most cultural primatologists, multispecies ethnographers, and science studies scholars can probably agree, namely that humans should exercise their political agency to save the great apes from extinction. In this respect, primatologists are no less morally engaged than their critics in the humanities and social sciences. As moral animals, we all are. But that should not keep us from occasionally climbing out of the trenches to think about our situation from a distance. What I am proposing is an ethical self-experiment: assuming that we won't be able to change the dire fate of most wild ape cultures (I would be more than happy to lose that bet), I propose to change our self-relation to escape the sense of self-loathing that pervades the Anthropocene discourse. While I do hope that my account of the chimpanzee culture debate helps to reanimate a constructive conversation between humanities scholars and primatologists, that can hardly count as a reason to self-censor contentious philosophical ideas.

38. Konersmann 2015.

39. See also Matsuzawa 2012, 302–5.

40. Saito et al. 2014.

REFERENCES

Abu-Lughod, Lila. 1991. "Writing against Culture." In *Recapturing Anthropology: Working in the Present*, edited by Richard G. Fox, 137–62. Santa Fe: School of American Research Press.

Adorno, Theodor W. 1967. *Prismen: Kulturkritik und Gesellschaft*. Frankfurt am Main, Germany: Suhrkamp.

———. 2005. *Minima Moralia: Reflections on a Damaged Life*. London: Verso.

Agar, Michael H. 1980. *The Professional Stranger: An Informal Introduction to Ethnography*. San Diego: Academic Press.

Alcayna-Stevens, Lys. 2016. "Habituating Field Scientists." *Social Studies of Science* 46 (6): 833–53.

Allen, Barry. 1997. "The Chimpanzee's Tool." *Common Knowledge* 6:34–51.

———. 2008. *Artifice and Design: Art and Technology in Human Experience*. Ithaca, NY: Cornell University Press.

Allen, Colin. 2002. "A Skeptic's Progress: Review of Daniel Povinelli's Folk Physics for Apes." *Biology and Philosophy* 17:695–702.

Allen, Elizabeth, Barbara Beckwith, Jon Beckwith, Steven Chorover, David Culver, Herb Schreier, Miriam Rosenthal, et al. 1975. "Against "Sociobiology." *New York Review of Books*, no. 182 (November): 184–86.

Altmann, Jeanne. 1974. "Observational Study of Behavior: Sampling Methods." *Behaviour* 49 (3/4): 227–67.

Anderson, Amanda, and Joseph Valente. 2002. "Introduction: Discipline and Freedom." In *Disciplinarity at the Fin de Siècle*, edited by Amanda Anderson and Joseph Valente, 1–18. Princeton, NJ: Princeton University Press.

Angier, Natalie. 1999. "Chimpanzees Doin' What Comes Culturally." *New York Times*, 17 June 1999. http://www.nytimes.com/1999/06/17/us/chimpanzees-doin-what-comes-culturally.html.

Arandjelovic, Mimi, Colleen R. Stephens, Maureen S. McCarthy, Paula Dieguez, Ammie K. Kalan, Nuria Maldonado, Christophe Boesch, and Hjalmar S. Kuehl. 2016. "Chimp&See: An Online Citizen Science Platform for Large-Scale, Remote Video Camera Trap Annotation of Chimpanzee Behaviour, Demography and Individual Identification." PeerJ Preprints, e1792v1. https://doi.org/10.7287/peerj.preprints.1792v1.

Aristotle. 1997. *Poetics*. Translated by George Whalley. Montreal: McGill-Queen's University Press.

Asano, Toshio, Tetsuya Kojima, Tetsuro Matsuzawa, Kisou Kubota, and Kiyoko Murofushi. 1982. "Object and Color Naming in Chimpanzees." *Proceedings of the Japan Academy, Series B* 58:118–22. https://doi.org/10.2183/pjab.58.118.

Asquith, Pamela J. 1981. "Some Aspects of Anthropomorphism in the Terminology and Philosophy Underlying Western and Japanese Studies of the Social Behaviour of Non-human Primates." PhD diss., University of Oxford. http://ora.ox.ac.uk/objects/uuid:ced23a88-1ca0-47a9-a3be-a72dbaea1788.

———. 1983. "The Monkey Memorial Service of Japanese Primatologists." *RAIN* 54:3–4.

———. 1986a. "Anthropomorphism and the Japanese and Western Traditions in Primatology." In *Primate Ontogeny, Cognition and Social Behavior*, edited by James G. Else and Phyllis C. Lee, 61–71. Cambridge: Cambridge University Press.

———. 1986b. "Imanishi's Impact in Japan." *Nature* 323 (23 October): 675–76.

———. 1989. "Provisioning and the Study of Free-Ranging Primates: History, Effects, and Prospects." *Yearbook of Physical Anthropology* 32:129–58.

———. 1991. "Primate Research Groups in Japan: Orientations and East–West Differences." In *The Monkeys of Arashiyama: Thirty-Five Years of Research in Japan and the West*, edited by Linda M. Fedigan and Pamela J. Asquith, 81–98. Albany, NY: SUNY Press.

———. 1999. "The 'World System' of Anthropology and 'Professional Others.'" In *Anthropological Theory in North America*, edited by E. L. Cerroni-Long, 33–46. Westport, CT: Bergin & Garvey.

———. 2000. "Negotiating Science: Internationalization and Japanese Primatology." In *Primate Encounters: Models of Science, Gender, and Society*, edited by Shirley C. Strum and Linda M. Fedigan, 165–83. Chicago: University of Chicago Press.

———. 2002. Introduction to *A Japanese View of Nature: The World of Living Things*, xxix–xliii. London: Routledge.

———. 2006. "Imanishi Kinji's Natural Ethic: The Contribution of a Twentieth-Century Japanese Scientist to Ideas of 'Being in the World.'" In *Historical Consciousness, Historiography, and Modern Japanese Values*, edited by James C. Baxter, 201–6. Kyoto: International Research Center for Japanese Studies. http://publications.nichibun.ac.jp/region/d/NSH/series/symp/2006-11-30-1/s001/s020/pdf/article.pdf.

———. 2007. "Sources for Imanishi Kinji's Views of Sociality and Evolutionary Outcomes." *Journal of Biosciences* 32 (4): 635–41. https://doi.org/10.1007/s12038-007-0063-7.

Assheuer, Thomas. 2005. "Hartz IV in der Synapse." *Die Zeit*, 31 May 2005.

Ayouba, Ahidjo, Chantal Akoua-Koffi, Sébastien Calvignac-Spencer, Amandine Esteban, Sabrina Locatelli, Hui Li, Yingying Li, et al. 2013. "Evidence for Continuing Cross-Species Transmission of SIVsmm to Humans: Characterization of a New HIV-2 Lineage in Rural Côte d'Ivoire." *AIDS* 27 (15). https://doi.org/10.1097/01.aids.0000432443.22684.50.

Balter, Michael. 2001. "Max Planck's Meeting of the Anthropological Minds." *Science* 293 (5533): 1246–49.

Battel, Andrew. 2009. *The Strange Adventures of Andrew Battell of Leigh, in Angola and the Adjoining Regions*. Edited by E. G. Ravenstein. London: BiblioBazaar.

Baxter, James C. 2006. Introduction to *Historical Consciousness, Historiography, and Modern Japanese Values*, v–xvi. Edited by James C. Baxter. Kyoto: International Research Center for

Japanese Studies. http://publications.nichibun.ac.jp/region/d/NSH/series/symp/2006-11-30-1/s001/s020/pdf/article.pdf.

Beatty, Harry. 1950. "A Note on the Behavior of the Chimpanzee." *Journal of Mammology* 32 (1): 118.

Beck, Benjamin B. 1982. "Chimpocentrism: Bias in Cognitive Ethology." *Journal of Human Evolution* 11:3–17.

Befu, Harumi. 2001. *Hegemony of Homogeneity: An Anthropological Analysis of Nihonjinron*. Melbourne: Trans Pacific Press.

Benedict, Ruth. 1946. *The Chrysanthemum and the Sword*. New York: Houghton Mifflin.

Bennett, Max, and Peter Hacker. 2003. *Philosophical Foundations of Neuroscience*. Malden, UK: Blackwell.

Berlin, Brent, and Paul Kay. 1969. *Basic Color Terms: Their Universality and Evolution*. Berkeley: University of California Press.

Biro, Dora. 2013. "The PrimateCast #12: An Interview with Dr. Dora Biro." Center for International Collaboration and Advanced Studies in Primatology. 27 March 2013. http://www.cicasp.pri.kyoto-u.ac.jp/news/podcasts/primatecast-12-interview-dr-dora-biro.

Biro, Dora, Susana Carvalho, and Tetsuro Matsuzawa. 2010. "Tools, Traditions, and Technologies: Interdisciplinary Approaches to Chimpanzee Nut Cracking." In *The Mind of the Chimpanzee: Ecological and Experimental Perspectives*, edited by Elizabeth V. Lonsdorf, Stephen R. Ross, and Tetsuro Matsuzawa, 142–55. Chicago: University of Chicago Press.

Biro, Dora, Noriko Inoue-Nakamura, Rikako Tonooka, Gen Yamakoshi, Claudia Sousa, and Tetsuro Matsuzawa. 2003. "Cultural Innovation and Transmission of Tool Use in Wild Chimpanzees: Evidence from Field Experiments." *Animal Cognition* 6 (4): 213–23. https://doi.org/10.1007/s10071-003-0183-x.

Blanckaert, Claude. 1993. "Buffon and the Natural History of Man: Writing History and the 'Foundational Myth' of Anthropology." *History of the Human Sciences* 61:13–50. https://doi.org/10.1177/095269519300600102.

Bloor, David. 1976. *Knowledge and Social Imagery*. Chicago: University of Chicago Press.

Bodley, John. 2008. *Anthropology and Contemporary Human Problems*. Lanham, MD: AltaMira Press.

Boellstorff, Tom. 2008. *Coming of Age in Second Life: An Anthropologist Explores the Virtually Human*. Princeton, NJ: Princeton University Press.

Boesch, Christophe. 1978. "Nouvelles observations sur les chimpanzés de la Forêt de Taï (Côte d'Ivoire)." *La terre et la vie* 32:195–201.

———. 1991a. "Handedness in Wild Chimpanzees." *International Journal of Primatology* 12 (6): 541–58. https://doi.org/10.1007/BF02547669.

———. 1991b. "Symbolic Communication in Wild Chimpanzees?" *Human Evolution* 6 (1): 81–89. https://doi.org/10.1007/BF02435610.

———. 1991c. "Teaching among Wild Chimpanzees." *Animal Behaviour* 41 (3): 530–32. https://doi.org/10.1016/S0003-3472(05)80857-7.

———. 1993. "Towards a New Image of Culture in Wild Chimpanzees?" *Behavioral and Brain Sciences* 16 (3): 514–15. https://doi.org/10.1017/S0140525X00031277.

———. 1994. "Cooperative Hunting in Wild Chimpanzees." *Animal Behaviour* 48 (3): 653–67. https://doi.org/10.1006/anbe.1994.1285.

———. 1996. "The Emergence of Cultures among Wild Chimpanzees." In *Evolution of Social Behaviour Patterns in Primates and Man*, 251–68. Proceedings of the British Academy 88. New York: Oxford University Press.

———. 2005. "Joint Cooperative Hunting among Wild Chimpanzees: Taking Natural Observations Seriously." *Behavioral and Brain Sciences* 28 (5): 692–93.

———. 2007. "What Makes Us Human (*Homo Sapiens*)? The Challenge of Cognitive Cross-Species Comparison." *Journal of Comparative Psychology* 121 (3): 227–40.

———. 2008. "Taking Development and Ecology Seriously When Comparing Cognition: Reply to Tomasello and Call (2008)." *Journal of Comparative Psychology* 122 (4): 453–55. https://doi.org/10.1037/0735-7036.122.4.453.

———. 2009. *The Real Chimpanzee: Sex Strategies in the Forest*. Cambridge: Cambridge University Press.

———. 2012a. "From Material to Symbolic Cultures: Culture in Primates." In *The Oxford Handbook of Culture and Psychology*, edited by Jan Valsiner, 677–94. Oxford: Oxford University Press.

———. 2012b. *Wild Cultures*. New York: Cambridge University Press.

Boesch, Christophe, and Hedwige Boesch. 1982. "Optimisation of Nut-Cracking with Natural Hammers by Wild Chimpanzees." *Behaviour* 83 (3/4): 265–86.

———. 1989. "Hunting Behavior of Wild Chimpanzees in the Taï National Park." *American Journal of Physical Anthropology* 78 (4): 547–73.

———. 1990. "Tool Use and Tool Making in Wild Chimpanzees." *Folia Primatologica* 54:86–99.

Boesch, Christophe, and Hedwige Boesch-Achermann. 2000. *The Chimpanzees of the Taï Forest: Behavioural Ecology and Evolution*. Oxford: Oxford University Press.

Boesch, Christophe, Daša Bombjaková, Adam Boyette, and Amelia Meier. 2017. "Technical Intelligence and Culture: Nut Cracking in Humans and Chimpanzees." *American Journal of Physical Anthropology* 163 (2): 339–55. https://doi.org/10.1002/ajpa.23211.

Boesch, Christophe, Josephine Head, and Martha M. Robbins. 2009. "Complex Tool Sets for Honey Extraction among Chimpanzees in Loango National Park, Gabon." *Journal of Human Evolution* 56 (6): 560–69.

Boesch, Christophe, Paul Marchesi, Nathalie Marchesi, Barbara Fruth, and Frédéric Joulian. 1994. "Is Nut Cracking in Wild Chimpanzees a Cultural Behaviour?" *Journal of Human Evolution* 26 (4): 325–38. https://doi.org/10.1006/jhev.1994.1020.

Boesch, Christophe, and Sanjida O'Connell. 2012. *Chimpanzee: The Making of the Film*. New York: Disney Editions.

Boesch, Christophe, and Michael Tomasello. 1998. "Chimpanzee and Human Cultures." *Current Anthropology* 39 (5): 591–614. https://doi.org/10.1086/204785.

Boesch-Achermann, Hedwige, and Christophe Boesch. 1993. "Tool Use in Wild Chimpanzees: New Light from Dark Forests." *Current Directions in Psychological Science* 2 (1): 18–21.

Bonner, John T. 1980. *The Evolution of Culture in Animals*. Princeton, NJ: Princeton University Press.

Bonnie, Kristin E., and Frans B. M. de Waal. 2006. "Affiliation Promotes the Transmission of a Social Custom: Handclasp Grooming among Captive Chimpanzees." *Primates* 47 (1): 27–34. https://doi.org/10.1007/s10329-005-0141-0.

Bourdieu, Pierre. 1977. *Outline of a Theory of Practice*. Cambridge: Cambridge University Press.

———. 1999. *The Weight of the World: Social Suffering in Contemporary Society*. Cambridge, UK: Polity.

———. 2000. *Pascalian Meditations*. Stanford, CA: Stanford University Press.

Bourdieu, Pierre, and Loïc Wacquant. 1992. *An Invitation to Reflexive Sociology*. Chicago: University of Chicago Press.

Boyd, Robert. 2018. *A Different Kind of Animal: How Culture Transformed Our Species*. Princeton, NJ: Princeton University Press.

Boyd, Robert, and Peter J. Richerson. 1985. *Culture and the Evolutionary Process*. Chicago: University of Chicago Press.

———. 2005. *The Origin and Evolution of Cultures*. Oxford: Oxford University Press.

Bradford, Phillips Verner, and Harvey Blume. 1992. *Ota Benga: The Pygmy in the Zoo*. New York: St. Martin's Press.

Brumann, Christoph. 1999. "Writing for Culture: Why a Successful Concept Should Not Be Discarded." *Current Anthropology* 40:S1–27.

Bryson-Morrison, Nicola, Joseph Tzanopoulos, Tetsuro Matsuzawa, and Tatyana Humle. 2017. "Activity and Habitat Use of Chimpanzees (*Pan troglodytes verus*) in the Anthropogenic Landscape of Bossou, Guinea, West Africa." *International Journal of Primatology*, January, 1–21. https://doi.org/10.1007/s10764-016-9947-4.

Burgess, John. 1986. "Nakasone Suggests Minorities Put U.S. Society behind Japan's." *Washington Post*, 24 September 1986. https://www.washingtonpost.com/archive/politics/1986/09/24/nakasone-suggests-minorities-put-us-society-behind-japans/aa67c979-4dff-42e7-9a13-ba93f4a07c1f/.

Byrne, Richard W. 2007. "Culture in Great Apes: Using Intricate Complexity in Feeding Skills to Trace the Evolutionary Origin of Human Technical Prowess." *Philosophical Transactions of the Royal Society B: Biological Sciences* 362 (1480): 577–85. https://doi.org/10.1098/rstb.2006.1996.

Cabral, Amilcar. 1973. "National Liberation and Culture." In *Return to the Source: Selected Speeches*, edited by Africa Information Service, 39–56. New York: Monthly Review Press.

Call, Josep, Brian Hare, Malinda Carpenter, and Michael Tomasello. 2004. "Unwilling or Unable? Chimpanzees' Understanding of Intentional Action." *Developmental Science* 7:488–98.

Callon, Michel, and Bruno Latour. 1981. "Unscrewing the Big Leviathan: How Actors Macro-Structure Reality and How Sociologists Help Them to Do So." In *Advances in Social Theory and Methodology: Toward an Integration of Micro- and Macro-Sociologies*, edited by Karin Knorr-Cetina and Aaron V. Cicourel, 277–303. London: Routledge.

———. 1992. "Don't Throw the Baby Out with the Bath School! A Reply to Collins and Yearley." In *Science as Practice and Culture*, edited by Andrew Pickering, 343–68. Chicago: University of Chicago Press.

Campbell, Geneviève, Hjalmar Kühl, Paul N'Goran Kouamé, and Christophe Boesch. 2008. "Alarming Decline of West African Chimpanzees in Côte d'Ivoire." *Current Biology* 18 (19): R903–4. https://doi.org/10.1016/j.cub.2008.08.015.

Campbell, Geneviève, Hjalmar S. Kühl, Abdoulaye Diarrassouba, K. Paul N'Goran, and Christophe Boesch. 2011. "Long-Term Research Sites as Refugia for Threatened and Over-Harvested Species." *Biology Letters* 7 (5): 723–26. https://doi.org/10.1098/rsbl.2011.0155.

Candea, Matei. 2010. "'I Fell in Love with Carlos the Meerkat': Engagement and Detachment in Human–Animal Relations." *American Ethnologist* 37 (2): 241–58.

———. 2014. "Objects Made Out of Action." In *Objects and Materials: A Routledge Companion*, edited by Penny Harvey, Eleanor Conlin Casella, Gillian Evans, Hannah Know, Christine McLean, Elizabeth B. Silva, Nicholas Thoburn, and Kath Woodward, 338–48. London: Routledge.

———. 2016. "De deux modalités de comparaison en anthropologie sociale." *L'Homme*, no. 218, 183–218.

Carpenter, Clarence Ray. 1942. "Sexual Behavior of Free Ranging Rhesus Monkeys (*Macaca mulatta*). I. Specimens, Procedures and Behavioral Characteristics of Estrus." *Journal of Comparative Psychology* 33 (1): 113–42.

———. 1960. "Comment on Imanishi's 'Social Organization of Subhuman Primates in Their Natural Habitat.'" *Current Anthropology* 15 (6): 402–3.

Cartmill, Matt. 1991. "Book Review of Donna Haraway's Primate Visions." *International Journal of Primatology* 12 (1): 67–75.

Carvalho, Susana, Dora Biro, William C. McGrew, and Tetsuro Matsuzawa. 2009. "Tool-Composite Reuse in Wild Chimpanzees (*Pan troglodytes*): Archaeologically Invisible Steps in the Technological Evolution of Early Hominins?" *Animal Cognition* 12 (1): S103–14. https://doi.org/10.1007/s10071-009-0271-7.

Chakrabarty, Dipesh. 2016. "Humanities in the Anthropocene: The Crisis of an Enduring Kantian Fable." *New Literary History* 47 (2): 377–97. https://doi.org/10.1353/nlh.2016.0019.

Chance, Michael R. A. 1960. "Comment on Imanishi's 'Social Organization of Subhuman Primates in Their Natural Habitat.'" *Current Anthropology* 15 (6): 403–4.

Chomsky, Noam. 1959. "A Review of B. F. Skinner's Verbal Behavior." *Language* 35 (1): 26–58.

———. 1965. *Aspects of the Theory of Syntax*. Cambridge, MA: MIT Press.

Churchland, Patricia S. 1986. *Neurophilosophy: Toward a Unified Science of the Mind-Brain*. Cambridge, MA: MIT Press.

Cibot, Marie, Sarah Bortolamiol, Andrew Seguya, and Sabrina Krief. 2015. "Chimpanzees Facing a Dangerous Situation: A High-Traffic Asphalted Road in the Sebitoli Area of Kibale National Park, Uganda." *American Journal of Primatology* 77 (8): 890–900. https://doi.org/10.1002/ajp.22417.

Clifford, James. 1988. *The Predicament of Culture: Twentieth-Century Ethnography, Literature, and Art*. Cambridge, MA: Harvard University Press.

———. 2000. "Taking Identity Politics Seriously: 'The Contradictory, Stony Ground . . .'" In *Without Guarantees: In Honour of Stuart Hall*, edited by Paul Gilroy, Lawrence Grossberg, and Angela McRobbie, 94–112. London: Verso. https://books.google.com/books?hl=en&lr=&id=MYcFKbUlozkC&oi=fnd&pg=PA94&dq=clifford+james+2000+identity&ots=M_LVpAQfG7&sig=jI6hlHl2aR8pf96rYVVzyEaodxo.

———. 2005. "Rearticulating Anthropology." In *Unwrapping the Sacred Bundle: Reflections on the Disciplining of Anthropology*, edited by Daniel Segal and Sylvia Yanagisako, 25–48. Durham, NC: Duke University Press.

Clifford, James, and George E. Marcus, eds. 1986. *Writing Culture: The Poetics and Politics of Ethnography*. Berkeley: University of California Press.

Cohen, Lawrence. 1998. *No Aging in India: Alzheimer's, the Bad Family, and Other Modern Things*. Berkeley: University of California Press.

Collins, Harry M. 1983. "The Sociology of Scientific Knowledge: Studies of Contemporary Science." *Annual Review of Sociology* 9:265–85.

———. 1985. *Changing Order: Replication and Induction in Scientific Practice*. Chicago: University of Chicago Press.

Corbey, Raymond H. A. 2005. *The Metaphysics of Apes: Negotiating the Animal-Human Boundary*. Cambridge: Cambridge University Press.

Cove, John. 1995. *What the Bones Say: Tasmanian Aborigines, Science and Domination*. Ottawa: Carleton University Press.

Crist, Eileen. 1999. *Images of Animals: Anthropocentrism and Animal Mind*. Philadelphia: Temple University Press.

Croft, Darren P., Rufus A. Johnstone, Samuel Ellis, Stuart Nattrass, Daniel W. Franks, Lauren J. N. Brent, Sonia Mazzi, Kenneth C. Balcomb, John K. B. Ford, and Michael A. Cant. 2017. "Reproductive Conflict and the Evolution of Menopause in Killer Whales." *Current Biology* 27 (2): 298–304. https://doi.org/10.1016/j.cub.2016.12.015.

Cross, Sherrie. 1996. "Prestige and Comfort: The Development of Social Darwinism in Early Meiji Japan, and the Role of Edward Sylvester Morse." *Annals of Science* 53 (4): 323–44.

Crunchant, Anne-Sophie, Monika Egerer, Alexander Loos, Tilo Burghardt, Klaus Zuberbühler, Katherine Corogenes, Vera Leinert, Lars Kulik, and Hjalmar S. Kühl. 2017. "Automated Face Detection for Occurrence and Occupancy Estimation in Chimpanzees." *American Journal of Primatology* 79 (3). https://doi.org/10.1002/ajp.22627.

Crutzen, Paul J. 2002. "Geology of Mankind." *Nature* 415 (6867): 23. https://doi.org/10.1038/415023a.

Dale, Peter N. 1986. *The Myth of Japanese Uniqueness*. New York: St. Martin's Press.

Dart, Raymond. 1925. "*Australopithecus africanus*: The Man-Ape of South Africa." *Nature* 115 (2884): 195–99.

Darwin, Charles. (1859) 2009. *On the Origin of Species*. Oxford: Oxford University Press.

———. (1871) 1981. *The Descent of Man, and Selection in Relation to Sex*. Princeton, NJ: Princeton University Press.

Daston, Lorraine. 1995. "The Moral Economy of Science." *Osiris* 10:2–24.

———. 2005. "Intelligences: Angelic, Animal, Human." In *Thinking with Animals: New Perspectives on Anthropomorphism*, edited by Lorraine Daston and Gregg Mitman, 37–58. New York: Columbia University Press.

———. 2011. "The Empire of Observation, 1600–1800." In *Histories of Scientific Observation*, edited by Lorraine Daston and Elizabeth Lunbeck, 81–114. Chicago: University of Chicago Press.

———. 2014. "Objectivity and Impartiality: Epistemic Virtues in the Humanities." In *The Making of the Humanities*, vol. 3, *The Modern Humanities*, edited by Rens Bod, J. Maat, and T. Weststeijn, 27–41. Amsterdam: Amsterdam University Press.

———. 2017a. "Epilogue: Time of the Archive." In *Science in the Archives: Pasts, Presents, Futures*, edited by Lorraine Daston, 329–32. Chicago: University of Chicago Press.

———. 2017b. "The Immortal Archive: Nineteenth-Century Science Imagines the Future." In *Science in the Archives: Pasts, Presents, Futures*, edited by Lorraine Daston, 159–84. Chicago: University of Chicago Press.

Daston, Lorraine, and Peter Galison. 2007. *Objectivity*. New York: Zone Books.

Daston, Lorraine, and Fernando Vidal. 2004. "Doing What Comes Naturally." In *The Moral Authority of Nature*, edited by Lorraine Daston and Fernando Vidal, 1–23. Chicago: University of Chicago Press.

Dawkins, Marian Stamp. 2007. *Observing Animal Behaviour: Design and Analysis of Quantitative Data*. Oxford: Oxford University Press.

Dawkins, Richard. 2004. *The Ancestor's Tale: A Pilgrimage to the Dawn of Life*. London: Weidenfeld and Nicolson.

de Melo Daly, Gabriela Bezerra. 2018. "Drawing and Blurring Boundaries between Species: An Etho-Ethnography of Human-Chimpanzee Social Relations at the Primate Research Institute of Kyoto University." PhD diss., Collège de France.

de Pracontal, Michel. 2010. *Kaluchua: Cultures, techniques et traditions des sociétés animales*. Paris: Seuil.

Descartes, René. (1637) 2003. "Discourse on Method." In *Discourse on Method and Meditations*, 1–52. Mineola, NY: Dover.

Descola, Philippe. 2013. *Beyond Nature and Culture*. Chicago: University of Chicago Press.

Despret, Vinciane. 2004. "The Body We Care For: Figures of Anthropo-Zoo-Genesis." *Body & Society* 10 (2–3): 111–34. https://doi.org/10.1177/1357034X04042938.

———. 2006. "Sheep Do Have Opinions." In *Making Things Public: Atmospheres of Democracy*, edited by Bruno Latour and Peter Weibel, 360–70. Cambridge, MA: MIT Press. http://orbi.ulg.ac.be/handle/2268/135590.

———. 2013. "Responding Bodies and Partial Affinities in Human-Animal Worlds." *Theory, Culture & Society* 30 (7–8): 51–76. https://doi.org/10.1177/0263276413496852.

DeVore, Irven. 1992. "An Interview with Sherwood Washburn." *Current Anthropology* 33 (4): 411–23.

DeVore, Irven, and K. R. L. Hall. 1965. "Baboon Ecology." In *Primate Behavior: Field Studies of Monkeys and Apes*, edited by Irven DeVore, 20–52. New York: Holt, Rinehart and Winston.

de Waal, Frans B. M. 1982. *Chimpanzee Politics: Power and Sex among Apes*. New York: Harper & Row.

———. 2001. *The Ape and the Sushi Master: Cultural Reflections by a Primatologist*. New York: Basic Books.

———. 2003. "Silent Invasion: Imanishi's Primatology and Cultural Bias in Science." *Animal Cognition* 6 (4): 293–99. https://doi.org/10.1007/s10071-003-0197-4.

———. 2005. *Our Inner Ape: A Leading Primatologist Explains Why We Are Who We Are*. New York: Riverhead Books.

———. 2010. *The Age of Empathy: Nature's Lessons for a Kinder Society*. New York: Broadway Books.

DeWalt, Kathleen M., and Billie R. DeWalt. 2011. *Participant Observation: A Guide for Fieldworkers*. 2nd ed. Lanham, MD: AltaMira Press.

Diderot, Denis. (1796) 2009. *Jacques the Fatalist*. Oxford: Oxford University Press.

Dobzhansky, Theodosius. 1955. *Evolution, Genetics and Man*. New York: Wiley.

Dolhinow, Phyllis, ed. 1968. *Primates: Studies in Adaptation and Variability*. New York: Holt, Rinehart and Winston.

Dupré, John. 2002. "Conversations with Apes: Reflections on the Scientific Study of Language." In *Humans and Other Animals*, 236–57. Oxford, UK: Clarendon Press.

Durkheim, Émile. 1895. *Les règles de la méthode sociologique*. Paris: Librairie Félix Alcan.

———. 1964. *The Division of Labor in Society*. New York: Free Press.

Eggan, Fred. 1954. "Social Anthropology and the Method of Controlled Comparison." *American Anthropologist* 56 (5): 743–63.

Elton, Charles Sutherland. 1930. *Animal Ecology and Evolution*. Oxford, UK: Claredon Press.

Emlen, John T. 1960. "Comment on Imanishi's 'Social Organization of Subhuman Primates in Their Natural Habitat.'" *Current Anthropology* 15 (6): 404–5.

Engelman, Robert. 2016. "Africa's Population Will Soar Dangerously Unless Women Are More Empowered." *Scientific American*, February 2016. https://www.scientificamerican.com/article/africa-s-population-will-soar-dangerously-unless-women-are-more-empowered/.

Engels, Friedrich. 1987. "Dialectics of Nature." In *Karl Marx. Frederick Engels: Collected Works*, vol. 25, 313–590. New York: International Publishers.

Estep, Daniel, and Suzanne Hetts. 1992. "Interactions, Relationships, and Bonds: The Conceptual Basis for Scientist-Animal Relations." In *The Inevitable Bond: Examining Scientist-Animal Interaction*, edited by Dianne Balfour and Hank Davis, 6–26. New York: Cambridge University Press.

Estrada, Alejandro, Paul A. Garber, Anthony B. Rylands, Christian Roos, Eduardo Fernandez-Duque, Anthony Di Fiore, K. Anne-Isola Nekaris, et al. 2017. "Impending Extinction Crisis of the World's Primates: Why Primates Matter." *Science Advances* 31:e1600946. https://doi.org/10.1126/sciadv.1600946.

Evans-Pritchard, Edward Evan. 1937. *Witchcraft, Oracles and Magic among the Azande*. Oxford: Oxford University Press.

Fabian, Johannes. 1983. *Time and the Other: How Anthropology Makes Its Object*. New York: Columbia University Press.

Fairet, Emilie, Sandra Bell, Kharl Remanda, and Joanna M. Setchell. 2014. "Rural Emptiness and Its Influence on Subsistence Farming in Contemporary Gabon: A Case Study in Loango National Park." *Society, Biology & Human Affairs* 78 (1–2): 39–59.

Faith, J. Tyler, John Rowan, Andrew Du, and Paul L. Koch. 2018. "Plio-Pleistocene Decline of African Megaherbivores: No Evidence for Ancient Hominin Impacts." *Science* 362 (6417): 938–41. https://doi.org/10.1126/science.aau2728.

Farman, Abou. 2014. "Misanthropology?" *Platypus: The Castac Blog*, 9 December 2014. http://blog.castac.org/2014/12/misanthropology/.

Fechner, Gustav Theodor. 1966. *Elements of Psychophysics*. Edited by Edwin Garrigues Boring and Davis H. Howes. Translated by Helmut E. Adler. New York: Holt, Rinehart and Winston.

Feyerabend, Paul. 1962. "Explanation, Reduction and Empiricism." In *Scientific Explanation, Space, and Time*, vol. 3, edited by Herbert Feigl and Grover Maxwell, 28–97. Minnesota Studies in the Philosophy of Science. Minneapolis: University of Minnesota Press.

Fischer, Joachim. 2015. "Michael Tomasello—Protagonist der philosophischen Anthropologie im 21. Jahrhundert?" In *Kritikfiguren: Festschrift für Gerard Raulet zum 65. Geburtstag / Figures de la critique: En hommage à Gerard Raulet*, edited by Olivier Agard, Manfred Gangl, Françoise Lartillot, and Gilbert Merlio, 321–42. Frankfurt am Main. Germany: Peter Lang.

Fisher, James, and Robert A. Hinde. 1949. "The Opening of Milk Bottles by Birds." *British Birds* 42:347–59.

Fisher, Philip. 2002. *The Vehement Passions*. Princeton, NJ: Princeton University Press.

Fisher, Ronald A. 1925. *Statistical Methods for Research Workers*. Edinburgh: Oliver and Boyd.

———. 1935. *The Design of Experiments*. Edinburgh: Oliver and Boyd.

Fisler, George F. 1967. "Nonbreeding Activities of Three Adult Males in a Band of Free-Ranging Rhesus Monkeys." *Journal of Mammalogy* 48 (1): 70–78. https://doi.org/10.2307/1378171.

Formenty, Pierre, Christophe Hatz, Bernard Le Guenno, Agnés Stoll, Philipp Rogenmoser, and Andreas Widmer. 1999. "Human Infection Due to Ebola Virus, Subtype Côte d'Ivoire: Clinical and Biologic Presentation." *Journal of Infectious Diseases* 179 (Supplement 1): S48–53. https://doi.org/10.1086/514285.

Fortun, Kim. 2015. "Figuring Out Theory: Ethnographic Sketches." In *Theory Can Be More Than It Used to Be: Learning Anthropology's Method in a Time of Transition*, edited by Dominic Boyer, James D. Faubion, and George E. Marcus, 147–67. Ithaca, NY: Cornell University Press.

Fossey, Dian. 1983. *Gorillas in the Mist*. Boston: Houghton Mifflin.

Fothergill, Alastair, and Mark Linfield. 2012. *Chimpanzee*. Documentary. Directed by Alastair Fothergill and Mark Linfield. Produced by Alastair Fothergill, Mark Linfield, and Alix Tidmarsh.

Foucault, Michel. 1998. "Pierre Boulez: Passing through the Screen." In *Aesthetics, Method, and Epistemology: Essential Works of Foucault, 1954–1984*, edited by James Faubion, 241–44. New York: New Press.

Fouts, Roger. 1997. *Next of Kin: What Chimpanzees Have Taught Me about Who We Are*. New York: Morrow, William.

Fouts, Roger, and Stephen Tukel Mills. 1997. *Next of Kin: My Conversations with Chimpanzees*. New York: Avon Books.

Fridland, Ellen, and Richard Moore. 2014. "Imitation Reconsidered." *Philosophical Psychology* 28 (6): 856–80.

Frisch, John E. 1959. "Research on Primate Behavior in Japan." *American Anthropologist* 61 (4): 584–96. https://doi.org/10.1525/aa.1959.61.4.02a00040.

Fuentes, Agustín. 2010. "Naturalcultural Encounters in Bali: Monkeys, Temples, Tourists, and Ethnoprimatology." *Cultural Anthropology* 25 (4): 600–624. https://doi.org/10.1111/j.1548-1360.2010.01071.x.

Fuentes, Agustín, and Kimberley J. Hockings. 2010. "The Ethnoprimatological Approach in Primatology." *American Journal of Primatology* 72 (10): 841–47.

Galef, Bennett G. 1990. "Tradition in Animals: Field Observations and Laboratory Analyses." In *Interpretation and Explanation in the Study of Animal Behavior*, vol. 1, *Interpretation, Intentionality, and Communication*, edited by Marc Bekoff and Dale Jamieson, 74. Boulder, CO: Westview Press.

———. 1992. "The Question of Animal Culture." *Human Nature* 3 (2): 157–78.

Gardner, Allen R., and Beatrix T. Gardner, eds. 1989. *Teaching Sign Language to Chimpanzees*. Albany, NY: SUNY Press.

Geertz, Clifford. 1973. "Thick Description: Toward an Interpretive Theory of Culture." In *The Interpretation of Cultures: Selected Essays*, 3–30. New York: Basic Books.

Gellner, Ernest. 1998. *Language and Solitude: Wittgenstein, Malinowski and the Habsburg Dilemma*. Cambridge: Cambridge University Press.

Genette, Gérard. 1997. *Paratexts: Thresholds of Interpretation*. Cambridge: Cambridge University Press.

Germain, Jacques. 1984. *Guinée: Peuples de la forêt*. Paris: Académie des sciences d'outre-mer. http://www.webguinee.net/bibliotheque/ethnographie/jGermain/tdm.html.

Geulen, Christian. 2010. "Menschen, die auf Tiere starren: Zu Michael Tomasellos Thesen über Kommunikation und Kooperation." In "Universität," special issue, *Nach Feierabend: Zürcher Jahrbuch für Wissensgeschichte* 6:155–62.

Geyer, Christian. 2004. *Hirnforschung und Willensfreiheit: Zur Deutung der neuesten Experimente*. Frankfurt am Main. Germany: Suhrkamp.

Giono, Jean. 1984. *The Man Who Planted Trees*. White River Junction, VT: Chelsea Green.

Goodall, Jane. 1962. "Nest Building Behavior in the Free Ranging Chimpanzee." *Annals of the New York Academy of Sciences* 102 (2): 455–67.

———. 1963. "Feeding Behaviour of Wild Chimpanzees: A Preliminary Report." *Symposium of the Zoological Society of London* 10:39–48.

———. 1971. *In the Shadow of Man*. New York: Houghton Mifflin Harcourt.

———. 1973. "Cultural Elements in a Chimpanzee Community." In *Precultural Primate Behavior*, edited by Emil W. Menzel, 144–84. Basel, Switzerland: S. Karger.

Goodwin, Reiko Matsuda, and Michael A. Huffman. 2017. "History of Primatology—Japan." In *The International Encyclopedia of Primatology*, edited by Agustín Fuentes, 248–59. Tokyo: Springer. https://doi.org/10.1002/9781119179313.wbprim0408.

Gould, Stephen Jay, and Richard C. Lewontin. 1979. "The Spandrels of San Marco and the Panglossian Paradigm: A Critique of the Adaptationist Programme." *Proceedings of the Royal Society of London, Series B, Biological Sciences* 205 (1161): 581–98.

Gray, John. 2002. *Straw Dogs: Thoughts on Humans and Other Animals*. London: Granta Books.

Green, Steven. 1975. "Dialects in Japanese Monkeys: Vocal Learning and Cultural Transmission of Locale-Specific Vocal Behavior?" *Zeitschrift für Tierpsychologie* 38 (3): 304–14. https://doi.org/10.1111/j.1439-0310.1975.tb02006.x.

Grether, W. F. 1940. "Chimpanzee Color Vision I. Hue Discrimination at Three Spectral Points." *Journal of Comparative Psychology* 29 (2): 167–77. https://doi.org/10.1037/h0056693.

Gruber, Jacob W. 1970. "Ethnographic Salvage and the Shaping of Anthropology." *American Anthropologist* 72 (6): 1289–99. https://doi.org/10.1525/aa.1970.72.6.02a00040.

Gruber, Thibaud, Klaus Zuberbühler, Fabrice Clément, and Carel van Schaik. 2015. "Apes Have Culture but May Not Know That They Do." *Comparative Psychology* 6:91. https://doi.org/10.3389/fpsyg.2015.00091.

Grützmacher, Kim, Verena Keil, Vera Leinert, Floraine Leguillon, Arthur Henlin, Emmanuel Couacy-Hymann, Sophie Köndgen, et al. 2018. "Human Quarantine: Toward Reducing Infectious Pressure on Chimpanzees at the Taï Chimpanzee Project, Côte d'Ivoire." *American Journal of Primatology* 80 (1): e22619. https://doi.org/10.1002/ajp.22619.

Guengant, Jean-Pierre, and John F. May. 2013. "African Demography." *Global Journal of Emerging Market Economies* 5 (3): 215–67.

Gumbrecht, Hans Ulrich. 2014. *Our Broad Present: Time and Contemporary Culture*. New York: Columbia University Press.

Gyger, Marcel, and Peter Marler. 1988. "Food Calling in Domestic Fowl (*Gallus gallus*): The Role of External Referents and Deception." *Animal Behaviour* 36 (2): 358–65.

Habermas, Jürgen. 1985. *The Theory of Communicative Action*. Vol. 2. Boston: Beacon Press.

———. 1997. *A Berlin Republic: Writings on Germany*. Cambridge, UK: Polity.

———. 2008. *Between Naturalism and Religion: Philosophical Essays*. Cambridge, UK: Polity.

———. 2009. "Es beginnt mit dem Zeigefinger: Review of Michael Tomasello's Origins of Human Communication." *Die Zeit*, no. 51 (October): 45.

———. 2013. "Bohrungen an der Quelle des objektiven Geistes: Hegel-Preis für Michael Tomasello." In *Im Sog der Technokratie: Kleine Politische Schriften XII*, 166–73. Frankfurt am Main, Germany: Suhrkamp.

———. 2015. *The Lure of Technocracy*. Malden, MA: Polity.

Hacking, Ian. 1983. *Representing and Intervening: Introductory Topics in the Philosophy of Natural Science*. Cambridge: Cambridge University Press.

———. 1990. *The Taming of Chance*. New York: Cambridge University Press.

———. 1999. *The Social Construction of What?* Cambridge, MA: Harvard University Press.

———. 2005. "The Cartesian Vision Fulfilled: Analogue Bodies and Digital Minds." *Interdisciplinary Science Reviews* 30 (2): 153–66.

———. 2007. "Our Neo-Cartesian Bodies in Parts." *Critical Inquiry* 34 (1): 78–105.

Hagner, Michael. 2015. *Zur Sache des Buches*. Göttingen, Germany: Wallstein Verlag.

Hallowell, A. Irving. 1961. "The Protocultural Foundations of Human Adaptation." In *Social Life of Early Man*, edited by Sherwood L. Washburn, 236–55. Chicago: Aldine.

Halstead, Beverly. 1985. "Anti-Darwinian Theory in Japan." *Nature* 317 (6038): 587–89.

———. 1987. "Imanishi's Influence on Evolution Theory in Japan." *Nature* 326 (March): 21.

Hamill, Pete. 2004. "The Alloy of New York." In *Reinventing the Melting Pot: The New Immigrants and What It Means to Be American*, edited by Tamar Jacoby, 167–82. New York: Basic Books.

Hanzel, Igor. 2012. "Der Spracherwerb bei Menschenaffen: Sue Savage-Rumbaughs methodologische Wende." *Deutsche Zeitschrift für Philosophie* 60 (5): 659–82. https://doi.org/10.1524/dzph.2012.0050.

Haraway, Donna. 1984. "Primatology Is Politics by Other Means." *PSA: Proceedings of the Biennial Meeting of the Philosophy of Science Association* 2:489–524.

———. 1989. *Primate Visions: Gender, Race, and Nature in the World of Modern Science*. New York: Routledge.

———. 1991. "The Biological Enterprise: Sex, Mind, and Profit from Human Engineering to Sociobiology." In *Simians, Cyborgs, and Women: The Reinvention of Nature*, 43–68. New York: Routledge.

———. 2003. *The Companion Species Manifesto: Dogs, People, and Significant Otherness*. Chicago: Prickly Paradigm Press.

———. 2008. *When Species Meet*. Minneapolis: University of Minnesota Press.

———. 2015. "Anthropocene, Capitalocene, Plantationocene, Chthulucene: Making Kin." *Environmental Humanities* 6 (1): 159–65. https://doi.org/10.1215/22011919-3615934.

———. 2016. *Staying with the Trouble: Making Kin in the Chthulucene*. Durham, NC: Duke University Press.

Haraway, Donna, Noboru Ishikawa, Scott F. Gilbert, Kenneth Olwig, Anna L. Tsing, and Nils Bubandt. 2016. "Anthropologists Are Talking—About the Anthropocene." *Ethnos* 81 (3): 535–64. https://doi.org/10.1080/00141844.2015.1105838.

Harding, Sandra. 2015. *Objectivity and Diversity: Another Logic of Scientific Research*. Chicago: University of Chicago Press.

Hart, Hornell, and Adele Pantzer. 1925. "Have Subhuman Animals Culture?" *American Journal of Sociology* 306:703–9.

Hartmann, Martin. 2011. "Against First Nature: Critical Theory and Neuroscience." In *Critical Neuroscience: A Handbook of the Social and Cultural Contexts of Neuroscience*, edited by Suparna Choudhury and Jan Slaby, 67–84. Malden, MA: Wiley-Blackwell.

Haslam, Michael, Adriana Hernandez-Aguilar, Victoria Ling, Susana Carvalho, Ignacio de la Torre, April DeStefano, Andrew Du, et al. 2009. "Primate Archaeology." *Nature* 460 (7253): 339–44. https://doi.org/10.1038/nature08188.

Hawkes, Kristen, J. F. O'Connell, N. G. Blurton Jones, H. Alvarez, and E. L. Charnov. 1998. "Grandmothering, Menopause, and the Evolution of Human Life Histories." *Proceedings of the National Academy of Sciences* 95 (3): 1336–39.

Hayashi, Misato, Yuu Mizuno, and Tetsuro Matsuzawa. 2005. "How Does Stone-Tool Use Emerge? Introduction of Stones and Nuts to Naïve Chimpanzees in Captivity." *Primates* 46 (2): 91–102. https://doi.org/10.1007/s10329-004-0110-z.

Hayes, Keith J., and Catherine Hayes. 1951. "The Intellectual Development of a Home-Raised Chimpanzee." *Proceedings of the American Philosophical Society* 95 (2): 105–9.

Hegel, Georg Wilhelm Friedrich. (1807) 1977. *Phenomenology of Spirit*. Translated by A. V. Miller. Rev. ed. Oxford: Oxford University Press.

Henrich, Joseph. 2016. *The Secret of Our Success: How Culture Is Driving Human Evolution, Domesticating Our Species, and Making Us Smarter*. Princeton, NJ: Princeton University Press.

Herzfeld, Chris. 2016. *Wattana: An Orangutan in Paris*. Chicago: University of Chicago Press.

Herzfeld, Chris, and Dominique Lestel. 2005. "Knot Tying in Great Apes: Etho-Ethnology of an Unusual Tool Behavior." *Social Science Information* 44 (4): 621–53. https://doi.org/10.1177/0539018405058205.

Hester, James J. 1968. "Pioneer Methods in Salvage Anthropology." *Anthropological Quarterly* 41 (3): 132–46. https://doi.org/10.2307/3316788.

Hewes, Gordon W. 1994. "The Baseline for Comparing Human and Nonhuman Primate Behavior." In *Hominid Culture in Primate Perspective*, edited by Duane Quiatt and Junichiro Itani, 59–94. Niwot: University Press of Colorado.

Hinde, Robert A. 1973. "On the Design of Check-Sheets." *Primates* 14 (4):393–406. https://doi.org/10.1007/BF01731360.

———. 2000. "Some Reflections on Primatology at Cambridge and the Science Studies Debate." In *Primate Encounters: Models of Science, Gender, and Society*, edited by Shirley C. Strum and Linda M. Fedigan, 104–15. Chicago: University of Chicago Press.

Hirata, Satoshi, and Maura L. Celli. 2003. "Role of Mothers in the Acquisition of Tool-Use Behaviours by Captive Infant Chimpanzees." *Animal Cognition* 6 (4): 235–44. https://doi.org/10.1007/s10071-003-0187-6.

Hirata, Satoshi, and Misato Hayashi. 2011. "The Emergence of Stone-Tool Use in Captive Chimpanzees." In *The Chimpanzees of Bossou and Nimba*, edited by Tetsuro Matsuzawa, Tatyana Humle, and Yukimaru Sugiyama, 183–90. Primatology Monographs. Tokyo: Springer Japan. https://doi.org/10.1007/978-4-431-53921-6_20.

Hirata, Satoshi, Naruki Morimura, and Chiharu Houki. 2009. "How to Crack Nuts: Acquisition Process in Captive Chimpanzees (*Pan troglodytes*) Observing a Model." *Animal Cognition* 12 (1): 87–101. https://doi.org/10.1007/s10071-009-0275-3.

Hirata, Satoshi, Kunio Watanabe, and Masao Kawai. 2001. "'Sweet-Potato Washing' Revisited." In *Primate Origins of Human Cognition and Behavior*, edited by Tetsuro Matsuzawa, 487–508. Tokyo: Springer.

Hockings, Kimberley J. 2007. "Human-Chimpanzee Coexistence at Bossou, the Republic of Guinea: A Chimpanzee Perspective." Stirling, UK: University of Stirling. https://dspace.stir.ac.uk/handle/1893/189.

———. 2011a. "Behavioral Flexibility and Division of Roles in Chimpanzee Road-Crossing." In *The Chimpanzees of Bossou and Nimba*, edited by Tetsuro Matsuzawa, Tatyana Humle, and Yukimaru Sugiyama, 221–29. Primatology Monographs. Tokyo: Springer Japan. https://doi.org/10.1007/978-4-431-53921-6_24.

———. 2011b. "The Tool Repertoire of Bossou Chimpanzees." In *The Chimpanzees of Bossou and Nimba*, edited by Tetsuro Matsuzawa, Tatyana Humle, and Yukimaru Sugiyama, 61–71. Primatology Monographs. Tokyo: Springer Japan. https://doi.org/10.1007/978-4-431-53921-6_24.

Hockings, Kimberley J., James R. Anderson, and Tetsuro Matsuzawa. 2006. "Road Crossing in Chimpanzees: A Risky Business." *Current Biology* 16 (17): R668–70. https://doi.org/10.1016/j.cub.2006.08.019.

Hockings, Kimberley J., Nicola Bryson-Morrison, Susana Carvalho, Michiko Fujisawa, Tatyana Humle, William C. McGrew, Miho Nakamura, et al. 2015. "Tools to Tipple: Ethanol Ingestion by Wild Chimpanzees Using Leaf-Sponges." *Royal Society Open Science* 2 (6): 1–6. https://doi.org/10.1098/rsos.150150.

Hockings, Kimberley J., Matthew R. McLennan, Susana Carvalho, Marc Ancrenaz, René Bobe, Richard W. Byrne, Robin I. M. Dunbar, et al. 2015. "Apes in the Anthropocene: Flexibility and Survival." *Trends in Ecology & Evolution* 30 (4): 215–22.

Hockings, Kimberley J., Gen Yamakoshi, Asami Kabasawa, and Tetsuro Matsuzawa. 2010. "Attacks on Local Persons by Chimpanzees in Bossou, Republic of Guinea: Long-Term Perspectives." *American Journal of Primatology* 72 (10): 887–96. https://doi.org/10.1002/ajp.20784.

Hokkyo, Noboru. 1987. "Comments on Anti-Darwinian Theory in Japan: Human Concerns beyond Natural Science." *Journal of Social and Biological Structures* 10:377–79.

Hollinger, David A. 1995. *Postethnic America: Beyond Multiculturalism*. New York: Basic Books.

Holmberg, Allan R. 1955. "Participant Intervention in the Field." *Human Organization* 14 (1):23–26.

Holmes, Douglas R. 2000. *Integral Europe: Fast Capitalism, Multiculturalism, Neofascism*. Princeton, NJ: Princeton University Press.

Honneth, Axel. 2008. *Reification: A New Look at an Old Idea*. Oxford: Oxford University Press.

Hoppitt, William, and Kevin N. Laland. 2012. *Social Learning: An Introduction to Mechanisms, Methods, and Models*. Princeton, NJ: Princeton University Press.

Horner, Victoria, and Andrew Whiten. 2005. "Causal Knowledge and Imitation/Emulation Switching in Chimpanzees (*Pan troglodytes*) and Children (*Homo sapiens*)." *Animal Cognition* 8 (3): 164–81. https://doi.org/10.1007/s10071-004-0239-6.

Horner, Victoria, Andrew Whiten, Emma Flynn, and Frans B. M. de Waal. 2006. "Faithful Replication of Foraging Techniques along Cultural Transmission Chains by Chimpanzees and Children." *Proceedings of the National Academy of Sciences* 103 (37):13878–83. https://doi.org/10.1073/pnas.0606015103.

Houdart, Sophie. 2007. *La cour des miracles: Ethnologie d'un laboratoire japonais*. Paris: CNRS Éditions.

Hrdy, Sarah Blaffer. 2008. *Mothers and Others: The Evolutionary Origins of Mutual Understanding*. Cambridge, MA: Belknap Press of Harvard University Press.

Huffman, Michael A. 2014. "Learning to Become a Monkey." In *Primate Ethnographies*, edited by Karen B. Strier, 57–68. Upper Saddle River, NJ: Pearson Education.

Hughes, J. Donald. 1994. *Pan's Travails: Environmental Problems of the Ancient Greeks and Romans*. Baltimore: Johns Hopkins University Press.

Humle, Tatyana. 2011. "Location and Ecology." In *The Chimpanzees of Bossou and Nimba*, edited by Tetsuro Matsuzawa, Tatyana Humle, and Yukimaru Sugiyama, 13–21. Primatology Monographs. Tokyo: Springer Japan. https://doi.org/10.1007/978-4-431-53921-6_3.

Hunter, James Davison. 1991. *Culture Wars: The Struggle to Control the Family, Art, Education, Law, and Politics in America*. New York: Basic Books.

Huntington, Samuel P. 1993. "The Clash of Civilizations?" *Foreign Affairs* 72 (3): 22–49.

———. 1996. *The Clash of Civilizations and the Remaking of World Order*. New York: Simon & Schuster.

Huxley, Aldous. 1948. *Ape and Essence*. London: Chatto & Windus.

Idani, Gen'ichi, and Satoshi Hirata. 2007. "Studies at the Great Ape Research Institute, Hayashibara." In *Primate Perspectives on Behavior and Cognition*, edited by David A. Washburn, 29–36. Washington, DC: American Psychological Association.

Iida, Kaori. 2015. "A Controversial Idea as a Cultural Resource: The Lysenko Controversy and Discussions of Genetics as a 'Democratic' Science in Postwar Japan." *Social Studies of Science*, August. https://doi.org/10.1177/0306312715596460.

Ikeda, Kiyohiko, and Atuhiro Sibatani. 1995. "Kinji Imanishi's Biological Thought." In *Speciation and the Recognition Concept: Theory and Application*, edited by David M. Lambert and Hamish G. Spencer, 71–89. Baltimore: Johns Hopkins University Press.

Imanishi, Kinji. (1941) 2002. *A Japanese View of Nature: The World of Living Things*. Edited by Pamela J. Asquith. London: Routledge.

———. 1957a. "Identification: A Process of Enculturation in the Subhuman Society of *Macaca fuscasta*." *Primates* 1:1–29.

———. 1957b. "Social Behavior in Japanese Monkeys, *Macaca fuscata*." *Psychologia* 1:47–54.

———. 1960. "Social Organization of Subhuman Primates in Their Natural Habitat." *Current Anthropology* 15 (6): 393–407.

———. 1984. "A Proposal for Shizengaku: The Conclusion to My Study of Evolutionary Theory." *Journal of Social and Biological Structures* 7 (4): 357–68. https://doi.org/10.1016/0140-1750(84)90008-3.

Ingold, Tim. 2000. *The Perception of the Environment: Essays on Livelihood, Dwelling and Skill*. London: Routledge.

———. 2001. "The Use and Abuse of Ethnography." *Behavioral and Brain Sciences* 24 (2): 337.

———. 2007. "The Trouble with 'Evolutionary Biology.'" *Anthropology Today* 23 (2): 13–17.

———. 2008. "Anthropology Is Not Ethnography." *Proceedings of the British Academy* 15 (4): 69–92.

———. 2011. *Being Alive: Essays on Movement, Knowledge and Description*. New York: Routledge.

Ingold, Tim, and Gisli Palsson, eds. 2013. *Biosocial Becomings: Integrating Social and Biological Anthropology*. New York: Cambridge University Press.

———. 2016. "Perspectives on the Intersection of Biology and Society: Response to Tehrani and Carrithers." *Journal of the Royal Anthropological Institute* 22 (2): 459–60. https://doi.org/10.1111/1467-9655.12439.

Inoue, Sana, and Tetsuro Matsuzawa. 2007. "Working Memory of Numerals in Chimpanzees." *Current Biology* 17 (23): R1004–5. https://doi.org/10.1016/j.cub.2007.10.027.

Inoue-Nakamura, Noriko, and Tetsuro Matsuzawa. 1997. "Development of Stone Tool Use by Wild Chimpanzees (*Pan troglodytes*)." *Journal of Comparative Psychology* 111 (2): 159–73. http://dx.doi.org/10.1037/0735-7036.111.2.159.

Itani, Junichiro. 1958. "On the Acquisition and Propagation of a New Food Habit in the Natural Group of the Japanese Monkey at Takasaki-Yama." *Primates* 1 (2): 84–98. https://doi.org/10.1007/BF01813697.

———. 1985. "The Evolution of Primate Social Structures." *Man* 20 (4): 593–611. https://doi.org/10.2307/2802752.

Itani, Junichiro, and Akisato Nishimura. 1973. "The Study of Infrahuman Culture in Japan." In *Precultural Primate Behavior*, edited by Emil W. Menzel, 26–50. Basel, Switzerland: S. Karger.

Itō, Yosiaki. 1991. "Development of Ecology in Japan, with Special Reference to the Role of Kinji Imanishi." *Ecological Research* 6:139–55.

Izumi, Hiroaki. 2006. *Towards the Neo-Kyoto School: History and Development of the Primatological Approach of the Kyoto School in Japanese Primatology and Ecological Anthropology*. Occasional Papers No. 101. Edinburgh: University of Edinburgh, Center of African Studies.

Jacobi, Friedrich Heinrich. 2017. *Über die Lehre des Spinoza in Briefen an den Herrn Moses Mendelssohn*. Berlin: Hofenberg.

Jampel, Barbara. 1993. *Among the Wild Chimpanzees*. National Geographic DVD.

Jay, Martin. 1973. *The Dialectical Imagination: A History of the Frankfurt School and the Institute of Social Research, 1923–1950*. Berkeley: University of California Press.

Jensen, Casper Bruun, and Anders Blok. 2013. "Techno-Animism in Japan: Shinto Cosmograms, Actor-Network Theory, and the Enabling Powers of Non-Human Agencies." *Theory, Culture & Society* 30 (2): 84–115. https://doi.org/10.1177/0263276412456564.

Jensen, Casper Bruun, and Atsuro Morita. 2017. "Introduction: Minor Traditions, Shizen Equivocations, and Sophisticated Conjunctions." *Social Analysis* 61 (2): 1–14. https://doi.org/10.3167/sa.2017.610201.

Johannssen, Dennis. 2013. "Toward a Negative Anthropology." *Anthropology & Materialism: A Journal of Social Research*, no. 1 (October). https://doi.org/10.4000/am.194.

Johnstone, Bob. 1987. "How Debate Could Shatter the Peace." *New Scientist*, September 1987, 74–75.

Jolly, Alison. 2000. "The Bad Old Days of Primatology?" In *Primate Encounters: Models of Science, Gender, and Society*, edited by Shirley C. Strum and Linda M. Fedigan, 71–84. Chicago: University of Chicago Press.

Joulian, Frédéric. 1996. "Comparing Chimpanzee and Early Hominid Techniques: Some Contributions to Cultural and Cognitive Questions." In *Modelling the Early Human Mind*, edited by Paul Mellars and Kathleen Gibson, 173–89. Cambridge, UK: McDonald Institute Monographs.

Jumonville, Neil. 2002. "The Cultural Politics of the Sociobiology Debate." *Journal of the History of Biology* 35 (3): 569–93. https://doi.org/10.1023/A:1021190227056.

Kalland, Arne, and Pamela J. Asquith. 1997. "Japanese Perceptions of Nature: Ideals and Illusions." In *Japanese Images of Nature: Cultural Perspectives*, edited by Pamela J. Asquith and Arne Kalland, 1–35. London: RoutledgeCurzon.

Kano, Fumihiro, Christopher Krupenye, Satoshi Hirata, and Josep Call. 2017. "Eye Tracking Uncovered Great Apes' Ability to Anticipate That Other Individuals Will Act According to False Beliefs." *Communicative & Integrative Biology* 10 (2): e1299836. https://doi.org/10.1080/19420889.2017.1299836.

Kano, Fumihiro, and Masaki Tomonaga. 2013. "Head-Mounted Eye Tracking of a Chimpanzee under Naturalistic Conditions." *PLOS ONE* 8 (3): e59785. https://doi.org/10.1371/journal.pone.0059785.

Kant, Immanuel. (1800) 1992. *Lectures on Logic*. Cambridge: Cambridge University Press.

Kappeler, Peter M., and David P. Watts. 2012. *Long-Term Field Studies of Primates*. Berlin: Springer.

Kawade, Yoshimi. 1998. "Imanishi Kinji's Biosociology as a Forerunner of the Semiosphere Concept." *Semiotica* 120 (3–4): 273–97.

Kawai, Kaori, ed. 2013. *Groups: The Evolution of Human Sociality*. Kyoto: Kyoto University Press.

———, ed. 2017. *Institutions: The Evolution of Human Sociality*. Kyoto: Kyoto University Press.

Kawai, Masao. 1957. "On the Rank System in a Natural Group of Japanese Monkey (I)—the Basic and Dependent Rank." *Primates* 1 (2): 111–30.

———. 1965. "Newly-Acquired Pre-Cultural Behavior of the Natural Troop of Japanese Monkeys on Koshima Islet." *Primates* 6 (1): 1–30. https://doi.org/10.1007/BF01794457.

Kawamura, Syunzo. 1959. "The Process of Sub-Culture Propagation among Japanese Macaques." *Primates* 2 (1): 43–60. https://doi.org/10.1007/BF01666110.

Keller, Evelyn Fox. 1983. *A Feeling for the Organism: The Life and Work of Barbara McClintock*. New York: A. W. H. Freeman/Owl Book.

Kellert, Stephen R. 1991. "Japanese Perceptions of Wildlife." *Conservation Biology* 5 (3): 297–308. https://doi.org/10.1111/j.1523-1739.1991.tb00141.x.

———. 1993. "Attitudes, Knowledge, and Behavior toward Wildlife among the Industrial Superpowers: United States, Japan, and Germany." *Journal of Social Issues* 49 (1): 53–69. https://doi.org/10.1111/j.1540-4560.1993.tb00908.x.

King, Barbara J. 2001. "Debating Culture." Review of *The Ape and the Sushi Master*, by Frans de Waal. *Current Anthropology* 42 (3): 441–43.

———. 2004. "Towards an Ethnography of African Great Apes." *Social Anthropology* 12 (2): 195–207.

Kirksey, Eben. 2015. *Emergent Ecologies*. Durham, NC: Duke University Press.

Kirksey, Eben, and Stefan Helmreich. 2010. "The Emergence of Multispecies Ethnography." *Cultural Anthropology* 25 (4): 545–76.

Kirksey, Eben, Craig Schuetze, and Stefan Helmreich. 2014. "Introduction: Tactics of Multispecies Ethnography." In *The Multispecies Salon*, edited by Eben Kirksey, 1–24. Durham, NC: Duke University Press.

Kitcher, Philip. 1998. "A Plea for Science Studies." In *A House Built on Sand: Exposing Postmodernist Myths about Science*, edited by Noretta Koertge, 32–50. Oxford: Oxford University Press.

Knight, John. 2011. *Herding Monkeys to Paradise: How Macaque Troops Are Managed for Tourism in Japan*. Leiden, Netherlands: Brill.

Kohler, Robert E. 2002. *Landscapes and Labscapes: Exploring the Lab-Field Border in Biology*. Chicago: University of Chicago Press.

Köhler, Wolfgang. 1921. *Intelligenzprüfungen an Menschenaffen*. Berlin: Springer.

Kolbert, Elizabeth. 2014. *The Sixth Extinction: An Unnatural History*. New York: Henry Holt.

Köndgen, Sophie, Hjalmar Kühl, Paul K. N'Goran, Peter D. Walsh, Svenja Schenk, Nancy Ernst, Roman Biek, et al. 2008. "Pandemic Human Viruses Cause Decline of Endangered Great Apes." *Current Biology* 18 (4): 260–64. https://doi.org/10.1016/j.cub.2008.01.012.

Konersmann, Ralf. 2015. *Die Unruhe der Welt*. Frankfurt am Main, Germany: Fischer.

Koops, Kathelijne. 2011. "Chimpanzees in the Seringbara Region of the Nimba Mountains." In *The Chimpanzees of Bossou and Nimba*, edited by Tetsuro Matsuzawa, Tatyana Humle, and Yukimaru Sugiyama, 277–87. Primatology Monographs. Tokyo: Springer. https://doi.org/10.1007/978-4-431-53921-6_29.

Koops, Kathelijne, William C. McGrew, and Tetsuro Matsuzawa. 2013. "Ecology of Culture: Do Environmental Factors Influence Foraging Tool Use in Wild Chimpanzees, *Pan troglodytes verus*?" *Animal Behaviour* 85 (1): 175–85. https://doi.org/10.1016/j.anbehav.2012.10.022.

Kortlandt, Adriaan. 1986. "The Use of Stone Tools by Wild-Living Chimpanzees and Earliest Hominids." *Journal of Human Evolution* 15 (2): 77–132. https://doi.org/10.1016/S0047-2484(86)80068-9.

Kortlandt, Adriaan, and M. Kooij. 1963. "Protohominid Behaviour in Primates." *Symposium of the Zoological Society of London* 10:61–88.

Koselleck, Reinhart. 1988. *Critique and Crisis: Enlightenment and the Pathogenesis of Modern Society*. Cambridge, MA: MIT Press.

Koyama, Naoki. (1980) 2012. "Touches of Humanity in Monkey Society." In *The Monkeys of Stormy Mountain: 60 Years of Primatological Research on the Japanese Macaques of Arashiyama*, edited by Jean-Baptiste Leca, Michael A. Huffman, and Paul L. Vasey, 42–50. Cambridge: Cambridge University Press.

Kroeber, Alfred L. 1928. "Sub-Human Cultural Beginnings." *Quarterly Review of Biology* 3:325–42.

———. 1939. *Cultural and Natural Areas of Native North America*. Berkeley: University of California Press.

Kroeber, Alfred L., and Clyde Kluckhohn. 1952. *Culture: A Critical Review of Concepts and Definitions*. Cambridge, MA: Peabody Museum of American Archaeology and Ethnology.

Kropotkin, Pyotr. 1902. *Mutual Aid: A Factor in Evolution*. London: Heinneman.

Krou, Patrick. 2013. "Site touristique de Soko: Les singes sont en train de disparaître." *L'intelligent d'Abidjan*, 31 May 2013. http://news.abidjan.net/h/460883.html.

Krupenye, Christopher, Fumihiro Kano, Satoshi Hirata, Josep Call, and Michael Tomasello. 2016. "Great Apes Anticipate That Other Individuals Will Act According to False Beliefs." *Science* 354 (6308): 110–14. https://doi.org/10.1126/science.aaf8110.

Krützen, Michael, Janet Mann, Michael R. Heithaus, Richard C. Connor, Lars Bejder, and William B. Sherwin. 2005. "Cultural Transmission of Tool Use in Bottlenose Dolphins." *Proceedings of the National Academy of Sciences of the United States of America* 102 (25): 8939–43. https://doi.org/10.1073/pnas.0500232102.

Krützen, Michael, Carel P. van Schaik, and Andrew Whiten. 2007. "The Animal Cultures Debate: Response to Laland and Janik." *Trends in Ecology & Evolution* 22 (1): 6. https://doi.org/10.1016/j.tree.2006.10.011.

Kühl, Hjalmar S., and Tilo Burghardt. 2013. "Animal Biometrics: Quantifying and Detecting Phenotypic Appearance." *Trends in Ecology & Evolution* 28 (7): 432–41. https://doi.org/10.1016/j.tree.2013.02.013.

Kühl, Hjalmar S., Ammie K. Kalan, Mimi Arandjelovic, Floris Aubert, Lucy D'Auvergne, Annemarie Goedmakers, Sorrel Jones, et al. 2016. "Chimpanzee Accumulative Stone Throwing." *Scientific Reports*, 6 February 2016, srep22219. https://doi.org/10.1038/srep22219.

Kühl, Hjalmar S., Tenekwetche Sop, Elizabeth A. Williamson, Roger Mundry, David Brugière, Genevieve Campbell, Heather Cohen, et al. 2017. "The Critically Endangered Western Chimpanzee Declines by 80%." *American Journal of Primatology* 79 (9): e22681. https://doi.org/10.1002/ajp.22681.

Kühl, Hjalmar S., Liz Williamson, Crickette Sanz, David Morgan, and Christophe Boesch. 2007. "Launch of A.P.E.S. Database." *Gorilla Journal: Journal of Berggorilla and Regenwald Direkthilfe*, no. 34, 20–21.

Kuhn, Thomas S. 1962. *The Structure of Scientific Revolutions*. Chicago: University of Chicago Press.

Kuklick, Henrika, and Robert E. Kohler. 1996. "Introduction to 'Science in the Field.'" *Osiris* 11:1–14.

Kummer, Hans. 1971. *Primate Societies: Group Techniques of Ecological Adaptation*. Chicago: Aldine.

Kuper, Adam. 1994. *The Chosen Primate: Human Nature and Cultural Diversity*. Cambridge, MA: Harvard University Press.

———. 1999. *Culture: The Anthropologists' Account*. Cambridge, MA: Harvard University Press.

Kusch, Martin. 1995. *Psychologism: A Case Study in the Sociology of Philosophical Knowledge*. London: Routledge.

Kutsukake, Nobuyuki. 2010. "Lost in Translation: Field Primatology, Culture, and Interdisciplinary Approaches." In *Centralizing Fieldwork: Critical Perspectives from Primatology, Biological and Social Anthropology*, edited by Jeremy MacClancy and Agustín Fuentes, 104–20. New York: Berghahn Books.

Kuwayama, Takami. 2004. "The 'World System' of Anthropology: Japan and Asia in the Global Community of Anthropologists." In *The Making of Anthropology in East and Southeast Asia*, edited by Shinji Yamashita, Joseph Bosco, and J. S. Eades, 35–56. New York: Berghahn Books.

Laërtius, Diogenes. 1901. *The Lives and Opinions of Eminent Philosophers*. Translated by C. D. Yonge. London: George Bell and Sons.

Laland, Kevin N. 2017. *Darwin's Unfinished Symphony: How Culture Made the Human Mind*. Princeton, NJ: Princeton University Press.

Laland, Kevin N., and Bennett G. Galef, eds. 2009. *The Question of Animal Culture*. Cambridge, MA: Harvard University Press.

Laland, Kevin N., and Vincent M. Janik. 2006. "The Animal Cultures Debate." *Trends in Ecology & Evolution* 21 (10): 542–47.

Laland, Kevin N., Jeremy R. Kendal, and Rachel L. Kendal. 2009. "Animal Culture: Problems and Solutions." In *The Question of Animal Culture*, edited by Kevin N. Laland and Bennett G. Galef, 184–96. Cambridge, MA: Harvard University Press.

Laland, Kevin N., Kim Sterelny, John Odling-Smee, William Hoppitt, and Tobias Uller. 2011. "Cause and Effect in Biology Revisited: Is Mayr's Proximate-Ultimate Dichotomy Still Useful?' *Science* 334 (6062): 1512–16. https://doi.org/10.1126/science.1210879.

Lamont, Michèle. 2009. *How Professors Think*. Cambridge, MA: Harvard University Press.

Landau, Misia. 1991. *Narratives of Human Evolution*. New Haven, CT: Yale University Press.

Langer, William L. 1963. "Europe's Initial Population Explosion." *American Historical Review* 69 (1): 1–17. https://doi.org/10.2307/1904410.

Langergraber, Kevin E., Christophe Boesch, Eiji Inoue, Miho Inoue-Murayama, John C. Mitani, Toshisada Nishida, Anne Pusey, et al. 2011. "Genetic and 'Cultural' Similarity in Wild Chimpanzees." *Proceedings of the Royal Society of London B: Biological Sciences* 27 (8): 408–16. https://doi.org/10.1098/rspb.2010.1112.

Langlitz, Nicolas. 2005. *Die Zeit der Psychoanalyse: Lacan und das Problem der Sitzungsdauer*. Frankfurt am Main, Germany: Suhrkamp.

———. 2007. "What First-Order Observers Can Learn from Second-Order Observations." Anthropology of the Contemporary Research Collaboratory. ARC Concept Note no. 3. http://anthropos-lab.net/wp/publications/2007/08/conceptnote03.pdf.

———. 2012. *Neuropsychedelia: The Revival of Hallucinogen Research since the Decade of the Brain*. Berkeley: University of California Press.

———. 2015a. "On a Not So Chance Encounter between Neurophilosophy and Science Studies in a Sleep Laboratory." *History of the Human Sciences* 28 (4): 3–24. https://doi.org/10.1177/0952695115581576.

———. 2015b. "Vatted Dreams: Neurophilosophy and the Politics of Phenomenal Internalism." *Journal of the Royal Anthropological Institute* 21 (4): 739–57. https://doi.org/10.1111/1467-9655.12285.

———. 2016a. "*Homo academicus* und *Papio anubis* in der Reagan-Thatcher-Ära." *Nach Feierabend: Zürcher Jahrbuch für Wissensgeschichte, Wissen, ca. 1980* 12:235–43.

———. 2016b. "Is There a Place for Psychedelics in Philosophy? Fieldwork in Neuro- and Perennial Philosophy." *Common Knowledge* 22 (3): 373–84. https://doi.org/10.1215/0961754X-3622224.

———. 2017a. "Fieldwork in Skepticism: How an Anthropologist Learns to Cultivate Doubt and Other Virtues in a French Neuroscience Laboratory." Review of *Plastic Reason: An Anthropology of Brain Science in Embryogenetic Terms*, by Tobias Rees. *Dialectical Anthropology*, May 2017, 1–7.

———. 2017b. "Synthetic Primatology: What Humans and Chimpanzees Do in a Japanese Laboratory and the African Field." *British Journal for the History of Science Themes* 2:101–25.

———. 2019. "Primatology of Science: On the Birth of Actor-Network Theory from Baboon Field Observations." *Theory, Culture & Society* 36 (1): 83–105.

———. Forthcoming. "Cooperative Primates and Competitive Primatologists: Prosociality and Polemics in a Nonhuman Social Science." In *The Social Sciences through the Looking Glass: Studies in the Production of Knowledge*, edited by George Steinmetz and Didier Fassin. Oxford: Oxford University Press.

Latour, Bruno. 1987. *Science in Action: How to Follow Scientists and Engineers through Society*. Cambridge, MA: Harvard University Press.

———. 1993. *We Have Never Been Modern*. Cambridge, MA: Harvard University Press.

———. 2000. "A Well-Articulated Primatology: Reflections of a Fellow Traveler." In *Primate Encounters: Models of Science, Gender, and Society*, edited by Shirley C. Strum and Linda M. Fedigan, 358–81. Chicago: University of Chicago Press.

Latour, Bruno, and Shirley C. Strum. 1986. "Human Social Origins: Oh Please, Tell Us Another Story." *Journal of Social and Biological Structures* 9 (2): 169–87.

Leakey, Richard E., and Roger Lewin. 1995. *The Sixth Extinction: Patterns of Life and the Future of Humankind*. New York: Doubleday.

Leblan, Vincent. 2013. "Introduction: Emerging Approaches in the Anthropology/Primatology Borderland." *Revue de Primatologie*, no. 5 (December): 1–16. https://doi.org/10.4000/primatologie.1831.

———. 2016. "Territorial and Land-Use Rights Perspectives on Human-Chimpanzee-Elephant Coexistence in West Africa (Guinea, Guinea-Bissau, Senegal, Nineteenth to Twenty-First Centuries)." *Primates*, April 2016, 1–8. https://doi.org/10.1007/s10329-016-0532-4.

———. 2017. *Aux frontières du Singe: Relations entre hommes et chimpanzés au Kakandé, Guinée (XIXe-XXIe siècle)*. Paris: Éditions de l'École des hautes études en sciences sociales.

Leblan, Vincent, and Blandine Bricka. 2013. "Genies or the Opacity of Human-Animal Relationships in Kakandé, Guinea." *African Study Monographs* 34 (2): 85–108. https://doi.org/10.14989/179135.

Le Guenno, B., P. Formenty, M. Wyers, P. Gounon, F. Walker, and C. Boesch. 1995. "Isolation and Partial Characterisation of a New Strain of Ebola Virus." *Lancet* 345 (8960): 1271–74. https://doi.org/10.1016/S0140-6736(95)90925-7.

Lemov, Rebecca. 2015. "Anthropological Data in Danger, c. 1941–1965." In *Endangerment, Biodiversity and Culture*, edited by Fernando Vidal and Nélia Dias, 87–112. New York: Routledge.

Lestel, Dominique. 2001. *Les origines animales de la culture*. Paris: Flammarion.

———. 2002. "The Biosemiotics and Phylogenesis of Culture." *Social Science Information* 41 (1): 35–68.

———. 2011. "What Capabilities for the Animal?" *Biosemiotics* 4:83–102.

———. 2014. "Hybrid Communities." *Angelaki: Journal of the Theoretical Humanities* 19 (3): 61–73.

Lestel, Dominique, Florence Brunois, and Florence Gaunet. 2006. "Etho-Ethnology and Ethno-Ethology." *Social Science Information* 45 (2): 155–77. https://doi.org/10.1177/0539018406063633.

Lévi-Strauss, Claude. (1955) 1974. *Tristes tropiques*. New York: Atheneum.

———. 1969. *The Elementary Structures of Kinship*. Boston: Beacon Press.

———. 1976. "Jean-Jacques Rousseau, Founder of the Sciences of Man." In *Structural Anthropology*, vol. 3, 33–43. New York: Basic Books.

———. 2013. *The Other Face of the Moon*. Translated by Jane Marie Todd. Cambridge, MA: Harvard University Press.

Lévi-Strauss, Claude, and Didier Eribon. 1991. *Conversations with Claude Lévi-Strauss*. Translated by Paula Wissing. Chicago: University of Chicago Press.

Lévy, Bernard-Henri. 2010. *De la guerre en philosophie*. Paris: Éditions Grasset & Fasquelle.

Liebenberg, Louis. 2013. "The CyberTracker Story." CyberTracker Conservation. http://www.cybertracker.org/background/our-story.

Lindeman, Eduard C. 1924. *Social Discovery: An Approach to the Study of Functional Groups*. New York: Republic. https://books.google.com/books/about/Social_discovery.html?id=k9mTtAEACAAJ.

Lock, Margaret. 1993. *Encounters with Aging: Mythologies of Menopause in Japan and North America*. Berkeley: University of California Press.

———. 2002. *Twice Dead: Organ Transplants and the Reinvention of Death*. Berkeley: University of California Press.

Locke, John. (1689) 2000. *An Essay concerning Human Understanding*. London: Routledge.

Lorenz, Konrad. 1950. "The Comparative Method in Studying Innate Behavior Patterns." *Symposia of the Society for Experimental Biology* 4 (Physiological Mechanisms in Animal Behavior): 221–54.

———. 1952. *King Solomon's Ring: New Light on Animal Ways*. London: Methuen.

Luhmann, Niklas. 1991. "Paradigm Lost: On the Ethical Reflection of Morality; Speech on the Occasion of the Award of the Hegel Prize 1988." *Thesis Eleven* 291:82–94.

———. 1995. "Kultur als historischer Begriff." In *Gesellschaftsstruktur und Semantik: Studien zur Wissenssoziologie der modernen Gesellschaft*, 31–54. Frankfurt am Main, Germany: Suhrkamp.

———. 1998. *Observations on Modernity*. Stanford, CA: Stanford University Press.

Luncz, Lydia V., and Christophe Boesch. 2015. "The Extent of Cultural Variation between Adjacent Chimpanzee (*Pan troglodytes verus*) Communities: A Microecological Approach." *American Journal of Physical Anthropology* 156 (1): 67–75. https://doi.org/10.1002/ajpa.22628.

Luncz, Lydia V., Roger Mundry, and Christophe Boesch. 2012. "Evidence for Cultural Differences between Neighboring Chimpanzee Communities." *Current Biology* 22 (10): 922–26. https://doi.org/10.1016/j.cub.2012.03.031.

Lycett, Stephen J., Mark Collard, and William C. McGrew. 2007. "Phylogenetic Analyses of Behavior Support Existence of Culture among Wild Chimpanzees." *Proceedings of the National Academy of Sciences* 104 (45): 17588–92.

MacClancy, Jeremy, and Agustín Fuentes. 2010. "Centralizing Fieldwork." In *Centralizing Fieldwork: Critical Perspectives from Primatology, Biological, and Social Anthropology*, edited by Jeremy MacClancy and Agustín Fuentes, 1–26. New York: Berghahn Books.

Magee, Glenn Alexander. 2010. "Quietism in German Mysticism and Philosophy." *Common Knowledge* 16 (3): 457–73.

Malinowski, Bronislaw. (1922) 1961. *Argonauts of the Western Pacific*. London: Routledge. Reprint, New York: E. P. Dutton.

Malm, Andreas, and Alf Hornborg. 2014. "The Geology of Mankind? A Critique of the Anthropocene Narrative." *Anthropocene Review* 1 (1): 62–69.
Malthus, Thomas Robert. 1798. *An Essay on the Principle of Population*. London: J. Johnson.
Marcon, Federico. 2015. *The Knowledge of Nature and the Nature of Knowledge in Early Modern Japan*. Chicago: University of Chicago Press.
Marcus, George E., and Michael M. J. Fischer. 1986. *Anthropology as Cultural Critique: An Experimental Moment in the Human Sciences*. Chicago: University of Chicago Press.
Marcus, George E., and Nicolas Langlitz. 2014. "What Could the Comedy of Things Be?" *Comedy of Things* (blog). http://comedyofthings.com/box-2/text-production-2-marcus-langlitz/.
Markus, Gyorgy. 1987. "Why Is There No Hermeneutics of Natural Sciences? Some Preliminary Theses." *Science in Context* 1 (1): 5–51.
Martin, Paul S. 1966. "Africa and Pleistocene Overkill." *Nature*, no. 5060 (October): 339–42.
———. 2005. *Twilight of the Mammoths: Ice Age Extinction and the Rewilding of America*. Berkeley: University of California Press.
Matsuzawa, Tetsuro. 1985a. "Colour Naming and Classification in a Chimpanzee (*Pan troglodytes*)." *Journal of Human Evolution* 14 (3): 283–91. https://doi.org/10.1016/S0047-2484(85)80069-5.
———. 1985b. "Use of Numbers by a Chimpanzee." *Nature* 315 (6014): 57–59.
———. 1990. "Form Perception and Visual Acuity in a Chimpanzee." *Folia Primatologica* 55 (1): 24–32.
———. 1994. "Field Experiments on Use of Stone Tools by Chimpanzees in the Wild." In *Chimpanzee Cultures*, edited by Richard W. Wrangham, William C. McGrew, Frans De Waal, and P. G. Heltne, 351–70. Cambridge, MA: Harvard University Press.
———. 1996. "Chimpanzee Intelligence in Nature and in Captivity: Isomorphism of Symbol Use and Tool Use." In *Great Ape Societies*, edited by William C. McGrew, Linda Marchant, and Toshisada Nishida, 196–210. Cambridge: Cambridge University Press.
———. 1999. "Communication and Tool Use in Chimpanzees: Cultural and Social Contexts." In *The Design of Animal Communication*, edited by Marc D. Hauser and Mark Konishi, 645–71. Cambridge, MA: Massachusetts Institute of Technology.
———. 2003. "The Ai Project: Historical and Ecological Contexts." *Animal Cognition* 6 (4): 199–211. https://doi.org/10.1007/s10071-003-0199-2.
———. 2006. "Sociocognitive Development in Chimpanzees: A Synthesis of Laboratory Work and Fieldwork." In *Cognitive Development in Chimpanzees*, edited by Tetsuro Matsuzawa, Masaki Tomonaga, and Masayuki Tanaka, 3–33. Tokyo: Springer.
———. 2011a. "Education by Master-Apprenticeship." In *The Chimpanzees of Bossou and Nimba*, edited by Tetsuro Matsuzawa, Tatyana Humle, and Yukimaru Sugiyama, 201–8. Primatology Monographs. Tokyo: Springer Japan. https://doi.org/10.1007/978-4-431-53921-6_18.
———. 2011b. "Field Experiments of Tool-Use." In *The Chimpanzees of Bossou and Nimba*, edited by Tetsuro Matsuzawa, Tatyana Humle, and Yukimaru Sugiyama, 157–64. Primatology Monographs. Tokyo: Springer. https://doi.org/10.1007/978-4-431-53921-6_17.
———. 2011c. "Stone Tools for Nut-Cracking." In *The Chimpanzees of Bossou and Nimba*, edited by Tetsuro Matsuzawa, Tatyana Humle, and Yukimaru Sugiyama, 73–83. Primatology Monographs. Tokyo: Springer Japan. https://doi.org/10.1007/978-4-431-53921-6_8.

———. 2012. "What Is Uniquely Human? A View from Comparative Development in Humans and Chimpanzees." In *The Primate Mind: Built to Connect with Other Minds*, edited by Frans de Waal and Pier Francesco Ferrari, 288–305. Cambridge, MA: Harvard University Press.
———. 2016. "Mountain Day: Isomorphism of Mountaineering and Science." *Primates* 57 (4): 441–44.
———. 2018. "Chimpanzee Velu: The Wild Chimpanzee Who Passed Away at the Estimated Age of 58." *Primates* 59 (2): 107–11. https://doi.org/10.1007/s10329-018-0654-y.
Matsuzawa, Tetsuro, Dora Biro, Tatyana Humle, Noriko Inoue-Nakamura, Rikako Tonooka, and Gen Yamakoshi. 2001. "Emergence of Culture in Wild Chimpanzees: Education by Master-Apprenticeship." In *Primate Origins of Human Cognition and Behavior*, edited by Tetsuro Matsuzawa, 557–74. Springer Japan. https://doi.org/10.1007/978-4-431-09423-4_28.
Matsuzawa, Tetsuro, and William C. McGrew. 2008. "Kinji Imanishi and 60 Years of Japanese Primatology." *Current Biology* 18 (14): R587–91. https://doi.org/10.1016/j.cub.2008.05.040.
Matsuzawa, Tetsuro, Gaku Ohashi, Tatyana Humle, Nicolas Granier, Makan Kourouma, and Aly Gaspard Soumah. 2011. "Green Corridor Project: Planting Trees in the Savanna between Bossou and Nimba." In *The Chimpanzees of Bossou and Nimba*, edited by Tetsuro Matsuzawa, Tatyana Humle, and Yukimaru Sugiyama, 361–70. Primatology Monographs. Tokyo: Springer Japan. https://doi.org/10.1007/978-4-431-53921-6_38.
Matsuzawa, Tetsuro, and Juichi Yamagiwa. 2018. "Primatology: The Beginning." *Primates* 59 (4): 313–26. https://doi.org/10.1007/s10329-018-0672-9.
Mayr, Ernst. 1961. "Cause and Effect in Biology." *Science* 134 (3489): 1501–6.
McElreath, Mary Brooke, Christophe Boesch, Hjalmar Kuehl, and Richard McElreath. 2017. "Complex Dynamics from Simple Cognition: The Primary Ratchet Effect in Animal Culture." Preprint, May 2017, BioRxiv. https://doi.org/10.1101/134247.
McGovern, Mike. 2011. *Making War in Côte d'Ivoire*. Chicago: University of Chicago Press.
———. 2012. *Unmasking the State: Making Guinea Modern*. Chicago: University of Chicago Press.
McGrew, William C. 1987. "Tools to Get Food: The Subsistants of Tasmanian Aborigines and Tanzanian Chimpanzees Compared." *Journal of Anthropological Research* 43 (3): 247–58.
———. 1992. *Chimpanzee Material Culture: Implications for Human Evolution*. Cambridge: Cambridge University Press.
———. 2003. "Ten Dispatches from the Chimpanzee Culture Wars." In *Animal Social Complexity: Intelligence, Culture, and Individualized Societies*, edited by Frans B. M. de Waal and Peter L. Tyack, 419–39. Cambridge, MA: Harvard University Press.
———. 2004. *The Cultured Chimpanzee: Reflections on Cultural Primatology*. Cambridge: Cambridge University Press.
———. 2007. "New Wine in New Bottles: Prospects and Pitfalls of Cultural Primatology." *Journal of Anthropological Research* 63 (2): 167–83.
———. 2010. "New Theaters of Conflict in the Animal Culture Wars: Recent Findings from Chimpanzees." In *The Mind of the Chimpanzee: Ecological and Experimental Perspectives*, edited by Elizabeth V. Lonsdorf, Stephen R. Ross, and Tetsuro Matsuzawa, 168–77. Chicago: University of Chicago Press.
———. 2015. "The Cultured Chimpanzee: Nonsense or Breakthrough." *Human Ethology Bulletin—Proceedings of the XXII ISHE Conference*, 41–52.

———. 2017. "Ourselves Explained." *Human Ethology Bulletin* 32 (3): 141–44.

McGrew, William C., and Caroline E. G. Tutin. 1978. "Evidence for a Social Custom in Wild Chimpanzees?' *Man* 13 (2): 234–51.

McGrew, William C., Caroline E. G. Tutin, and P. J. Baldwin. 1979. "Chimpanzees, Tools, and Termites: Cross-Cultural Comparisons of Senegal, Tanzania, and Rio Muni." *Man* 14 (2): 185–214.

McLennan, Matthew R., and Catherine M. Hill. 2010. "Chimpanzee Responses to Researchers in a Disturbed Forest–Farm Mosaic at Bulindi, Western Uganda." *American Journal of Primatology* 72 (10): 907–18. https://doi.org/10.1002/ajp.20839.

Mead, Margaret. 1928. *Coming of Age in Samoa: A Psychological Study of Primitive Youth for Western Civilisation*. New York: Morrow.

Menzel, Emil W., ed. 1973. *Precultural Primate Behavior*. Basel, Switzerland: S. Karger.

Mesoudi, Alex, and Andrew Whiten. 2008. "The Multiple Roles of Cultural Transmission Experiments in Understanding Human Cultural Evolution." *Philosophical Transactions of the Royal Society B: Biological Sciences* 363 (1509): 3489–501. https://doi.org/10.1098/rstb.2008.0129.

Mesoudi, Alex, Andrew Whiten, and Kevin N. Laland. 2006. "Towards a Unified Science of Cultural Evolution." *Behavioral and Brain Sciences* 29 (4): 329–47.

Mitman, Gregg. 1992. *The State of Nature: Ecology, Community, and American Social Thought, 1900–1950*. Chicago: University of Chicago Press.

Moderlak, Tom. 2016. *Intersubjektivität als Philosophisch-Anthropologische Kategorie: Arnold Gehlen und Michael Tomasello*. Hamburg, Germany: Verlag Dr. Kovač.

Montgomery, Georgina M. 2005. "Place, Practice and Primatology: Clarence Ray Carpenter, Primate Communication and the Development of Field Methodology, 1931–1945." *Journal of the History of Biology* 38 (3): 495–533.

———. 2015. *Primates in the Real World: Escaping Primate Folklore and Creating Primate Science*. Charlottesville: University of Virginia Press.

Moore, Jason W. 2017. "The Capitalocene Part I: On the Nature and Origins of Our Ecological Crisis." *Journal of Peasant Studies* 44 (3): 594–630.

———. 2018. "The Capitalocene Part II: Accumulation by Appropriation and the Centrality of Unpaid Work/Energy." *Journal of Peasant Studies* 45 (2): 237–79.

Moore, Richard. 2013. "Imitation and Conventional Communication." *Biology & Philosophy* 28 (3): 481–500. https://doi.org/10.1007/s10539-012-9349-8.

Moore, Richard, Josep Call, and Michael Tomasello. 2015. "Production and Comprehension of Gestures between Orang-Utans (*Pongo pygmaeus*) in a Referential Communication Game." *PLOS ONE* 106:e0129726. https://doi.org/10.1371/journal.pone.0129726.

Moore, Richard, Bettina Mueller, Juliane Kaminski, and Michael Tomasello. 2015. "Two-Year-Old Children but Not Domestic Dogs Understand Communicative Intentions without Language, Gestures, or Gaze." *Developmental Science* 18 (2): 232–42. https://doi.org/10.1111/desc.12206.

Morgan, Bethan J., and Ekwoge E. Abwe. 2006. "Chimpanzees Use Stone Hammers in Cameroon." *Current Biology* 16 (16): R632–33.

Morgan, C. Lloyd. 1894. *An Introduction to Comparative Psychology*. London: W. Scott.

Morimura, Naruki, Gen'ichi Idani, and Tetsuro Matsuzawa. 2010. "The First Chimpanzee Sanctuary in Japan: An Attempt to Care for the 'Surplus' of Biomedical Research." *American Journal of Primatology* 73 (3): 226–32. https://doi.org/10.1002/ajp.20887.

Morris, Desmond. 1979. *Animal Days*. London: Jonathan Cape.

Morris-Suzuki, Tessa. 1995. "The Invention and Reinvention of 'Japanese Culture.'" *Journal of Asian Studies* 54 (3): 759–80. https://doi.org/10.2307/2059450.

———. 1998. *Re-Inventing Japan: Time, Space, Nation*. Armonk, NY: M. E. Sharpe.

Musil, Robert. 1995. "On the Essay." In *Precision and Soul: Essays and Addresses*, edited and translated by Burton Pike and David S. Luft, 48–51. Chicago: University of Chicago Press.

Myowa-Yamakoshi, M., and T. Matsuzawa. 1999. "Factors Influencing Imitation of Manipulatory Actions in Chimpanzees (*Pan troglodytes*)." *Journal of Comparative Psychology* 113 (2): 128–36.

———. 2000. "Imitation of Intentional Manipulatory Actions in Chimpanzees (*Pan troglodytes*)." *Journal of Comparative Psychology* 114 (4): 381–91.

Naam, Ramez. 2013. *The Infinite Resource: The Power of Ideas on a Finite Planet*. Lebanon, NH: University Press of New England.

Nakamura, Michio. 2009. "Interaction Studies in Japanese Primatology: Their Scope, Uniqueness, and the Future." *Primates* 50 (2): 142–52. https://doi.org/10.1007/s10329-009-0133-6.

———. 2010. "Ubiquity of Culture and Possible Social Inheritance of Sociality among Wild Chimpanzees." In *The Mind of the Chimpanzee: Ecological and Experimental Perspectives*, edited by Elizabeth V. Lonsdorf, Stephen R. Ross, and Tetsuro Matsuzawa, 156–67. Chicago: University of Chicago Press.

Nakamura, Michio, and Toshisada Nishida. 2006. "Subtle Behavioral Variation in Wild Chimpanzees, with Special Reference to Imanishi's Concept of Kaluchua." *Primates* 47 (1): 35–42. https://doi.org/10.1007/s10329-005-0142-z.

Neumann, Ralfs. 2012. "Animal Behaviour Research: Publication Analysis 1999–2010." *Lab Times*, no. 3, 34–36.

Nietzsche, Friedrich. (1882) 2001. *The Joyous Science*. Cambridge: Cambridge University Press.

Nishida, Kitaro. 1992. *An Inquiry into the Good*. New Haven, CT: Yale University Press.

Nishida, Toshisada. 1976. "The Bark-Eating Habits in Primates, with Special Reference to Their Status in the Diet of Wild Chimpanzees." *Folia Primatologica* 25 (4): 277–87. https://doi.org/10.1159/000155720.

———. 1987. "Local Traditions and Cultural Transmission." In *Primate Societies*, edited by Barbara B. Smuts, Dorothy L. Cheney, Robert M. Seyfarth, and Richard W. Wrangham, 462–74. Chicago: University of Chicago Press.

———. 2012. *Chimpanzees of the Lakeshore: Natural History and Culture at Mahale*. Cambridge: Cambridge University Press.

Nishida, Toshisada, Koichiro Zamma, Takahisa Matsusaka, Agumi Inaba, and William C. McGrew. 2010. *Chimpanzee Behavior in the Wild: An Audio-Visual Encyclopedia*. Tokyo: Springer.

Nungesser, Frithjof. 2016. "Die intrinsische Sozialität rücksichtslosen Handelns: Über Michael Tomasello und die dunklen Seiten humanspezfischer Kooperation." In "Kooperation, Sozialität und Kultur: Michael Tomasellos Arbeiten in der soziologischen Diskussion," edited by Gert Albert, Jens Greve, and Rainer Schützeichel, special issue, *Zeitschrift für Theoretische Soziologie* 3:128–62. Weinheim, Germany: Julius Beltz.

Oden, David L., Roger K. Thompson, and David Premack. 1988. "Spontaneous Transfer of Matching by Infant Chimpanzees (*Pan troglodytes*)." *Journal of Experimental Psychology: Animal Behavior Processes* 14 (2): 140–45. https://doi.org/10.1037/0097-7403.14.2.140.

Oelze, Vicky M., Geraldine Fahy, Gottfried Hohmann, Martha M. Robbins, Vera Leinert, Kevin Lee, Henk Eshuis, et al. 2016. "Comparative Isotope Ecology of African Great Apes." *Journal of Human Evolution* 101 (December): 1–16. https://doi.org/10.1016/j.jhevol.2016.08.007.

Ohnuki-Tierney, Emiko. 1987. *The Monkey as Mirror: Symbolic Transformations in Japanese History and Ritual*. Princeton, NJ: Princeton University Press.

Ouattara, Karim, Alban Lemasson, and Klaus Zuberbühler. 2009. "Campbell's Monkeys Use Affixation to Alter Call Meaning." *PLOS ONE* 4 (11): e7808. https://doi.org/10.1371/journal.pone.0007808.

Pääbo, Svante. 2014. *Neanderthal Man: In Search of Lost Genomes*. New York: Basic Books.

Perpeet, Wilhelm. 1976. "Kultur, Kulturphilosophie." In *Historisches Wörterbuch der Philosophie*, vol. 4, edited by Joachim Ritter and Karlfried Gründer, 1309–24. Basel, Switzerland: Schwabe.

Perry, George H., and Brian F. Codding. 2017. "Monkeys Overharvest Shellfish." *eLife*, 6 September 2017, e30865. https://doi.org/10.7554/eLife.30865.

Perry, Susan, Mary Baker, Linda Fedigan, Julie Gros-Louis, Katherine Jack, Katherine C. MacKinnon, Joseph H. Manson, Melissa Panger, Kendra Pyle, and Lisa Rose. 2003. "Social Conventions in Wild White-Faced Capuchin Monkeys: Evidence for Traditions in a Neotropical Primate." *Current Anthropology* 44 (2): 241–68. https://doi.org/10.1086/345825.

Perry, Susan E. 2006. "What Cultural Primatology Can Tell Anthropologists about the Evolution of Culture." *Annual Review of Anthropology* 35 (1): 171–90. https://doi.org/10.1146/annurev.anthro.35.081705.123312.

Peterson, Dale. 2014. *Jane Goodall: The Woman Who Redefined Man*. Boston: Houghton Mifflin.

Petherbridge, Danielle. 2013. *The Critical Theory of Axel Honneth*. Lanham, MD: Lexington Books.

Pickering, Andrew. 1995. *The Mangle of Practice: Time, Agency, and Science*. Chicago: University of Chicago Press.

Pina-Cabral, João de. 2000. "The Ethnographic Present Revisited." *Social Anthropology* 8 (3): 341–48.

Pinker, Steven. 2002. *The Blank Slate: The Modern Denial of Human Nature*. New York: Penguin.

Popper, Karl. 1957. *The Poverty of Historicism*. London: Routledge.

Porter, Theodore M. 2012. "Thin Description: Surface and Depth in Science and Science Studies." *Osiris* 27 (1): 209–26. https://doi.org/10.1086/667828.

Povinelli, Daniel J. 2000. *Folk Physics for Apes: The Chimpanzee's Theory of How the World Works*. Oxford: Oxford University Press.

Povinelli, Daniel J., and Theodore J. Povinelli. 2001. Review of *The Chimpanzees of the Taï Forest*, by Christophe Boesch and Hedwige Boesch-Achermann. *Ethology* 107 (5): 463–64.

Premack, David, and Ann J. Premack. 1983. *The Mind of an Ape*. New York: W. W. Norton.

Price, Derek J. de Solla. 1970. "Citation Measures of Hard Science, Soft Science, Technology and Nonscience." In *Communication among Scientists and Engineers*, edited by Carnot E. Nelson and Donald K. Pollock, 3–22. Lexington, MA: Heath.

Prinz, Jesse. 2008. "Empirical Philosophy and Experimental Philosophy." In *Experimental Philosophy*, edited by Shaun Nichols and Joshua Knobe, 189–208. Oxford: Oxford University Press.

Quammen, David. 2012. *Spillover: Animal Infections and the Next Human Pandemic*. New York: W. W. Norton.

Quiatt, Duane. 1997. "Silent Partners? Observations on Some Systematic Relations among Observer Perspective, Theory, and Behavior." In *Anthropomorphism, Anecdotes, and Animals*, edited by R. Mitchell, Nicholas S. Thompson, and H. L. Miles, 220–36. New York: SUNY Press.

Quine, Willard Van Orman. 1960. *Word and Object*. Cambridge, MA: MIT Press.

Rabinow, Paul. 2003. *Anthropos Today: Reflections on Modern Equipment*. Princeton, NJ: Princeton University Press.

———. 2008. *Marking Time: On the Anthropology of the Contemporary*. Princeton, NJ: Princeton University Press.

———. 2011. *The Accompaniment: Assembling the Contemporary*. Chicago: University of Chicago Press.

Radcliffe-Brown, Alfred R. 1940. "On Social Structure." *Journal of the Royal Anthropological Institute of Great Britain and Ireland* 70 (1): 1–12.

Radick, Gregory. 2007. *The Simian Tongue: The Long Debate about Animal Language*. Chicago: University of Chicago Press.

———. 2016. "The Unmaking of a Modern Synthesis: Noam Chomsky, Charles Hockett, and the Politics of Behaviorism, 1955–1965." *Isis* 107 (1): 49–73. https://doi.org/10.1086/686177.

Rakoczy, Hannes, and Michael Tomasello. 2008. "Kollektive Intentionalität und kulturelle Entwicklung." *Deutsche Zeitschrift für Philosophie* 56 (3): 401–10. https://doi.org/10.1524/dzph.2008.0031.

Rawlings, Bruce, Marina Davila-Ross, and Sarah T. Boysen. 2014. "Semi-Wild Chimpanzees Open Hard-Shelled Fruits Differently across Communities." *Animal Cognition* 17 (4): 891–99. https://doi.org/10.1007/s10071-013-0722-z.

Rees, Amanda. 2006. "A Place That Answers Questions: Primatological Field Sites and the Making of Authentic Observations." *Studies in History and Philosophy of Biological and Biomedical Sciences* 37 (2): 311–33.

———. 2007. "Reflections on the Field: Primatology, Popular Science and the Politics of Personhood." *Social Studies of Science* 37 (6): 881–907.

———. 2009a. *The Infanticide Controversy: Primatology and the Art of Field Science*. Chicago: University of Chicago Press.

———. 2009b. "The Undead Darwin: Iconic Narrative, Scientific Controversy and the History of Science." *History of Science* 47 (4): 445–57. https://doi.org/10.1177/007327530904700406.

Rees, Tobias. 2016. *Plastic Reason: An Anthropology of Brain Science in Embryogenetic Terms*. Berkeley: University of California Press.

Rehberg, Karl-Siegbert. 2016. "Sonderstellung oder ökologische Nische? Wolfgang Köhler und Michael Tomasello aus der Sicht der philosophischen Anthropologie." In "Kooperation, Sozialität und Kultur: Michael Tomasellos Arbeiten in der soziologischen Diskussion," edited by Gert Albert, Jens Greve, and Rainer Schützeichel, special issue, *Zeitschrift für Theoretische Soziologie* 3:28–44. Weinheim, Germany: Julius Beltz.

Reichenbach, Hans. 1947. *Elements of Symbolic Logic*. New York: Macmillan.

Rendell, Luke, and Hal Whitehead. 2001. "Culture in Whales and Dolphins." *Behavioral and Brain Sciences* 24 (2): 309–24. https://doi.org/10.1017/S0140525X0100396X.

Reynolds, Vernon. 1975. "How Wild Are the Gombe Chimpanzees?" *Man* 10 (1): 123–25.
———. 1991. Review of *Primate Visions*, by Donna Haraway. *Man* 26 (1): 167–68.
———. 2005. *The Chimpanzees of Budongo Forest: Ecology, Behaviour, and Conservation*. Oxford: Oxford University Press.
Rheinberger, Hans-Jörg. 2015. *Natur und Kultur im Spiegel des Wissens*. Heidelberg, Germany: Universitätsverlag Winter.
Richards, Paul. 1993. "Natural Symbols and Natural History: Chimpanzees, Elephants and Experiments in Mende Thought." In *Environmentalism: The View from Anthropology*, edited by Kay Milton, 143–56. London: Routledge.
Riesen, A. H. 1970. "Chimpanzee Visual Perception." In *The Chimpanzee*, vol. 2, edited by Geoffrey H. Bourne, 93–119. Baltimore: University Park Press.
Rodman, Peter S. 1990. "Flawed Vision: Deconstruction of Primatology and Primatologists." *Current Anthropology* 31 (4): 484–86.
Roitman, Janet. 2013. *Anti-crisis*. Durham, NC: Duke University Press.
Rorty, Richard. 1999. *Achieving Our Country: Leftist Thought in Twentieth-Century America*. Cambridge, MA: Harvard University Press.
Rosaldo, Renato. 1989. "Imperialist Nostalgia." *Representations* 26:107–22.
Rousseau, Jean-Jacques. 1997. *"The Discourses" and Other Early Political Writings*. Cambridge: Cambridge University Press.
Rowell, Thelma E. 1967. "Variability in the Social Organization of Primates." In *Primate Ethology*, edited by Desmond Morris, 219–35. Chicago: Aldine.
Rudwick, Martin J. S. 1997. *Georges Cuvier, Fossil Bones, and Geological Catastrophes: New Translations & Interpretations of the Primary Texts*. Chicago: University of Chicago Press.
Ruhnau, Jürgen. 2018. "Fatalismus." In *Historisches Wörterbuch der Philosophie*. Basel, Switzerland: Schwabe. https://www.schwabeonline.ch/schwabe-xaveropp/elibrary/start.xav#__elibrary__%2F%2F*%5B%40attr_id%3D%27verw.fatalismus%27%5D__1533575667348.
Rumbaugh, Duane M., ed. 1977. *Language Learning by a Chimpanzee: The Lana Project*. Illustrated ed. New York: Academic Press.
Russon, Anne, and Kristin Andrews. 2010. "Orangutan Pantomime: Elaborating the Message." *Biology Letters*, August 2010, rsbl20100564. https://doi.org/10.1098/rsbl.2010.0564.
Sahlins, Marshall. 1976. *The Use and Abuse of Biology: An Anthropological Critique of Sociobiology*. Ann Arbor: University of Michigan Press.
———. 1999. "Two or Three Things That I Know about Culture." *Journal of the Royal Anthropological Institute* 5 (3): 399–421. https://doi.org/10.2307/2661275.
———. 2000. "'Sentimental Pessimism' and Ethnographic Experience; or, Why Culture Is Not a Disappearing 'Object.'" In *Biographies of Scientific Objects*, edited by Lorraine Daston, 158–202. Chicago: University of Chicago Press.
Sahlins, Marshall D. 2008. *The Western Illusion of Human Nature: With Reflections on the Long History of Hierarchy, Equality and the Sublimation of Anarchy in the West, and Comparative Notes on Other Conceptions of the Human Condition*. Chicago: Prickly Paradigm Press.
Said, Edward W. 1978. *Orientalism*. New York: Pantheon Books.
Saito, Aya, Misato Hayashi, Hideko Takeshita, and Tetsuro Matsuzawa. 2014. "The Origin of Representational Drawing: A Comparison of Human Children and Chimpanzees." *Child Development* 85 (6): 2232–46. https://doi.org/10.1111/cdev.12319.

Sakai, Tomoko, Satoshi Hirata, Kohki Fuwa, Keiko Sugama, Kiyo Kusunoki, Haruyuki Makishima, Tatsuya Eguchi, Shigehito Yamada, Naomichi Ogihara, and Hideko Takeshita. 2012. "Fetal Brain Development in Chimpanzees versus Humans." *Current Biology* 22 (18): R791–92. https://doi.org/10.1016/j.cub.2012.06.062.

Sakura, Osamu. 1995. "What Is This Thing Called 'Group of Animals'? A Proposal of the Pluralistic Terminology Implicated from Non-human Primate Ecology." *Annals of the Japan Association for Philosophy of Science* 8 (5): 237–52.

———. 1998. "Similarities and Varieties: A Brief Sketch on the Reception of Darwinism and Sociobiology in Japan." *Biology and Philosophy* 13 (3): 341–57. https://doi.org/10.1023/A:1006504623820.

———. 2005. Review of *A Japanese View of Nature*, by Kinji Imanishi. *Primates* 46 (4): 287–89. https://doi.org/10.1007/s10329-005-0133-0.

Sakura, Osamu, Toshiyuki Sawaguchi, Hiroko Kudo, and Shin'ichi Yoshikubo. 1986. "Declining Support for Imanishi." *Nature* 323 (586) (16 October): 586.

Samuni, Liran, Roger Mundry, Joseph Terkel, Klaus Zuberbühler, and Catherine Hobaiter. 2014. "Socially Learned Habituation to Human Observers in Wild Chimpanzees." *Animal Cognition* 17 (4): 997–1005.

Sanjek, Roger. 1991. "The Ethnographic Present." *Man* 26 (4): 609–28. https://doi.org/10.2307/2803772.

Sapir, Edward. 1921. *Language: An Introduction to the Study of Speech*. New York: Harcourt, Brace and World.

Sasaki, Takao, and Dora Biro. 2017. "Cumulative Culture Can Emerge from Collective Intelligence in Animal Groups." *Nature Communications*, 8 April 2017, 15049. https://doi.org/10.1038/ncomms15049.

Savage, Thomas S., and Jeffries Wyman. 1843. "Observations on the External Characteristics, Habits and Organisation of the *Troglodytes niger*, Geoff." *Boston Journal of Natural History* 4:362–86.

Savage-Rumbaugh, Sue. 1999. "Ape Language: Between a Rock and a Hard Place." In *The Origin of Language: What Nonhuman Primates Can Tell Us*, edited by Barbara J. King, 115–89. Santa Fe: School for Advanced Research Press.

Savage-Rumbaugh, Sue, William M. Fields, Pär Segerdahl, and Duane Rumbaugh. 2005. "Culture Prefigures Cognition in *Pan/Homo* Bonobos." *Theoria: Revista de teoría, historia y fundamentos de la ciencia* 20 (3): 311–28.

Schaller, George B. 1964. *The Year of the Gorilla*. Chicago: University of Chicago Press.

Schiavenato, Stephanie, Esther Rottenburg, and Nicolas Langlitz. 2016. "Time and the Other Primates." *Anthropology Now* 8 (3): 135–42. https://doi.org/10.1080/19428200.2016.1242928.

Schickore, Jutta. 2017. *About Method: Experimenters, Snake Venom, and the History of Writing Scientifically*. Chicago: University of Chicago Press.

Schmuhl, Hans-Walter. 2008. *The Kaiser Wilhelm Institute for Anthropology, Human Heredity and Eugenics, 1927–1945: Crossing Boundaries*. Dordrecht, Netherlands: Springer.

Schnädelbach, Herbert. 1984. *Philosophy in Germany, 1831–1933*. Cambridge: Cambridge University Press.

Schoener, Thomas W. 1983. "Field Experiments on Interspecific Competition." *American Naturalist* 122 (2): 240–85.

Schultz, Adolph H. 1960. "Comment on Imanishi's 'Social Organization of Subhuman Primates in Their Natural Habitat.'" *Current Anthropology* 15 (6): 405.

Sebastiani, Silvia. 2015. "Challenging Boundaries: Apes and Savages in Enlightenment." In *Simianization: Apes, Gender, Class, and Race*, edited by Wulf D. Hund, Charles W. Mills, and Silvia Sebastiani, 105–37. Berlin: LIT Verlag.

———. 2019. "A 'Monster with a Human Visage': The Orangutan, Savagery and the Borders of Humanity in the Global Enlightenment." *History of the Human Sciences* 32 (4): 80–99. https://doi.org/10.1177/0952695119836619.

Segerdahl, Pär, William M. Fields, and Sue Savage-Rumbaugh. 2005. *Kanzi's Primal Language: The Cultural Initiation of Primates into Language*. Houndmills, UK: Palgrave Macmillan.

Segerstråle, Ullica. 2000. *Defenders of the Truth: The Battle for Science in the Sociobiology Debate and Beyond*. Oxford: Oxford University Press.

Sepkoski, David. 2015. "Extinction, Diversity, and Endangerment." In *Endangerment, Biodiversity and Culture*, edited by Fernando Vidal and Nélia Dias, 62–86. New York: Routledge.

Shapin, Steven. 1994. *A Social History of Truth: Civility and Science in Seventeenth-Century England*. Chicago: University of Chicago Press.

———. 1996. *The Scientific Revolution*. Chicago: University of Chicago Press.

Shapin, Steven, and Simon Schaffer. 1985. *Leviathan and the Air-Pump: Hobbes, Boyle, and the Experimental Life*. Princeton, NJ: Princeton University Press.

Shimao, Eikoh. 1981. "Darwinism in Japan, 1877–1927." *Annals of Science* 38 (1): 93–102.

Shipley, Bill. 2000. *Cause and Correlation in Biology: A User's Guide to Path Analysis, Structural Equations and Causal Inference with R*. Cambridge: Cambridge University Press.

Shipman, Pat. 2015. *The Invaders: How Humans and Their Dogs Drove Neanderthals to Extinction*. 3rd ed. Cambridge, MA: Belknap Press of Harvard University Press.

Sibatani, Atuhiro. 1972. "An Exile's View of the Contemporary Scene." *Nature* 240 (24 November): 191–93.

———. 1983. "The Anti-Selectionism of Kinji Imanishi and Social Anti-Darwinism in Japan." *Journal of Social and Biological Structures* 6 (4): 335–43.

Silberberg, Alan, and David Kearns. 2009. "Memory for the Order of Briefly Presented Numerals in Humans as a Function of Practice." *Animal Cognition* 12 (2): 405–7. https://doi.org/10.1007/s10071-008-0206-8.

Skinner, Burrhus Frederic. 1957. *Verbal Behavior*. Upper Saddle River, NJ: Prentice-Hall.

Sleeboom, Margaret. 2004. *Academic Nations in China and Japan: Framed by Concepts of Nature, Culture and the Universal*. London: Routledge.

Smuts, Barbara B. 2001. "Encounters with Wild Minds." *Journal of Consciousness Studies* 8 (5–7): 293–309.

Snow, C. P. 1964. *The Two Cultures*. Cambridge: Cambridge University Press.

Sommer, Volker. 2009. "Kein Wir-Gefühl im Pongoland." Review of *The Origins of Human Communication*, by Michael Tomasello. *Frankfurter Rundschau*, 27 September 2009. http://www.fr.de/kultur/literatur/anthropologie-kein-wir-gefuehl-im-pongoland-a-1077163.

Sperling, Susan. 1991. "Baboons with Briefcases vs. Langurs in Lipstick: Feminism and Functionalism in Primate Studies." In *Gender at the Crossroads of Knowledge: Feminist Anthropology in the Postmodern Era*, edited by Micaela di Leonardo, 204–34. Berkeley: University of California Press.

Sponsel, Leslie E. 1997. "The Human Niche in Amazonia: Explorations in Ethnoprimatology." In *New World Primates: Ecology, Evolution, and Behavior*, edited by Warren G. Kinzey, 143–65. New York: Aldine de Gruyter. https://books.google.com/books?hl=en&lr=&id=1AdneqhwvCwC&oi=fnd&pg=PA143&dq=sponsel+1997+ethnoprimatology&ots=qmkVWFF7Cj&sig=nooHnVEwl1ElsQLhoPmtNWcIcUw.

———. 2012. *Spiritual Ecology: A Quiet Revolution*. Santa Barbara, CA: Praeger.

Spradley, James P. 1980. *Participant Observation*. Long Grove, IL: Harcourt Brace Jovanovich.

Sprague, David S. 2004. Review of *A Japanese View of Nature*, by Kinji Imanishi. *Journal of Japanese Studies* 30 (1): 211–15.

Stallard, Brian. 2015. "How Do Chimpanzees Cross the Road? With All the Right Precautions." *Nature World News*, 20 April 2015. https://www.natureworldnews.com/articles/14193/20150420/chimpanzees-cross-road-right-precautions.htm.

Stanford, Craig B. 1991. Review of *Primate Visions*, by Donna Haraway. *American Anthropologist* 93 (4): 1031–32.

Sterelny, Kim. 2003. *Thought in a Hostile World: The Evolution of Human Cognition*. Oxford, UK: Blackwell.

———. 2012. *The Evolved Apprentice*. Cambridge, MA: MIT Press.

Stocking, George W. 1968. *Race, Culture, and Evolution: Essays in the History of Anthropology*. Chicago: University of Chicago Press.

———. 1992. *The Ethnographer's Magic and Other Essays in the History of Anthropology*. Madison: University of Wisconsin Press.

Strathern, Marilyn. 1980. "No Nature, No Culture: The Hagen Case." In *Nature, Culture and Gender*, edited by Carol MacCormack and Marilyn Strathern, 174–222. Cambridge: Cambridge University Press.

———. 1988. *The Gender of the Gift: Problems with Women and Problems with Society in Melanesia*. Berkeley: University of California Press.

———. 1992. *After Nature: English Kinship in the Late Twentieth Century*. Cambridge: Cambridge University Press.

———. 2004a. *Commons and Borderlands: Working Papers on Interdisciplinarity, Accountability and the Flow of Knowledge*. Herefordshire, UK: Sean Kingston Publishing.

———. 2004b. "The Whole Person and Its Artifacts." *Annual Review of Anthropology* 33 (1): 1–19.

Strier, Karen B. 2003. "Primate Behavioral Ecology: From Ethnography to Ethology and Back." *American Anthropologist* 105 (1): 16–27.

Struhsaker, Thomas T., and Pierre Hunkeler. 1971. "Evidence of Tool-Using by Chimpanzees in the Ivory Coast." *Folia Primatologica* 15 (3–4): 212–19. https://doi.org/10.1159/000155380.

Strum, Shirley C. 1987. *Almost Human: A Journey into the World of Baboons*. Chicago: University of Chicago Press.

———. 2012. "Darwin's Monkey: Why Baboons Can't Become Human." *American Journal of Physical Anthropology* 149 (S55): 3–23.

———. 2017. "Baboons and the Origins of Actor-Network Theory: An Interview with Shirley Strum about the Shared History of Primate and Science Studies (by Nicolas Langlitz)." *BioSocieties* 12 (1): 158–67.

Strum, Shirley C., and Linda M. Fedigan. 2000. *Primate Encounters: Models of Science, Gender, and Society*. Chicago: University of Chicago Press.

Strum, Shirley C., and Bruno Latour. 1987. "Redefining the Social Link: From Baboons to Humans." *Social Science Information* 26 (4): 783–802.

Sugiyama, Yukimaru. 1965. "Short History of the Ecological and Sociological Studies on Non-human Primates in Japan." *Primates* 6 (3–4): 457–60.

———. 1999. "Socioecological Factors of Male Chimpanzee Migration at Bossou, Guinea." *Primates* 40 (1): 61–68. https://doi.org/10.1007/BF02557702.

Sugiyama, Yukimaru, and Jeremy Koman. 1979. "Tool-Using and -Making Behavior in Wild Chimpanzees at Bossou, Guinea." *Primates* 20 (4): 513–24. https://doi.org/10.1007/BF02373433.

Takada, Akira. 2013. "Mutual Coordination of Behaviors in Human–Chimpanzee Interactions: A Case Study in a Laboratory Setting." *Revue de primatologie*, no. 5. https://doi.org/10.4000/primatologie.1902.

Takasaki, Hiroyuki. 2000. "Traditions of the Kyoto School of Field Primatology in Japan." In *Primate Encounters: Models of Science, Gender, and Society*, edited by Shirley C. Strum and Linda M. Fedigan, 151–64. Chicago: University of Chicago Press.

Takeshita, Hideko, Masako Myowa-Yamakoshi, and Satoshi Hirata. 2006. "A New Comparative Perspective on Prenatal Motor Behaviors: Preliminary Research with Four-Dimensional Ultrasonography." In *Cognitive Development in Chimpanzees*, edited by Tetsuro Matsuzawa, Masaki Tomonaga, and Masayuki Tanaka, 37–47. Tokyo: Springer.

Tanner, Jakob. 2004. *Historische Anthropologie zur Einführung*. Hamburg, Germany: Junius.

Taylor, Bron. 2009. *Dark Green Religion: Nature Spirituality and the Planetary Future*. Berkeley: University of California Press.

Taylor, Richard. 1962. "Fatalism." *Philosophical Review* 71 (1): 56–66.

Tehrani, Jamshid J., and Michael Carrithers. 2015. "Perspectives on the Intersection of Biology and Society." *Journal of the Royal Anthropological Institute* 21 (2): 470–72. https://doi.org/10.1111/1467-9655.12216.

Terrace, Herbert S. 1980. *Nim: A Chimpanzee Who Learned Sign Language*. London: Eyre Methuen.

Thies, Christian. 2017. "Michael Tomasello und die philosophische Anthropologie." *Philosophische Rundschau* 64 (2): 107–21.

Thomas, Julia Adeney. 2001. *Reconfiguring Modernity: Concepts of Nature in Japanese Political Ideology*. Berkeley: University of California Press.

———. 2015. "Who Is the 'We' Endangered by Climate Change?" In *Endangerment, Biodiversity and Culture*, edited by Fernando Vidal and Nélia Dias, 241–60. New York: Routledge.

Thomas, Marion. 2005. "Are Animals Just Noisy Machines?: Louis Boutan and the Co-invention of Animal and Child Psychology in the French Third Republic." *Journal of the History of Biology* 38 (3): 425–60. https://doi.org/10.1007/s10739-005-0555-y.

———. 2006. "Yerkes, Hamilton and the Experimental Study of the Ape Mind: From Evolutionary Psychiatry to Eugenic Politics." *Studies in History and Philosophy of Science Part C: Studies in History and Philosophy of Biological and Biomedical Sciences* 37 (2): 273–94. https://doi.org/10.1016/j.shpsc.2006.03.011.

———. 2016. "Between Biomedical and Psychological Experiments: The Unexpected Connections between the Pasteur Institutes and the Study of Animal Mind in the Second Quarter of Twentieth-Century France." *Studies in History and Philosophy of Science Part C: Studies in History and Philosophy of Biological and Biomedical Sciences* 55 (February): 29–40. https://doi.org/10.1016/j.shpsc.2015.10.010.

Thompson, Jo A. Myers. 1994. "Cultural Diversity in the Behavior of Pan." In *Hominid Culture in Primate Perspective*, edited by Duane Quiatt and Junichiro Itani, 95–115. Niwot: University Press of Colorado.

Thompson, Melissa Emery, James H. Jones, Anne E. Pusey, Stella Brewer-Marsden, Jane Goodall, David Marsden, Tetsuro Matsuzawa, et al. 2007. "Aging and Fertility Patterns in Wild Chimpanzees Provide Insights into the Evolution of Menopause." *Current Biology* 17 (24): 2150–56. https://doi.org/10.1016/j.cub.2007.11.033.

Tomasello, Michael. 1990. "Cultural Transmission in Chimpanzee Tool Use and Signaling?" In *"Language" and Intelligence in Monkeys and Apes: Comparative Developmental Perspectives*, edited by Sue Taylor Parker and Kathleen Rita Gibson, 274–311. Cambridge: Cambridge University Press.

———. 1999. *The Cultural Origins of Human Cognition*. Cambridge, MA: Harvard University Press.

———. 2001. "'Sie zeigen nur auf Dinge, die sie haben wollen': Kognition bei Menschen und Menschenaffen (Interview)." *Philokles: Zeitschrift für populäre Philosophie*, no. 2, 3–11.

———. 2006. *Die kulturelle Entwicklung des menschlichen Denkens: Zur Evolution der Kognition*. Frankfurt am Main, Germany: Suhrkamp.

———. 2008. *Origins of Human Communication*. Cambridge, MA: MIT Press.

———. 2009a. "Rede, gehalten Anlässlich der Verleihung des Hegel-Preises 2009 [English]." Stuttgart, Germany. http://www.stuttgart.de/img/mdb/item/383875/51641.pdf.

———. 2009b. *Why We Cooperate*. Cambridge, MA: MIT Press.

———. 2014. *A Natural History of Human Thinking*. Cambridge, MA: Harvard University Press.

Tomasello, Michael, and Josep Call. 2008. "Assessing the Validity of Ape-Human Comparisons: A Reply to Boesch (2007)." *Journal of Comparative Psychology* 122 (4): 449–52. https://doi.org/10.1037/0735-7036.122.4.449.

Tomasello, Michael, Maryann Davis-Dasilva, Lael Camak, and Kim Bard. 1987. "Observational Learning of Tool-Use by Young Chimpanzees." *Human Evolution* 2 (2): 175–83.

Tomasello, Michael, Ann Cale Kruger, and Hilary Horn Ratner. 1993. "Cultural Learning." *Behavioral and Brain Sciences* 16 (3): 495–511. https://doi.org/10.1017/S0140525X0003123X.

Tomasello, Michael, Sue Savage-Rumbaugh, and Ann Cale Kruger. 1993. "Imitative Learning of Actions on Objects by Children, Chimpanzees, and Enculturated Chimpanzees." *Child Development* 64 (6): 1688–705. https://doi.org/10.1111/1467-8624.ep9406130028.

Traweek, Sharon. 1988. *Beamtimes and Lifetimes: The World of High Energy Physicists*. Cambridge, MA: Harvard University Press.

Trigger, Bruce G. 1981. "Archaeology and the Ethnographic Present." *Anthropologica* 23 (1): 3–17. https://doi.org/10.2307/25605060.

Tsunoda, Tadanobu. 1985. *The Japanese Brain: Uniqueness and Universality*. Tokyo: Taishukan.

Turner, Terrence. 1991. "Representing, Resisting, Rethinking: Historical Transformations of Kayapo Culture and Anthropological Consciousness." In *Colonial Situations: Essays on the*

Contextualization of Ethnographic Knowledge, edited by George W. Stocking, 285–313. Madison: University of Wisconsin Press.

Tylor, Edward Burnett. 1871. *Primitive Culture: Researches into the Development of Mythology, Philosophy, Religion, Art, and Custom*. London: J. Murray.

Ueno, Ari, Satoshi Hirata, Kohki Fuwa, Keiko Sugama, Kiyo Kusunoki, Goh Matsuda, Hirokata Fukushima, Kazuo Hiraki, Masaki Tomonaga, and Toshikazu Hasegawa. 2008. "Auditory ERPs to Stimulus Deviance in an Awake Chimpanzee (*Pan troglodytes*): Towards Hominid Cognitive Neurosciences." *PLOS ONE* 3 (1): e1442.

———. 2010. "Brain Activity in an Awake Chimpanzee in Response to the Sound of Her Own Name." *Biology Letters* 6 (3): 311–13. https://doi.org/10.1098/rsbl.2009.0864.

UNESCO. 1952. *The Race Concept: Results of an Inquiry*. Paris: UNESCO.

Unoura, Hiroshi. 1999. "Samurai Darwinism: Hiroyuki Kato and the Reception of Darwin's Theory in Modern Japan from the 1880s to the 1900s." *History and Anthropology* 11 (2–3): 235–55.

van Leeuwen, Edwin J. C., Katherine A. Cronin, and Daniel B. M. Haun. 2014. "A Group-Specific Arbitrary Tradition in Chimpanzees (*Pan troglodytes*)." *Animal Cognition* 17 (6): 1421–25. https://doi.org/10.1007/s10071-014-0766-8.

van Leeuwen, Edwin J. C., Katherine A. Cronin, Daniel B. M. Haun, Roger Mundry, and Mark D. Bodamer. 2012. "Neighbouring Chimpanzee Communities Show Different Preferences in Social Grooming Behaviour." *Proceedings of the Royal Society B: Biological Sciences* 279 (1746): 4362–67. https://doi.org/10.1098/rspb.2012.1543.

van Schaik, Carel P., Marc Ancrenaz, Gwendolyn Borgen, Birute Galdikas, Cheryl D. Knott, Ian Singleton, Akira Suzuki, Sri Suci Utami, and Michelle Merrill. 2003. "Orangutan Cultures and the Evolution of Material Culture." *Science* 299 (5603): 102–5. https://doi.org/10.1126/science.1078004.

van Schaik, Carel P., and Judith M. Burkart. 2011. "Social Learning and Evolution: The Cultural Intelligence Hypothesis." *Philosophical Transactions of the Royal Society B: Biological Sciences* 366 (1567): 1008–16. https://doi.org/10.1098/rstb.2010.0304.

van Schaik, Carel P., and Kai Michel. 2016. *The Good Book of Human Nature: An Evolutionary Reading of the Bible*. New York: Basic Books.

Venturini, Tommaso, and Bruno Latour. 2010. "The Social Fabric: Digital Traces and Quali-Quantitative Methods." In *Proceedings of Future en Seine 2009*, 87–101. Paris: Éditions Future en Seine.

Verroux, Vincent. 2003. "L'homme et la biosphère dans la réserve des Monts Nimba (République de Guinée): Savoirs naturalistes locaux et gestion de l'environnement." Master's thesis (mémoire de Diplôme d'études approfondies). Paris: Muséum national d'histoire naturelle.

Vetter, Jeremy, ed. 2011. *Knowing Global Environments: New Historical Perspectives on the Field Sciences*. New Brunswick, NJ: Rutgers University Press.

Vidal, Fernando, and Nélia Dias, eds. 2015. *Endangerment, Biodiversity and Culture*. New York: Routledge.

Visalberghi, Elisabetta, and Dorothy Fragaszy. 1990. "Food-Washing Behaviour in Tufted Capuchin Monkeys, *Cebus apella*, and Crabeating Macaques, *Macaca fascicularis*." *Animal Behaviour* 40 (5): 829–36.

Viveiros de Castro, Eduardo. 1998. "Cosmological Deixis and Amerindian Perspectivism." *Journal of the Royal Anthropological Institute* 4 (3): 469–88.

———. 2003. "(Anthropology) AND (Science)." *Manchester Papers in Social Anthropology* 7. http://nansi.abaetenet.net/abaetextos/anthropology-and-science-e-viveiros-de-castro.

———. 2004. "Perspectival Anthropology and the Method of Controlled Equivocation." *Tipití: Journal of the Society for the Anthropology of Lowland South America* 21. http://digitalcommons.trinity.edu/tipiti/vol2/iss1/1.

Voltaire. 1901. *A Philosophical Dictionary*. Vol. 6. New York: E. R. DuMont.

———. 2006. *Candide*. Oxford: Oxford University Press.

von Herder, Johann Gottfried. 2002. "On the Change of Taste." In *Herder: Philosophical Writings*, edited by Michael N. Forster, 247–56. Cambridge: Cambridge University Press.

von Uexküll, Jakob. (1934) 2010. *A Foray into the Worlds of Animals and Humans*. Minneapolis: University of Minnesota Press.

Wagner, Roy. 1981. *The Invention of Culture*. Chicago: University of Chicago Press.

Wallman, Joel. 1992. *Aping Language*. Cambridge: Cambridge University Press.

Washburn, Sherwood L. 1944. "Thinking about Race." *Science Education* 28 (2): 65–76.

———. 1951. "The New Physical Anthropology." *Transactions of the New York Academy of Sciences* 13 (7): 298–304.

———. 1963. "The Study of Race." *American Anthropologist* 65 (3): 521–31.

———. 1973. "The Promise of Primatology." *American Journal of Physical Anthropology* 38 (2): 177–82. https://doi.org/10.1002/ajpa.1330380206.

———. 1983. "Evolution of a Teacher." *Annual Review of Anthropology* 12:1–24.

Washburn, Sherwood L., and Burton Benedict. 1979. "Non-Human Primate Culture." *Man* 14 (1): 163–64.

Washburn, Sherwood L., and Ruth Moore. 1973. *Ape into Man: Study of Human Evolution*. Boston: Little, Brown.

Watson, Claire F. I., Hannah M. Buchanan-Smith, and Christine A. Caldwell. 2014. "Call Playback Artificially Generates a Temporary Cultural Style of High Affiliation in Marmosets." *Animal Behaviour* 93 (July): 163–71. https://doi.org/10.1016/j.anbehav.2014.04.027.

Watts, David P., and Sylvia J. Amsler. 2013. "Chimpanzee-Red Colobus Encounter Rates Show a Red Colobus Population Decline Associated with Predation by Chimpanzees at Ngogo." *American Journal of Primatology* 75 (9): 927–37.

Weber, Max. (1909) 1978. "Urbanisation and Social Structure in the Ancient World." In *Max Weber: Selections in Translation*, 290–314. New York: Cambridge University Press.

———. 1958. "Science as a Vocation [1919]." In *From Max Weber: Essays in Sociology*, edited by H. H. Gerth and C. Wright Mills, 129–56. New York: Oxford University Press.

Weisman, Alan. 2007. *The World without Us*. New York: St. Martin's Press.

Welsch, Wolfgang. 2007. "Just What Is It That Makes *Homo Sapiens* So Different, So Appealing?" *Deutsche Zeitschrift für Philosophie* 55 (5): 751–60. https://doi.org/10.1524/dzph.2007.55.5.751.

White, Hayden. 1973. *Metahistory: The Historical Imagination in Nineteenth-Century Europe*. Baltimore: Johns Hopkins University Press.

White, Leslie A. 1949. *The Science of Culture: A Study of Man and Civilization*. Oxford, UK: Farrar, Straus.

Whiten, Andrew. 2002. "From the Field to the Laboratory and Back Again: Culture and 'Social Mind' in Primates." In *The Cognitive Animal*, edited by Marc Bekoff, Colin Allen, and Gordon M. Burghardt, 385–92. Cambridge, MA: MIT Press.

———. 2010. "A Coming of Age for Cultural Panthropology." In *The Mind of the Chimpanzee: Ecological and Experimental Perspectives*, edited by Elizabeth V. Lonsdorf, Stephen R. Ross, and Tetsuro Matsuzawa, 87–100. Chicago: University of Chicago Press.

———. 2017. "Social Learning and Culture in Child and Chimpanzee." *Annual Review of Psychology* 68 (1): 129–54. https://doi.org/10.1146/annurev-psych-010416-044108.

Whiten, Andrew, and Robert A. Barton. 1988. "Demise of the Checksheet: Using Off-the-Shelf Miniature Hand-Held Computers for Remote Fieldwork Applications." *Trends in Ecology & Evolution* 3 (6): 146–48.

Whiten, Andrew, Deborah M. Custance, Juan-Carlos Gomez, Patricia Teixidor, and Kim A. Bard. 1996. "Imitative Learning of Artificial Fruit Processing in Children (*Homo sapiens*) and Chimpanzees (*Pan troglodytes*)." *Journal of Comparative Psychology* 110 (1): 3–14. https://doi.org/10.1037/0735-7036.110.1.3.

Whiten, Andrew, Jane Goodall, William C. McGrew, Toshisada Nishida, Vernon Reynolds, Yukimaru Sugiyama, Caroline E. G. Tutin, Richard W. Wrangham, and Christophe Boesch. 1999. "Cultures in Chimpanzees." *Nature* 399 (6737): 682–85. https://doi.org/10.1038/21415.

Whiten, Andrew, Robert A. Hinde, Christopher B. Stringer, and Kevin N. Laland, eds. 2012. *Culture Evolves*. Oxford: Oxford University Press.

Whiten, Andrew, Victoria Horner, and Sarah Marshall-Pescini. 2003. "Cultural Panthropology." *Evolutionary Anthropology* 12 (2): 92–105.

Whiten, Andrew, and Carel P. van Schaik. 2007. "The Evolution of Animal 'Cultures' and Social Intelligence." *Philosophical Transactions of the Royal Society B: Biological Sciences* 362 (1480): 603–20. https://doi.org/10.1098/rstb.2006.1998.

Whorf, Benjamin Lee. 1956. *Language, Thought, and Reality: Selected Writings of Benjamin Lee Whorf*. Edited by John B. Carroll. Cambridge, MA: MIT Press.

Wieder, D. Lawrence. 1980. "Behavioristic Operationalism and the Life-World: Chimpanzees and Chimpanzee Researchers in Face-to-Face Interaction." *Sociological Inquiry* 50 (3–4): 75–103. https://doi.org/10.1111/j.1475-682X.1980.tb00017.x.

"Wild Chimps Look Both Ways before Crossing Roads." 2015. *New Scientist*, 17 April 2015. https://www.newscientist.com/article/dn27370-wild-chimps-look-both-ways-before-crossing-roads/.

Williams, Raymond. 1960. *Culture and Society: 1780–1950*. Garden City, NY: Anchor Books.

———. 1983. *Keywords: A Vocabulary of Culture and Society*. London: Fontana.

Williamson, Elizabeth A., and Anna T. C. Feistner. 2003. "Habituating Primates: Processes, Techniques, Variables and Ethics." In *Field and Laboratory Methods in Primatology: A Practical Guide*, edited by Joanna M. Setchell and Deborah J. Curtis, 25–39. Cambridge: Cambridge University Press.

Wilson, Edward O. 1975. *Sociobiology: The New Synthesis*. Cambridge, MA: Belknap Press of Harvard University Press.

———. 1978. *On Human Nature*. Cambridge, MA: Harvard University Press.

———. 1980. *Sociobiology*. Abridged ed. Cambridge, MA: Belknap Press of Harvard University Press.

———. 1994. *Naturalist*. Washington, DC: Island Press.

———. 1998. *Consilience: The Unity of Knowledge*. New York: Alfred A. Knopf.

———. 2015. *The Meaning of Human Existence*. New York: Liveright.

Wilson, Michael L., Christophe Boesch, Barbara Fruth, Takeshi Furuichi, Ian C. Gilby, Chie Hashimoto, Catherine L. Hobaiter, et al. 2014. "Lethal Aggression in *Pan* Is Better Explained by Adaptive Strategies Than Human Impacts." *Nature* 513 (7518): 414–17. https://doi.org/10.1038/nature13727.

Winch, Peter. 1990. *The Idea of a Social Science and Its Relation to Philosophy*. Atlantic Highlands, NJ: Humanities Press International.

Windelband, Wilhelm. 1904. *Geschichte und Naturwissenschaft: Rede zum Antritt des Rektorats*. Strassbourg, France: Heitz & Mündel.

Winkler, Heinrich August. 2007. *Germany: The Long Road West*. Oxford: Oxford University Press.

Wittig, Roman M. 2011. "A Comprehensive Guide to Chimpanzee Behavior." *American Journal of Primatology* 73 (5): 498–500. https://doi.org/10.1002/ajp.20923.

Wokler, Robert. 1978. "Perfectible Apes in Decadent Cultures: Rousseau's Anthropology Revisited." *Daedalus* 107 (3): 107–34.

Woods, Vanessa. 2007. *It's Every Monkey for Themselves: A True Story of Sex, Love, and Lies in the Jungle*. Crows Nest, Australia: Allen & Unwin.

Wrangham, Richard W. 2006. "Chimpanzees: The Culture-Zone Concept Becomes Untidy." *Current Biology* 16 (16): R634–35.

———. 2009. *Catching Fire: How Cooking Made Us Human*. New York: Basic Books.

Wrangham, Richard W., William C. McGrew, Frans B. M. de Waal, and Paul G. Heltne, eds. 1996. *Chimpanzee Cultures*. Cambridge, MA: Harvard University Press.

Wulf, Andrea. 2016. *The Invention of Nature: Alexander von Humboldt's New World*. New York: Vintage.

Wundt, Wilhelm. 1916. *Elements of Folk Psychology*. London: George Allen and Unwin.

Wunsch, Matthias. 2015. "Was macht menschliches Denken einzigartig? Zum Forschungsprogramm Michael Tomasellos." In *Interdisziplinäre Anthropologie, Jahrbuch 3/2015: Religion und Ritual*, edited by Gerald Hartung and Matthias Herrgen, 259–88. Wiesbaden, Germany: Springer.

Wynes, Seth, and Kimberly A. Nicholas. 2017. "The Climate Mitigation Gap: Education and Government Recommendations Miss the Most Effective Individual Actions." *Environmental Research Letters* 12 (7): 074024. https://doi.org/10.1088/1748-9326/aa7541.

Yagi, Shusuke. 2002. Foreword to *A Japanese View of Nature: The World of Living Things*, by Kinji Imanishi, xiii–xvi. London: Routledge.

Yamakoshi, Gen. 2002. "The Village in Guinea Where Chimpanzees Live alongside Humans." *Occasional Report: The Toyota Foundation*, no. 32, 1–4.

———. 2011. "The 'Prehistory' before 1976: Looking Back on Three Decades of Research on Bossou Chimpanzees." In *The Chimpanzees of Bossou and Nimba*, edited by Tetsuro Matsuzawa, Tatyana Humle, and Yukimaru Sugiyama, 35–44. Primatology Monographs. Tokyo: Springer Japan. https://doi.org/10.1007/978-4-431-53921-6_5.

Yamakoshi, Gen, and Vincent Leblan. 2013. "Conflicts between Indigenous and Scientific Concepts of Landscape Management for Wildlife Conservation: Human-Chimpanzee Politics of Coexistence at Bossou, Guinea." *Revue de primatologie*, no. 5 (December). https://doi.org/10.4000/primatologie.1762.

Yamamoto, Shinya, Tatyana Humle, and Masayuki Tanaka. 2013. "Basis for Cumulative Cultural Evolution in Chimpanzees: Social Learning of a More Efficient Tool-Use Technique." *PLOS ONE* 81:e55768. https://doi.org/10.1371/journal.pone.0055768.

Yerkes, Robert M. 1943. *Chimpanzees: A Laboratory Colony*. New Haven, CT: Yale University Press.

Yoshino, Kosaku. 1992. *Cultural Nationalism in Contemporary Japan*. London: Routledge.

Zammito, John H. 2004. *A Nice Derangement of Epistemes: Post-positivism in the Study of Science from Quine to Latour*. Chicago: University of Chicago Press.

Zimmerman, Andrew. 2001. *Anthropology and Antihumanism in Imperial Germany*. Chicago: University of Chicago Press.

Zuberbühler, Klaus. 2014. "Experimental Field Studies with Non-human Primates." *Current Opinion in Neurobiology* 28 (October): 150–56. https://doi.org/10.1016/j.conb.2014.07.012.

INDEX

Page numbers in *italics* refer to figures.

A NOTE ON THE TYPE

This book has been composed in Arno, an Old-style serif typeface in the classic Venetian tradition, designed by Robert Slimbach at Adobe.

CPSIA information can be obtained
at www.ICGtesting.com
Printed in the USA
LVHW090042080820
662410LV00004B/4